Liquid Crystal Displays

Wiley-SID Series in Display Technology

Editor:
Anthony C. Lowe

Display Systems:
Design and Applications
Lindsay W. MacDonald and **Anthony C. Lowe** (Eds)

Electronic Display Measurement:
Concepts, Techniques, and Instrumentation
Peter A. Keller

Projection Displays
Edward H. Stupp and **Matthew S. Brennesholz**

Reflective Liquid Crystal Displays
Shin-Tson Wu and **Deng-Ke Yang**

Liquid Crystal Displays: Addressing Schemes & Electro-Optical Effects
Ernst Lueder

Liquid Crystal Displays

ADDRESSING SCHEMES AND ELECTRO-OPTICAL EFFECTS

Ernst Lueder
University of Stuttgart, Germany

JOHN WILEY & SONS, LTD
Chichester • New York • Weinheim • Brisbane • Singapore • Toronto

Other Wiley Editorial Offices

John Wiley & Sons, Inc., 605 Third Avenue,
New York, NY 10158-0012, USA

Wiley-VCH Verlag GmbH
Pappelallee 3, D-69469 Weinheim, Germany

John Wiley, Australia 33 Park Road, Milton,
Queensland 4064, Australia

John Wiley & Sons (Canada) Ltd, 22 Worcester Road
Rexdale, Ontario, M9W 1L1, Canada

John Wiley & Sons (Asia) Pte Ltd, 2 Clementi Loop #02-01,
Jin Xing Distripark, Singapore 129809

Library of Congress Cataloguing in Publication Data

British Library Cataloguing in Publication Data

A catalogue record for this book is available from the British Library

ISBN **0 471 49029 6**

Typeset in 10/12pt Times by Thomson Press (India) Ltd., Chennai
Printed and bound in Great Britain by Antony Rowe Ltd., Chipppenham, Wilts
This book is printed on acid-free paper responsibly manufactured from sustainable forestry,
in which at least two trees are planted for each one used for paper production.

Contents

Foreword   xi

Preface   xiii

About the Author   xv

1 Introduction   1

2 Liquid Crystal Materials and Liquid Crystal Cells   3
  2.1   Properties of Liquid Crystals   3
        2.1.1   Shape and phases of liquid crystals   3
        2.1.2   Material properties of anisotropic liquid crystals   5
  2.2   The Operation of a Twisted Nematic LCD   11
        2.2.1   The electro-optical effects in transmissive twisted nematic LC-cells   11
        2.2.2   The addressing of LCDs by TFTs   17

3 Electro-optic Effects in Untwisted Nematic Liquid Crystals   21
  3.1   The Planar and Harmonic Wave of Light   21
  3.2   Propagation of Polarized Light in Birefringent
        Untwisted Nematic Liquid Crystal Cells   26
        3.2.1   The propagation of light in a Fréedericksz cell   26
        3.2.2   The transmissive Fréedericksz cell   31
        3.2.3   The reflective Fréedericksz cell   37
        3.2.4   The Fréedericksz cell as a phase-only modulator   39
        3.2.5   The DAP cell or the vertically aligned cell   43
        3.2.6   The HAN cell   45
        3.2.7   The π cell   46
        3.2.8   Switching dynamics of untwisted nematic LCDs   49

4   Electro-optic Effects in Twisted Nematic Liquid Crystals   55

4.1   The Propagation of Polarized Light in Twisted Nematic Liquid Crystal Cells   55

4.2   The Various Types of TN Cells   64

4.2.1   The regular TN cell   64

4.2.2   The supertwisted nematic LC cell (STN-LCD)   67

4.2.3   The mixed mode twisted nematic cell (MTN cell)   72

4.2.4   Reflective TN cells   74

4.3   Electronically Controlled Birefringence for the Generation of Colour   78

5   Descriptions of Polarization   81

5.1   The Characterizations of Polarization   81

5.2   A Differential Equation for the Propagation of Polarized
Light through Anisotropic Media   89

5.3   Special Cases for Propagation of Light   93

5.3.1   Incidence of linearly polarized light   93

5.3.2   Incident light is circularly polarized   95

**6   Propagation of Light with an Arbitrary Incident
Angle through Anisotropic Media   97**

6.1   Basic Equations for the Propagation of Light   97

6.2   Enhancement of the Performance of LC Cells   105

6.2.1   The degradation of picture quality   105

6.2.2   Optical compensation foils for the enhancement of picture quality   106

The enhancement of contrast   106

Compensation foils for LC molecules with different optical axis   108

6.2.3   Suppression of grey shade inversion and the preservation
of grey shade stability   113

6.2.4   Fabrication of compensation foils   114

6.3   Electro-optic Effects with Wide Viewing Angle   114

6.3.1   Multidomain pixels   114

6.3.2   In-Plane switching   116

6.3.3   Optically compensated bend cells   117

6.4   Polarizers with Increased Luminous Output   119

6.4.1   A reflective linear polarizer   119

6.4.2   A reflective polarizer working with circularly polarized light   120

6.5   Two Non-birefringent Foils   121

7   Modified Nematic Liquid Crystal Displays   123

7.1   Polymer Dispersed LCDs (PDLCDs)   123

7.1.1   The operation of a PDLCD   123

7.1.2   Applications of PDLCDs   127

7.2   Guest-Host Displays   128

7.2.1   The operation of Guest-Host displays   128

7.2.2   Reflective Guest-Host displays   131

8   Bistable Liquid Crystal Displays   137

8.1   Ferroelectric Liquid Crystal Displays (FLCDs)   137

8.2   Chiral Nematic Liquid Crystal Displays   145

8.3   Bistable Nematic Liquid Crystal Displays   152

8.3.1   Bistable twist cells   152

8.3.2   Grating aligned nematic devices   153

8.3.3   Monostable surface anchoring switching   153

9 **Continuously Light Modulating Ferroelectric Displays** **155**
   9.1   Deformed Helix Ferroelectric Devices   155
   9.2   Antiferroelectric LCDs   157

10 **Addressing Schemes for Liquid Crystal Displays** **161**

11 **Direct Addressing** **165**

12 **Passive Matrix Addressing of TN Displays** **167**
   12.1   The Basic Addressing Scheme and the Law of Alt and Pleshko   167
   12.2   Implementation of PM Addressing   172
   12.3   Multiple Line Addressing   176
       12.3.1   The basic equations   176
       12.3.2   Waveforms for the row selection   179
       12.3.3   Column voltage for MLA   181
       12.3.4   Implementation of multi-line addressing   182
       12.3.5   Modified PM addressing of STN cells   185
              Decreased levels of addressing voltages   185
              Contrast and grey shades for MLA   188
   12.4   Two Frequency Driving of PMLCDs   194

13 **Passive Matrix Addressing of Bistable Displays** **197**
   13.1   Addressing of Ferroelectric LCDs   197
       13.1.1   The $V - \tau_{min}$ addressing scheme   198
       13.1.2   The $V - 1/\tau$ addressing scheme   199
       13.1.3   Reducing crosstalk in FLCDs   201
       13.1.4   Ionic effects during addressing   202
   13.2   Addressing of Chiral Nematic Liquid Crystal Displays   205

14 **Addressing of Liquid Crystal Displays**
     with a-Si Thin Film Transistors (a-Si-TFTs) **211**
   14.1   Properties of a-Si Thin Film Transistors   211
   14.2   Static Operation of TFTs in an LCD   216
   14.3   The Dynamics of Switching by TFTs   224
   14.4   Bias-Temperature Stress Test of TFTs   230
   14.5   Drivers for AMLCDs   231
   14.6   The Entire Addressing System   238
   14.7   Layouts of Pixels with TFT Switches   241
   14.8   Fabrication Processes of a-Si TFTs   245

15 **Addressing of LCDs with Poly Si-TFTs** **249**
   15.1   Fabrication Steps for Top-Gate and Bottom-Gate Poly-Si TFTs   250
   15.2   Laser Crystallization by Scanning or Large Area Anneal   254
   15.3   Lightly Doped Drains for Poly-Si TFTs   256
   15.4   The Kink Effect and its Suppression   258
   15.5   Circuits with Poly-Si TFTs   259

16 **Liquid Crystal Displays on Silicon** **263**
   16.1   Fabrication of LCOS with DRAM-Type Analog Addressing   263
   16.2   SRAM-Type Digital Addressing of LCOS   265
   16.3   Microdisplays Using LCOS Technology   270

17 Addressing of Liquid Crystal Displays with Metal-Insulator-Metal Pixel Switches   **271**

18 Addressing of LCDs with Two-Terminal Devices and Optical, Plasma, Laser and e-beam Techniques   **281**

19 Colour Filters and Cell Assembly   **287**

19.1   Additive Colours Generated by Absorptive Photosensitive Pigmented Colour Filters   289

19.2   Additive and Subtractive Colours Generated by Reflective Dichroic Colour Filters   292

19.3   Colour Generation by Three Stacked Displays   294

19.4   Cell Assembly   294

20 Projectors with Liquid Crystal Light Valves   **295**

20.1   Single Transmissive Light Valve Systems   295

    20.1.1   The basic single light valve system   295

    20.1.2   The field sequential colour projector   296

    20.1.3   A single panel scrolling projector   297

    20.1.4   Single light valve projector with angular colour separation   298

    20.1.5   Single light valve projectors with a colour grating   298

20.2   Systems with Three Light Valves   300

    20.2.1   Projectors with three transmissive light valves   300

    20.2.2   Projectors with three reflective light valves   301

    20.2.3   Projectors with three LCOS light valves   301

20.3   Projectors with Two LC Light Valves   301

20.4   A Rear Projector with One or Three Light Valves   304

20.5   A Projector with Three Optically Addressed Light Valves   304

21 Liquid Crystal Displays with Plastic Substrates   **307**

21.1   Advantages of Plastic Substrates   307

21.2   Plastic Substrates and their Properties   307

21.3   Barrier Layers for Plastic Substrates   309

21.4   Thermo-Mechanical Problems with Plastics   310

21.5   Fabrication of TFTs and MIMs at Low Process Temperatures   314

    21.5.1   Fabrication of a-Si:H TFTs at low temperature   314

    21.5.2   Fabrication of low temperature poly-Si TFTs   316

    21.5.3   Fabrication of MIMs at low temperature   317

    21.5.4   Conductors and transparent electrodes for plastic substrates   317

22 Printing of Layers for LC-Cells   **319**

22.1   Printing Technologies   319

    22.1.1   Flexographic printing   319

    22.1.2   Knife coating   319

    22.1.3   Ink jet printing   320

    22.1.4   Silk screen printing   320

22.2   Printing of Layers for LCDs   322

22.3   Cell Building by Lamination   324

Appendix 1: Formats of Flat Panel Displays   **325**

Appendix 2: Optical Units of Displays   **327**

Appendix 3: Properties of Polarized Light   **329**

References   **335**

Index   **345**

Foreword

Electro-Optic Effects and Addressing of Liquid Crystal Displays by Ernst Lüder is the fourth volume in this Display Technology series published by John Wiley & Sons, Ltd., in association with the Society for Information Display.

It is our objective to publish books that are of a sufficiently advanced standard to satisfy graduate needs, but which also contain the information of a practical nature that is so much required by scientists and engineers working in the field.

Over the past thirty years, liquid crystal displays have developed from a research curiosity to become a dominant display technology, second only to the ubiquitous CRT. Yet, although there exist a number of acclaimed books that deal with the theory of liquid crystals, a practitioner in the display field must still search mainly through published research papers and conference proceedings to discover detailed, relevant and advanced information on the addressing of liquid crystal displays.

This volume has been written by an acknowledged leader in the field, who turned theory into practice by founding a world class institute, the Labor für Bildschirmtechnik at the University of Stuttgart. It is a major work, which successfully fills the gap in the literature on LCD addressing.

Of course, a book that described *only* addressing techniques would be incomplete because an informed discussion of the subtleties of addressing requires knowledge of how light propagates through anisotropic media. This volume therefore begins with a description of liquid crystal materials, goes on to describe the basic electro-optic effects that are of relevance to LCDs and then develops, from first principles, the theory of propagation of light through liquid crystalline media. Only then does it discuss the entire range of active and passive addressing techniques. The book concludes with some brief, but relevant chapters on cell assembly, new substrate materials and manufacturing techniques.

This is a self-contained and comprehensive work, which readers will find both intellectually rewarding and enormously useful.

Anthony C Lowe
Braishfield, UK, 2000

Preface

The overriding purpose of this book, as further outlined in the Introduction (Chapter 1), is to condense in one single volume all the basic information that is needed to understand the operation and the building of liquid crystal displays. This requires a treatment of a wealth of electro-optical effects as well as a description of the rich variety of addressing schemes. The latter has not been done for more than a decade.

In the pursuit of this ambitious goal I was very fortunate to have a number of experts at my side who offered advice and assistance for writing this book. Dr. Tony Lowe, the editor of this SID- series, lent his experience in selecting the contents of this book and in focusing on special topics. His most valuable assistance is gratefully appreciated. Dr. Mike Lee from the Imperial College in London enriched the chapters on addressing techniques with some most helpful suggestions and englightening discussions. I am very grateful for his support. I am also indebted to my coworker at Stuttgart University, Dr. Christoph Zeile, who contributed to the sections about electro-optical effects by numerous discussions and his helpful observations. I thank Mrs. Heidi Schuehle very much for typing the manuscript with competence and patience and for alerting me to various inconsistencies. Mr. Rene Troeger has skillfully drawn the figures for which I am very grateful.

Finally, I wish to thank John Wiley for their always pleasant cooperation as well as for the attractive production of the book.

Ernst Lueder
Scottsdale, Arizona, 2000

About the Author

Ernst Lueder was born in 1932. At his graduation from High School he was awarded the 'Scheffel'-prize for literary achievements.

In 1962 he received his doctor's degree in electrical engineering, and in 1966 his Habilitation, which qualified him to teach in the area of theoretical electrical engineering.

From 1968 to 1971 he worked for Bell Telephone Laboratories in Holmdel, New Jersey, USA, undertaking research into the design of miniaturized filters and communication systems, especially in thin film technology. He established laws for optimizing the dynamic range and the signal-to-noise ratio of two-ports.

In 1971 he was appointed a full professor at the Department of Electrical Communications, and named Director of the Institute of Network- and System's Theory at Stuttgart University. He specializes in the design of hybrid thin and thick film circuits, the development of sensors, thin film transistors and flat panel liquid crystal displays, in the synthesis of circuits, in the theory of communication systems and in the optimization of systems.

Since spring 1991 he has also headed a new DM 80 million laboratory for the fabrication of flat panel displays. Research activities in this laboratory include TFT- and MIM-addressed TN-, PDLC- and GH-displays, as well as bistable FLC- and PSCT-displays.

He retired in 1999.

He was a member of the IEEE, and became an IEEE Fellow in 1985. As a member of the German Society for Information Technology, ITG, he was for two years a member of the society's board of directors. He served in the Scientific Advisory group for the Heinrich-Hertz Institute in Berlin, and was chairman of this group for four years. Starting in 1994 he participated as a member of SID board of directors, as a director of the Mid-Europe chapter, and as vice-president for Europe. Further, he was a member of the SPIE, ISHM, FKTG, the German society for broadcast and television technology, and the New York Academy of Sciences (NYAS).

In 1991 he was awarded the order of merit 1st Class of the Federal Republic of Germany, and in 1998 he became a Fellow of SID.

1

Introduction

Liquid Crystal Displays (LCDs) have established a firm foothold on the market as flat panel displays for computers, transportation, communication (especially in its mobile version), instrumentation and, in the future, with increasing importance for television. The understanding of LCDs requires knowledge about the various electro-optical effects of liquid crystal cells, and about the control of the grey shades and colours in the picture elements (pixels, or pels) by addressing circuits.

The electro-optical effects are based either on the propagation of polarized light through anisotropic liquid crystal cells, or on the propagation of unpolarized light through scattering cells.

The grey shade controlling voltage across each pixel is provided mainly by either passive matrix or active matrix addressing. In passive matrix addressing, the voltage in each pixel is generated by voltages at the end of the rows (or lines) and the columns of the display, whereas active matrix addressing uses Thin Film Transistors (TFTs) or Metal Insulator Metal (MIMs) devices as switches in each pixel.

Further topics are the fabrication of conductors, transparent electrodes, TFTs and MIMs with thin film technology, the generation of colour filters, and the assembly and bonding of liquid crystal cells.

It is not only manufacturers of LCDs but also the vast community of users which need to grasp the essence of the physics and engineering of LCDs. The understanding of these topics enables users to select the appropriate LCD for their application, to tailor the optic performance to their needs (e.g. by optimizing the waveform at the addressing circuit by the addition of performance-enhancing sheets, or by selecting the appropriate location of the external ICs for signal processing), as well as for storing and feeding in of the picture information.

Further, manufacturers and users should be enabled to judge the suitability of future developments and trends for their purposes.

The first section of this book presents an overview of the properties of liquid crystal materials and a phenomenological description of the most frequently used type of LCDs, the

TFT-addressed twisted nematic (TN) LCD. The aim is to familiarize readers with the main aspects of LCDs, to introduce most of the terminology, to establish an understanding without doing calculations, and to alleviate the subsequent more detailed discussion without losing the overall picture in which the details are embedded.

The remaining portions of the book are devoted to an analytical investigation of the electro-optic effects, and to an elaboration of the addressing schemes complemented by the manufacture of the thin film components.

Applications are centred around transmissive and reflective displays and light valves for projectors. Plastic substrates and printing of layers replacing vacuum processes are examples of an emerging new display technology.

2

Liquid Crystal Materials and Liquid Crystal Cells

2.1 Properties of Liquid Crystals

2.1.1 Shape and phases of liquid crystals

Most liquid crystals consist of molecules shaped like the rod in Figure 2.1(a). The direction of the long axis is called the *director*, given by the vector $\vec{n}$, which is an apolar vector as $\vec{n}$ and $-\vec{n}$ are equivalent. Rod-shaped molecules are also termed *calamitic*. Other shapes of molecules are disc-like or discotic, as in Figure 2.1(b), and lath-like.

We focus on calamitic (Bahadur, 1990; Demus *et al.*, 1998a,b) liquid crystals as they are the most important for applications. Below the melting point T_m they are solid, crystalline and anisotropic, whereas above the clearing point with temperature $T_c > T_m$ they are a clear isotropic liquid. In the mesophase in Figure 2.2 in between T_m and T_c, the material has the appearance of a milky liquid, but still exhibits the ordered phases shown in Figure 2.2. These phases are now described in the sequence given by increasing temperature. The first phase above T_m is the smectic C phase (smectic is derived from the Greek word for soap). As all smectic phases, it is ordered in two dimensions. The molecules are arranged with random deviations tilted to the plane of the layer. In the smectic A phase the directors of the molecules are again with random deviations perpendicular to the plane of the layer. Next to the clearing point, the nematic phase appears with only a one-dimensional order (nematic in Greek means a thread, indicating the thread-like defects in the material). All members of the mesophase are anisotropic, as is the solid phase.

Some more phases of minor importance for display applications are below the smectic C phase, the smectic B_{hex} phase (hexatic B phase), with the same layers as smectic C but a short range close packed hexagonal structure, in Figure 2.3(a) seen against the director $\vec{n}$; in this direction, the smectic C phase exhibits the irregular structure in Figure 2.3(b). The phases J, G, E, K and H are located above T_m, and are smectic-like soft crystals with a long range order.

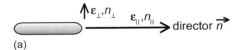

(a)

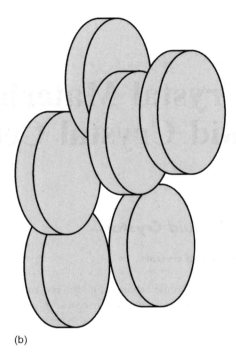

(b)

Figure 2.1 (a) Road-like or calamitic liquid crystal molecule with director n; (b) disclike or discotic liquid crystal molecules

The smectic C^* phase (chiral smectic C phase) in Figure 2.4 possesses a layered smectic structure in which the parallel directors of the molecules are rotated from layer to layer on the surface of a cone, resulting in a helix.

If chiral compounds such as cholesterol esters are added, the nematic phase changes to the cholesteric phase in Figure 2.5, which exhibits a helical structure in which, again, the director is rotated from layer to layer.

An as yet poorly understood peculiarity are the blue phases which occur in a small temperature range between the cholesteric and solid anisotropic phase.

More than 20 000 calamitic compounds are known.

Liquid crystals, the phases of which change with temperature, are called *thermotropic*. Those which change with the concentration of solvents and temperature are *lyotropic*. Calamitic and thermotropic liquid crystals are important for LCDs. Their nematic phase is the basis for both the most widely used Twisted Nematic (TN) cell with active matrix addressing, and for the SuperTwist Nematic (STN) cell with passive matrix addressing. Further LCDs based on calamitic and thermotropic nematic phases are Polymer Dispersed

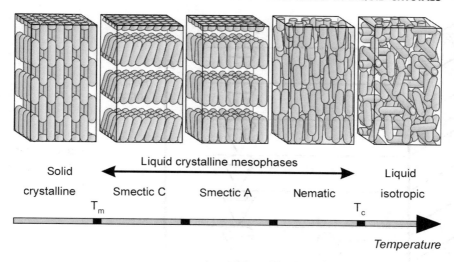

Figure 2.2   Phases of LC materials versus temperature

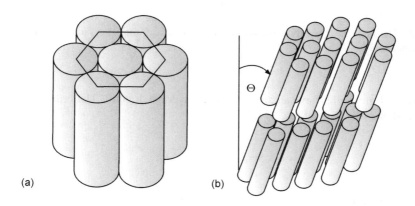

Figure 2.3   Top view of (a) the close packed hexagonal structure of the smectic B_{hex} phase, and (b) of the smectic C phase

Liquid Crystals (PDLC) and guest-host-LCDs. The smectic A and smectic C* phases provide bistable ferro-electric LCDs with passive matrix addressing. The cholesteric phase gave rise to the Stabilized Cholesteric Texture (SCT) with bistability at zero field. LCDs based on these phases will be discussed later.

To better understand electro-optical effects and electronic addressing, some materials properties have to be presented (Bahadur, 1990; Demus *et al.*, 1998a,b).

2.1.2 *Material properties of anisotropic liquid crystals*

The rod-like molecules have a head and a tail, which is, however, not taken into account by the direction of $\vec{n}$. Molecules in an unordered alignment exhibit an average director.

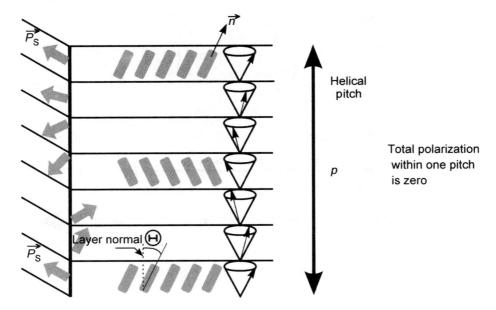

Figure 2.4 The helix in a layered structure of chiral smectic C liquid crystals with polarization $\vec{P}_s$ perpendicular to $\vec{n}$

The individual molecules have an angle Θ to this average director. The order parameter S of a phase is defined by (Tsvetkov, 1942)

$$S = \frac{1}{2}\langle 3\cos^2\Theta - 1\rangle, \tag{2.1}$$

where the bracket indicates that the average over a large number of molecules with angles Θ is taken. In a perfectly ordered state, $\Theta = 0$, and hence $S = 1$. A completely unordered phase has $S = 0$. In typical nematic phases, S lies in the region of 0.4 to 0.7, indicating that the molecules are rather disordered.

The energy needed for a phase transition, e.g., from smectic A to smectic C, is characterized by a transition enthalpy in kJ/mol. Extensive investigations of phase transitions have revealed the temperature dependance of physical parameters such as the helical pitch, the viscosity or the elastic coefficients.

Due to the ordered structure, all phases between T_m and T_c are anisotropic, meaning that all dielectric, optical and mechanical properties depend upon the direction.

The dielectric constant is $\varepsilon = \varepsilon_r\varepsilon_0$, where $\varepsilon_0 = 8.854 \cdot 10^{-14}$ F/m stands for the permittivity in vacuum and ε_r for the relative dielectric constant. This means, as shown in Figure 2.1, $\varepsilon_r = \varepsilon_\parallel$ in the direction parallel to the director and $\varepsilon_r = \varepsilon_\perp$ perpendicular to the director, leading to the dielectric anisotropy

$$\Delta\varepsilon = \varepsilon_\parallel - \varepsilon_\perp. \tag{2.2}$$

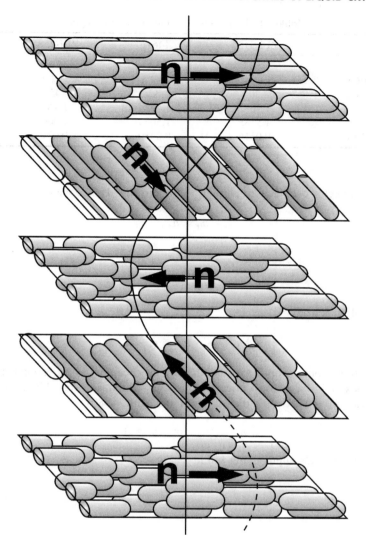

Figure 2.5 Helix of the cholesteric phase

Materials with $\Delta\varepsilon > 0$ are called p-type; their molecules align with the director parallel to the electric field, whereas in n-type materials with $\Delta\varepsilon < 0$, they align perpendicular to the field. This holds independent of the direction of the field vector. Values for $\Delta\varepsilon$ are found in the range from -0.8 to -6 and from 2 to 20. The addition of cyanogroups enlarges $\Delta\varepsilon$, whereas fluorine atoms in materials with $\Delta\varepsilon < 0$ lower $\Delta\varepsilon$ even further. Values for four materials are listed in Table 2.1.

The optical anisotropy Δn concerns the refractive indices n_0 for the ordinary beam of light, where the vector of the electrical field oscillates perpendicular to the optical axis that is perpendicular to the director and the refractive index n_e for the extraordinary beam of light,

Table 2.1 Properties of liquid crystal materials

	ZLI-3125	14616	ZLI-2585	14627
$T_C[°C]$	63	54	70	48
$\Delta\varepsilon$ (1kHz, 20°C)	+2.4	+2.3	−4.4	−3.5
η [mm^2/s] (20°C)	20	32	45	45
$n_0 = n_\perp$	1.4672	1.4554	1.469	1.4551
$n_e = n_\parallel$	1.5188	1.5034	1.506	1.4893
Δn (589 nm, 20°C)	0.0516	0.0480	0.037	0.0342

where the field vector oscillates in parallel to the director. Hence we obtain

$$n_0 = n_\perp, \tag{2.3}$$

and

$$n_e = n_\parallel, \tag{2.4}$$

and the optical anisotropy

$$\Delta n = n_\parallel - n_\perp = n_e - n_0. \tag{2.5}$$

More explanation about the optic axis and the ordinary beam will be given in Chapter 6. The refractive index n is based on optical frequencies which are very high. Therefore, the equation known from Maxwell's theory (Born and Wolf, 1980)

$$n = \sqrt{\varepsilon_r} \quad \text{for} \quad \mu_r = 1 \tag{2.6}$$

provides for frequencies approaching infinity:

$$\varepsilon_{r\parallel\infty} = n_\parallel^2, \tag{2.7}$$

$$\varepsilon_{r\perp\infty} = n_\perp^2 \tag{2.8}$$

and

$$\Delta\varepsilon_{r\infty} = n_\parallel^2 - n_\perp^2. \tag{2.9}$$

The refractive indices depend upon the wavelength λ. Values for Δn lie in the range $\Delta n \in [0.04, 0.45]$; some values are listed in Table 2.1. As a rule, materials with a high Δn are not stable to UV light. Due to the optical anisotropy, the material is birefringent. The speed of light is (Born and Wolf, 1980)

$$v = \frac{c}{\sqrt{n(\lambda)}}, \tag{2.10}$$

where c is the speed of light in vacuum. The speeds of light

$$v_\| = \frac{c}{\sqrt{n_\|(\lambda)}},$$
(2.11)

and

$$v_\perp = \frac{c}{\sqrt{n_\perp(\lambda)}},$$
(2.12)

where the E-vector oscillates parallel and perpendicular to the director, are different and dependent on the wavelength.

This is the key for the electro-optical effects in liquid crystal cells.

The direction with the larger refraction index $n_\|$ exhibits the smaller speed, and hence is called the slow axis, whereas $n_\perp$ defines the fast axis.

The dynamic behaviour of LC materials is affected greatly by the viscosity. Too high viscosities at lower temperatures slow down the movement of the molecules and yield the lower temperature limit of LC cells. The proximity to T_c provides the upper temperature limit. The dynamic viscosity η_d is defined as

$$\eta_d = \frac{F\,d}{A\,v} \quad \text{in} \quad \frac{Ns}{m^2} = Pa\,s,$$
(2.13)

where F is the force needed to shift a body with the area A with the velocity v over a viscous layer with a thickness d. For displays, the kinematic viscosity

$$\eta = \frac{\eta_d}{\delta} \quad \text{in} \quad \frac{mm^2}{s}$$
(2.14)

is used, where δ is the density of the viscous material. As for most LC materials, δ is around $1 Ns^2/mm^4$; the values for η_d and η do not differ much. The viscosity depends upon the orientation of the directors. For a random orientation, the bulk or turbid kinematic viscosity is given in Table 2.1. The rotational viscosity is measured according to Figure 2.6, where the vector of the rotation is perpendicular to the director. Values for dynamic rotational viscosities of LCs are 0.02 Pa s to about 0.5 Pa s. This viscosity is important for the movement of the director in an electric field.

The elastic constants belong to restoring torques if the field of directors is deformed. The three deformations from the equilibrium are splay, twist and bend, with the elastic constants K_{11}, K_{22} and K_{33}, as shown in Figure 2.7. The dimension is a force. The values are very small in the range of $10 \cdot 10^{-2} N$. These elastic forces determine the equilibrium in the presence of electric and magnetic fields.

A large variety of chemical compounds exhibit the properties of liquid crystals. The basic structure with rings, linking groups and terminal groups is shown in Figure 2.8. Rings can be cyclohexyl, pyridine, dioxane, phenylcyclohexane or phenyldioxane. Fluorinated compounds have a high specific resistance $\rho = 5 \cdot 10^{15}\,\Omega\,cm$. The characteristic temperatures of LC compounds can be shifted by additive ingredients. By this means, Merck's nematic compounds reached the wide temperature range of operation, from $-40°C$ to $120°C$, which is very suitable for automotive application. In Table 2.2 properties of LC materials with this wide temperature range are listed.

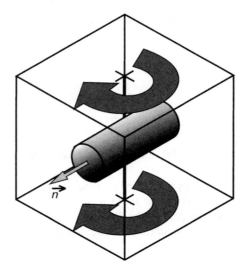

Figure 2.6 The rotational viscosity for rotation of a molecule perpendicular to the director

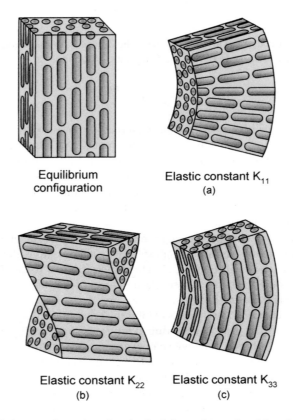

Figure 2.7 Equilibrium configuration; the elastic deformations splay (a), twist (b) and bend (c)

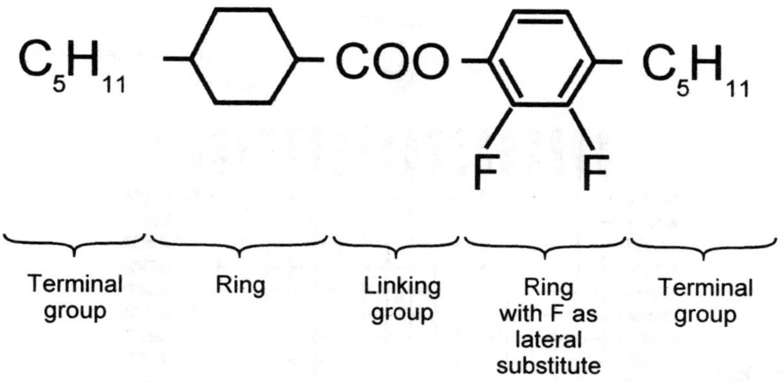

Terminal group — Ring — Linking group — Ring with F as lateral substitute — Terminal group

Figure 2.8 The basic structure of a calamitic LC molecule

Table 2.2 Properties of nematic LC materials with a wide temperature range

	MLC-1380000	MLC-13800100	MLC-1390000	MLC-13900100
Transition temp.				
smectic-nematic	$< -40°C$	$< -40°C$	$< -40°C$	$< -40°C$
Clearing pt T_c	110°C	111°C	110.5°C	110.5°C
Rotational				
viscosity, 20°C	228 mPas	151 mPas	235 mPas	167 mPas
$\Delta\varepsilon$ 1 kHz, 20°C	+8.9	+5.0	+8.3	+5.2
$n_0 = n_\perp$	1.4720	1.4832	1.4816	1.4906
$n_e = n_\parallel$	1.5622	1.5735	1.5888	1.5987
Δn	+0.0902	+0.0903	+0.1073	+0.1081

2.2 The Operation of a Twisted Nematic LCD

The liquid crystals used are calamitic and thermotropic in the nematic phase. The operation of this most widely applied LCD will be phenomenologically described in order to give an overview over the entire flat panel display system, including the addressing scheme (Demus et al., 1998a; Kaneko, 1987; Lueder, 1998a). This alleviates the more analytical and detailed treatments which follow.

2.2.1 The electro-optical effects in transmissive twisted nematic LC-cells

Figure 2.9 depicts the top view of a display panel with the conducting rows and columns terminating in the contact pads. The rectangular pixels can only be electrically addressed from those contact pads.

A colour VGA display, as used in laptops, has 480 rows and 3×320 columns forming triple dots for the three colours red, green and blue. An NTSC TV display has 484 rows and 3×450 columns corresponding to 653 400 pixels, whereas an HDTV display has

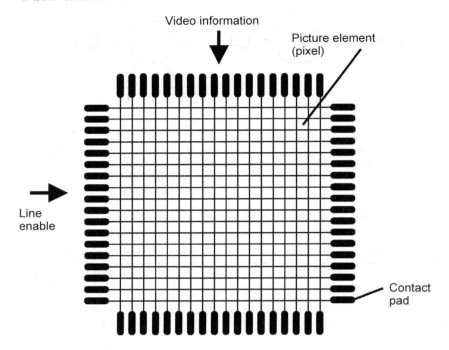

Figure 2.9 Top view of the rows, columns, pixels and contact pads of a display panel

$1080 \cdot 3 \cdot 1920 = 7\,320\,800$ pixels. For more standardized formats, see the table in Appendix 1.

Figure 2.10 shows a pixel of a transmissive twisted nematic LC-cell with no voltage applied. The white back light f passes the polarizer a. The light leaves it linearly polarized in the direction of the lines in the polarizer, and passes the glass substrate b, the transparent electrode c out of Indium-Tin-Oxide (ITO) and the transparent orientation layer g. This layer, made of an organic material such as polyimide, 100 nm thick, is rubbed to generate grooves in the direction of the plane of the polarized light. In these grooves the rod-like LC molecules are all anchored in parallel, but, as shown in Figure 2.11, with a pretilt angle α_0 to the surface of the orientation layer. The sequence of layers is the same on the second glass plate. A typical thickness of the cell in Figure 2.10 is $d = 3.5\,\mu$ to $4.5\,\mu$. The grooves on the second plate are perpendicular to those on the first plate. This forces the liquid crystal molecules to twist on a helix by $\beta = 90°$ from one plate to the other without the addition of chiral compounds. All twist angles are called β.

Due to the birefringence, the components of the electric field vector of the light in parallel and perpendicular to the directors travel with different speeds, which depend up-on the wavelength. They superimpose along their path between the two glass plates first to elliptically polarized light, in the distance $d/2$ from the input to circularly polarized light, then again to an elliptic polarization, and if

$$d = \frac{\sqrt{3}}{2} \frac{\lambda}{\Delta n} \qquad (2.15)$$

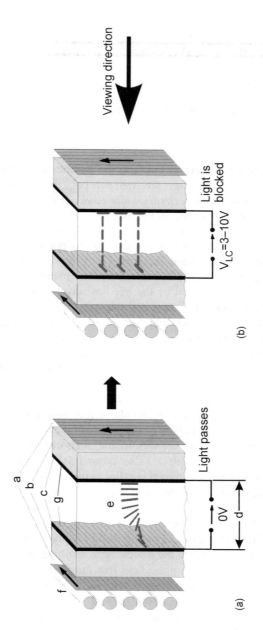

Figure 2.10 The structure of a TN-LCD (a) while light is passing, and (b) while light is blocked. a: polarizer; b: glass substrate; c: transparent electrode; g: orientation layer; e: liquid crystal; f: illumination

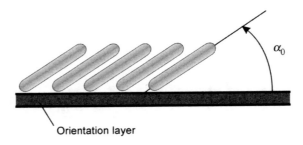

Figure 2.11   LC molecules with pretilt angle α_0 on top of the orientation layer

they reach the analyser again linearly polarized, but with the polarization plane rotated by 90°. If the analyser is crossed with the polarizer, the light can pass the analyser. The pixel appears white. This operation is termed the *normally white mode*. If the analyser is rotated by 90°, a parallel analyser, the light is blocked in the analyser. The pixel is black. This is called the *normally black mode*. A useful visualization of what happens to the light while travelling through the cell is as follows: the planes of the various polarizations follow the twist of the helix. This is, however, only true if Equation (2.15) holds. The explanation is also only true for light travelling and viewed perpendicular to the plane of the substrate. If viewed under a different angle, light perceived by the eye has travelled in a different path with different angles to the director and a different cell thickness d.

If a voltage V_{LC} of the order of 2 V is applied across the cell, as shown in Figure 2.10(b), using the two transparent ITO-electrodes 100 nm thick, the resulting electric field attempts to align the molecules for $\Delta\varepsilon > 0$ parallel to the field. This holds independent of the sign of the vector of the electrical field, as already pointed out in Section 2.1.2. Hence, the following effects are not dependent on the polarity of V_{LC}. Due to the anchoring forces, a thin LC layer on top of the orientation layers maintains its position almost parallel to the surfaces. A threshold voltage V_{th} is needed to overcome intermolecular forces before the twisted molecules start to rotate. A uniform start over the plane of the panel is favoured by a pretilt angle around 3°, which seems to avoid strong differences in the anchoring forces. Only at a saturation voltage V_{max} several times V_{th} with a value around 10 V have all molecules besides those on top of the orientation layers aligned parallel to the electric field, as depicted in Figure 2.10(b). In this state the vector of the electrical field of the incoming light oscillates perpendicular to the directors, and encounters only the refractive index $n_\perp$. Hence, no bi-refringence takes place and the wave reaches the crossed analyser in the same linearly polarized form as at the input. The analyser blocks the light and the pixel appears black. This is an excellent black state as it is independent of the wavelength, resulting in a blocking of the light. This black state is gradually reached from the field-free initial state by increasing the voltage V_{LC} from OV over an intermediate voltage up to V_{max}, which is also gradually rotating the molecules in Figure 2.12 from the initial twisted state with directors parallel to the surfaces (Figure 2.10(a)) over an intermediate state with the director already tilted down with tilt angle α (Figure 2.12(b)) to the final state with directors parallel ($\alpha = 90°$) to the electric field. The transmitted luminance, also termed *transmittance*, of the light is shown in Figure 2.13 for the normally white mode discussed so far. In the normally black mode, the analyser is parallel to the polarizer and allows the light to pass at the voltage

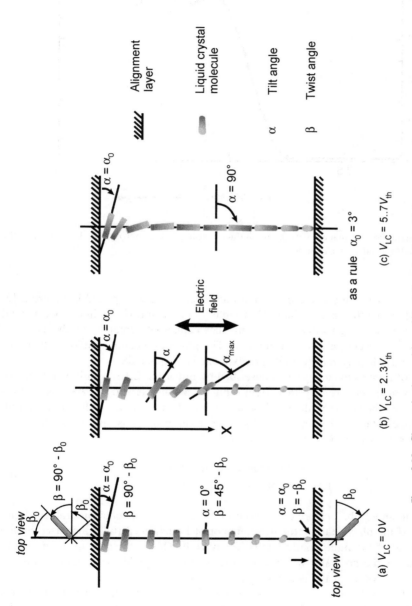

Figure 2.12 Change in the position of the LC molecules with increasing voltage

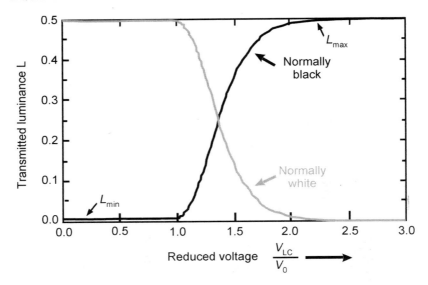

Figure 2.13 Transmitted luminance versus the reduced voltage V_{LC} across the LC cell for the normally white and normally black modes

$V_{th} \le V_{LC} \le V_{max}$. For this mode the transmitted luminance is also depicted in Figure 2.13. Only in this mode is the threshold voltage V_{th} visible, as in the normally white mode a small change in luminance at a high value of the luminance cannot be perceived by the eye.

Luminance is the correct term for 'brightness'. The physical meaning and dimensions of luminance and other display-related units are explained in Appendix 2.

The blocking of light in the analyser as described by Equation (2.15) is only valid for one wavelength for which, as a rule, yellow light with $\lambda = 505$ nm is chosen. As other wavelengths can still pass the analyser, the black state is not perfect. As a rule, it has a bluish tint. The imperfect black state can be improved by compensation foils, as discussed later.

The Contrast Ratio CR of a display is defined by

$$CR = \frac{\text{highest luminance in the pixels}, L_{max}}{\text{lowest luminance in the pixels}, L_{min}}. \tag{2.16}$$

The measurement should be performed without the interference of reflected ambient light, i.e. in darkness. If the black state in the denominator of Equation (2.16) is increased by the imperfect blocking of the light, contrast falls in any case. This is the case in the normally black state, whereas the normally white state described above yields an excellent contrast due to a much lower value of the denominator in Equation (2.16).

Grey shades of a pixel are controlled by the voltage V_{LC} in Figure 2.13, which modulates the luminance from a full but imperfect black up to a full white. Luminance differs when the display is viewed under angles different from perpendicular to the glass plates. Contrast decreases the more oblique the angles become.

The TFT addressing circuit will be placed on the glass next to the backlight in Figure 2.10. In a colour display, the glass plate facing the viewer carries the pixellized colour filter,

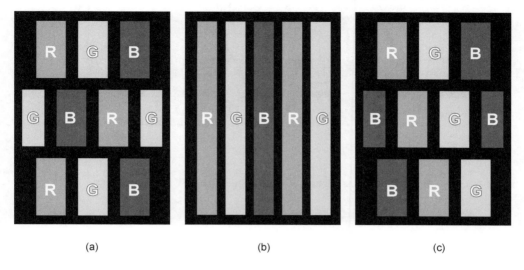

Figure 2.14 The geometrical arrangement of colour pixels for red R, green G, and blue B (a) in triangles, (b) in stripes and (c) in diagonal form

shown in Figure 2.14. The pixels for red, green and blue are covered with a compound which absorbs all wavelengths originating from the white backlight besides red, green and blue, respectively. The saturation of the colours is individually controlled for each pixel by the voltage V_{LC} in the same way as for grey shades.

The geometrical arrangement of the colour pixels in triangles in Figure 2.14(a) and along diagonals in Figure 2.14(c) are recommended for moving TV pictures, whereas the colour stripes in Figure 2.14(b) are preferred for computer displays often presenting rectangular graphs.

The cross-section of a colour filter in Figure 2.15 contains the colour materials for R, G and B, an absorptive layer with a low reflection in between the pixels, a so-called black matrix, an overcoat layer and, in the case of TFT addressing, an unpixellized ITO electrode over the entire display area. For other addressing schemes, the ITO layer is no longer unstructured. The ITO electrode on the TFT-carrying plate is pixellized. The black matrix prevents light between the pixels, which is neither controlled by the voltage V_{LC} at the ITO electrodes nor exhibits the desired colour, from seeping through the cell. This light would lighten up the black state, and would thus degrade contrast and the saturation of colour. A suitable material for a black matrix is an organic material with carbon particles exhibiting a reflectivity of only 4 percent, whereas the previously used Cr-oxide has a reflectivity of 40 percent. The overcoat layer (e.g. out of a methacrylate resin solution) equalizes the different heights of the colour pixels and protects them.

2.2.2   The addressing of LCDs by TFTs

So far we know that we have to control the grey shade individually in each pixel by applying the appropriate pixel voltage V_{LC}, but by only using the external contact pads in Figure 2.9.

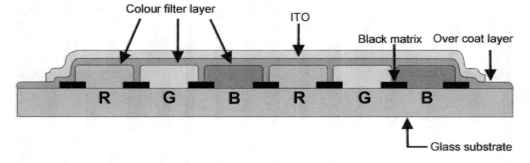

Figure 2.15   Cross-section of a colour filter for TFT addressed LCDs

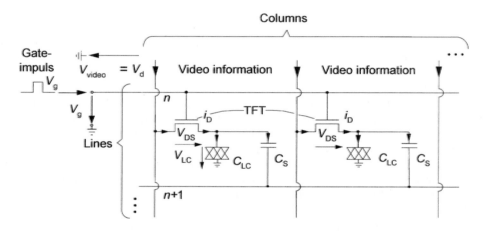

Figure 2.16   TFT addressing of the pixels in a row

The TFT-addressed LCD, usually called an Active Matrix LCD (AMLCD), solves this task as depicted in Figure 2.16. It shows two pixels of a row of pixels, with the row- and column-conductors and ground represented by the unstructured ITO electrode on the colour plate in Figure 2.15. The TFTs are n-channel Field Effect Transistors (FETs) fabricated with thin film technology. They operate as switches in the pixels. All TFTs in a row are rendered conductive by a positive gate impulse V_g. TFTs in other rows are blocked by grounding the rows. The video information is fed in through the columns and the conducting TFTs into all the pixels of a row simultaneously. More specifically, the video voltage V_d corresponding to a desired grey shade charges the LC-capacitor C_{LC} and an additional thin-film storage capacitor C_s up to the voltage V_d. This constitutes an amplitude modulation. The operation addresses one line at a time, as opposed to one pixel at a time, of the e-beam in CRTs. During the charging time, the storage capacitor connected to the succeeding line $n+1$ is grounded, and hence connected in parallel to C_{LC}. As this is no more true during other phases of the operation, degradations of the addressing waveform are introduced, as discussed later.

The pixel switches have to charge N rows in the frame time T_f in which a picture is written. Hence, the row-address time is

$$T_r = T_f/N. \tag{2.17}$$

The waveform of the pixel-voltage V_{LC} is depicted in Figure 2.17. In the time T_r, the storage capacitors are charged with the time constant

$$T_{on} = (C_{LC} + C_s)R_{on} \le 0.1 T_r = 0.1 \frac{T_f}{N}, \tag{2.18}$$

where R_{on} is the on-resistance of the TFT. The inequality guarantees that at the end of T_r, the voltage V_{LC} is only 1 percent below the desired voltage V_d in Figure 2.16. The TFTs need to be fast enough to make sure that even if their properties fluctuate, as indicated by dashed lines in Figure 2.17, they still charge the capacitors to the voltage V_d. After the time T_r, the transistor is blocked, but still has a finite off-resistance R_{off}. After T_f the row is addressed again and, the new picture information is fed in. During this time, the discharge of the capacitors should be small to provide a luminance as constant as possible. This yields an almost flicker-free picture, again as opposed to the CRT, where in the absence of storage the luminance of the phosphor decays after having shortly been hit by the e-beam. The constraint for the time constant T_{off} of the discharge is

$$T_{off} = (C_{LC} + C_s)R_{off} \ge 200 T_f, \tag{2.19}$$

which ensures a voltage drop of only 1 percent at T_f. From Equations (2.18) and (2.19), we obtain

$$\frac{R_{off}}{R_{on}} = \frac{I_{on}}{I_{off}} \ge 2000 N. \tag{2.20}$$

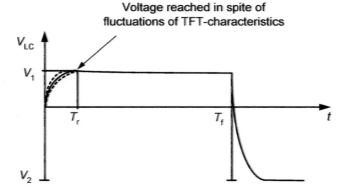

Figure 2.17 = Voltage reached in spite of fluctuations of TFT-characteristics

Figure 2.17 Waveform of the voltage across a pixel during charging and discharge of the storage capacitor

For an NTSC display with $N=484$, we require $R_{off}/R_{on} \geq 968 \cdot 10^3$. With the practically achievable value for the off-current $I_{off}=1\,pA$, the value for the on-current $I_{on} \geq 1\,\mu A$ meets the constraint in inequality (2.20).

The voltage across the pixels has to be free of dc in order to avoid dissociation of the ingredients in the LC material. Therefore, the voltage V_{LC} applied in the next frame time has the alternate sign as indicated in Figure 2.17. As we have observed in Section 2.2.1, the operation of the TN cell is independent of the sign of the electrical field vector. The alteration in sign from frame to frame is mandatory for a sufficient life of the LC cell. The transmitted luminance of an LC cell in Figure 2.13 exhibits a linear dependence on V_{LC} in a range of about 3 V. For 256 grey shades, this results in steps of $11.7\,mV/$grey shade. Special care will be taken to ensure the accuracy of these steps, as the voltages are altered by parasitic capacitive couplings. Remedies against this damage will be the alternating signs of V_{LC} from row to row, column to columns, frame to frame, and combinations of these measures.

TFT-addressed LCDs offer an appealing picture quality, mainly due to the absence of flicker. TFTs are fabricated together with the storage capacitors, ITO electrodes and conductors in thin film technology on glass or plastics as substrates. Glass substrates used to be 1.2 mm thick, and are presently down to 0.7 mm, whereas plastic materials excel in a thickness of only $100\,\mu$ to $200\,\mu$. As monocrystalline layers are unfeasible in thin film technology, the semiconductor in the TFTs is either amorphous or polycrystalline. Materials are a-Si, poly-Si and, more seldom, CdSe. Alternative pixel switches are MIMs or thin film diodes.

A more economical method of addressing does not require the above-mentioned pixel switches, and generates the voltages across the pixels by a superposition of voltages at the contact pads. This is called Passive Matrix (PM) addressing which, however, exhibits shortcomings.

A more detailed discussion of all addressing schemes starts with Chapter 10.

3

Electro-optic Effects in Untwisted Nematic Liquid Crystals

3.1 The Planar and Harmonic Wave of Light

A planar and harmonic wave of light (Lueder, 1998a; Born and Wolf, 1980; Yeh, 1988) is assumed to propagate through anisotropic liquid crystal layers with refractive indices $n_\parallel$ and $n_\perp$. Together with a polarizer and an analyser, they generate grey shades controlled by a voltage. For the derivation of these electro-optic effects, we first need to consider the basic equations for planar harmonic waves. A harmonic electric field $E(t)$ with amplitude E_0 is given by the scalar

$$E(t) = E_0 \cos(\omega t + \varphi), \tag{3.1}$$

where t is the time, ω the angular frequency and φ a phase angle. A travelling planar wave is a harmonic oscillation in a plane A in Figure 3.1 in the distance $z = \vec{r} \cdot \vec{s} = r \cos\Theta$ propagating in the direction of the unit vector $\vec{s}$ with the speed c_1; $\vec{r}$ is the vector from the origin to any location in the plane. Hence, replacing t by $t - ((\vec{r} \cdot \vec{s})/c_1)$ yields the equation for the planar harmonic wave in space and time as

$$E(t, r) = E_0 \cos\left(\omega\left(t - \frac{\vec{r}\,\vec{s}}{c_1}\right) + \varphi\right) = E_0 \cos\left(\omega t - \frac{\omega \vec{r}\,\vec{s}}{c_1} + \varphi\right). \tag{3.2}$$

Maxwell's equations provide for the speed of light in vacuum

$$c = \left(\sqrt{\mu_0 \varepsilon_0}\right)^{-1} \tag{3.3}$$

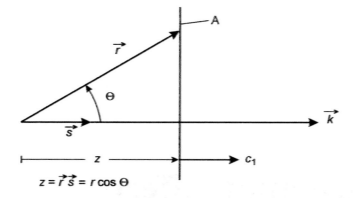

Figure 3.1 The plane A in which a planar wave travels with speed c_1 and wave vector $\vec{k}$ parallel to the normal $\vec{s}$

and in materials

$$c_1 = c \cdot \left(\sqrt{\mu_r \varepsilon_r}\right)^{-1},\qquad(3.4)$$

where μ_0, ε_0 are the permeability and permittivity in vacuum, and μ_r and ε_r are the pertinent relative constants in materials. Further, for $\mu_r = 1$ the refractive index is $n = \sqrt{\varepsilon_r}$ for $\mu_r = 1$, as known from Equation (2.6).

The wavelength λ in vacuum is given by

$$c = \lambda f,\qquad(3.5)$$

where $f = \omega/2\pi$ is the frequency. Equations (3.4), (2.6) and (3.5) provide for $\mu_r = 1$:

$$c_1 = \frac{c}{\sqrt{\varepsilon_r}} = \frac{c}{n} = \frac{\lambda f}{n}\qquad(3.6)$$

which results in the wave vector $\vec{k}$ from Equation (3.2) with Equation (3.6) as

$$\frac{\omega \vec{s}}{c_1} = \frac{2\pi}{\lambda} n \cdot \vec{s} = \vec{k}\qquad(3.7)$$

with

$$k = \frac{\omega}{c_1},\qquad(3.8)$$

and for vacuum

$$k_{\text{vac}} = \frac{\omega}{c}.\qquad(3.9)$$

The wave vector $\vec{k}$ parallel to $\vec{s}$ indicates the direction of the propagating wave. Its insertion in Equation (3.2) provides

$$E(t,r) = E_0 \cos(\omega t - \vec{k}\vec{r} + \varphi) \tag{3.10}$$

or

$$E(t,r) = E_0 \, Re \, e^{i(\omega t - \vec{k}\vec{r} + \varphi)} = E_0 \, Re \, e^{i\omega t} \, e^{i\varphi} \, e^{-i\vec{k}\vec{r}} \tag{3.11}$$

and with

$$A_0 = e^{i\omega t} \quad \text{and} \quad A_1 = e^{i\varphi} \tag{3.12}$$

$$E(t,r) = E_0 \, Re \, A_0 A_1 \, e^{-i\vec{k}\vec{r}}. \tag{3.13}$$

Equation (3.11) reveals that $\Phi = -\vec{k}\vec{r} + \varphi$ is the phase angle at distance r. The locus of constant phase Φ is determined by $\Phi = -\vec{k}\vec{r} + \varphi = const$, or as φ is a constant, by

$$-\vec{k}\vec{r} = const. \tag{3.14}$$

According to Figure 3.1, this is a plane perpendicular to $\vec{k}$, which is called the *surface of constant phase*, or the *wave surface* for short. The vector $\vec{E}$ lies in this plane, as assumed after Equation (3.1). Hence, $\vec{E}$ is perpendicular to $\vec{k}$. A more analytical statement follows after Equation (6.12).

In further calculations dealing with the propagation of the wave for varying r, the quantities A_0 and A_1 in Equation (3.13) can be omitted, as they do not change with r and have to be added again in the result, as given by Equation (3.13). Therefore, it is sufficient to deal only with the complex phasor

$$P_0 = E_0 e^{-i\vec{k}\vec{r}}, \tag{3.15}$$

which contains the amplitude E_0 and the change of phase $-\vec{k}\vec{r}$ in the distance r. Those phasors are widely used in optics (Born and Wolf, 1980), and in electric circuits fed with sinusoidal voltages in electrical engineering.

In the ξ-η-plane in Figure 3.2, the vector E of the electrical field is represented by the complex phasor P_0, with the complex components P_ξ and P_η as

$$P_0 = \xi_0 P_\xi + \eta_0 P_\eta, \tag{3.16}$$

where ξ_0 and η_0 are the unit vectors in the axes of the coordinates. The components are

$$P_\xi = P_0 \cos \beta \tag{3.17}$$

and

$$P_\eta = P_0 \sin \beta. \tag{3.18}$$

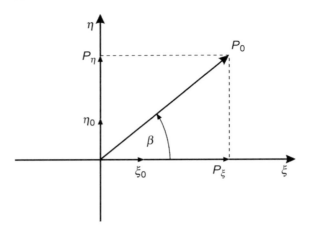

Figure 3.2 The phasor P_0 representing the vector E of an electrical field

In an anisotropic medium, the wave vector $\vec{k}$ in Equation (3.7) is different in the ξ- and η-direction, with the unit vectors $\vec{s}_\xi$ and $\vec{s}_\eta$, as the refraction indices n_ξ and n_η differ. Hence, we obtain from Equation (3.7) the wave vectors

$$\vec{k}_\xi = \frac{2\pi}{\lambda} n_\xi \vec{s}_\xi \tag{3.19}$$

and

$$\vec{k}_\eta = \frac{2\pi}{\lambda} n_\eta \vec{s}_\eta, \tag{3.20}$$

yielding with Equations (3.15), (3.16) (3.17) and (3.18)

$$P_\xi = E_0 \cos\beta \, e^{-i\vec{k}_\xi\vec{r}} = E_{\xi0} \, e^{-i\vec{k}_\xi\vec{r}} \tag{3.21}$$

and

$$P_\eta = E_0 \sin\beta \, e^{-i\vec{k}_\eta\vec{r}} = E_{\eta0} e^{-i\vec{k}_\eta\vec{r}}. \tag{3.22}$$

Note that $\vec{P}_\xi$ and $\vec{P}_\eta$ are perpendicular to $\vec{k}_\xi$ and $\vec{k}_\eta$, respectively.

The two components can be represented by a column vector, which is called the *Jones vector* (Jones, 1941):

$$J = \begin{pmatrix} J_\xi \\ J_\eta \end{pmatrix} = \begin{pmatrix} E_{\xi0} \, e^{-i\vec{k}_\xi\vec{r}} \\ E_{\eta0} \, e^{-i\vec{k}_\eta\vec{r}} \end{pmatrix}. \tag{3.23}$$

In a physical sense, J is not a vector as so far it consists only of two components and the vector product does not apply. It can also be termed a Jones matrix[1]. According to

[1] Usually only a 2×2 matrix is called a Jones matrix.

Equation (3.10), the scalars associated with the Jones vector are

$$E_\xi(t, r) = E_{\xi 0} \cos(\omega t - \vec{k}_\xi \vec{r} + \varphi) \tag{3.24}$$

and

$$E_\eta(t, r) = E_{\eta 0} \cos(\omega t - \vec{k}_\eta \vec{r} + \varphi). \tag{3.25}$$

In later investigations, the rectangular coordinates ξ and η will have to be rotated by an angle α into the new rectangular coordinates x and y, as shown in Figure 3.3. The vector E with the components E_ξ and E_η is transformed into the components E_x and E_y in the x-y-plane according to

$$E_x = \overbrace{E_\xi \cos\alpha}^{\text{I}} + \overbrace{E_\eta \sin\alpha}^{\text{II}} \tag{3.26}$$

$$E_y = -\overbrace{E_\xi \sin\alpha}^{\text{III}} + \overbrace{E_\eta \cos\alpha}^{\text{IV}} \tag{3.27}$$

if rotation by α is positive in the counter-clockwise direction. The roman numbers in Equations (3.26) and (3.27) indicate the sections marked in Figure 3.3, thus providing

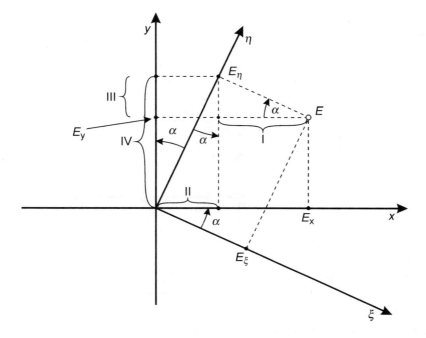

Figure 3.3 Rotation of the $\xi-\eta$ coordinates by α into the x-y coordinates

Equations (3.26) and (3.27). The matrix equation for (3.26) and (3.27) is

$$\begin{pmatrix} E_x \\ E_y \end{pmatrix} = \begin{pmatrix} \cos\alpha & \sin\alpha \\ -\sin\alpha & \cos\alpha \end{pmatrix} \begin{pmatrix} E_\xi \\ E_\eta \end{pmatrix} = R(\alpha) \begin{pmatrix} E_\xi \\ E_\eta \end{pmatrix}. \tag{3.28}$$

$R(\alpha)$ is the rotation matrix, and $R(-\alpha)$ stands for a rotation in clockwise direction.

3.2 Propagation of Polarized Light in Birefringent Untwisted Nematic Liquid Crystal Cells

3.2.1 The propagation of light in a Fréedericksz cell

We investigate the liquid crystal cell according to Fréedericksz (Fréedericksz and Zolina, 1933; Yeh and Gu, 1999) with top view in Figure 3.4(a) and the cross-section perpendicular to the top plane A in Figure 3.4(b). The vector E of the electric field in the plane A defined in the cartesian ξ-η-coordinates has the components using the notations in Equation (3.12)

$$E_\xi = E_{\xi 0} \cos(\omega t + \varphi) = E_{\xi 0}\, Re\, A_0 A_1 \tag{3.29}$$

and

$$E_\eta = 0 \tag{3.30}$$

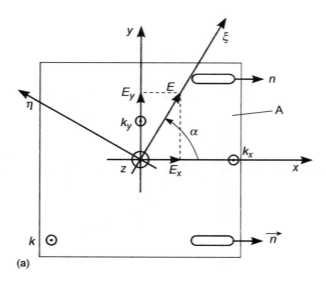

(a)

Figure 3.4 (a) Top view of Fréedericksz cell with direction of LC molecules and vector E of electric field; (b) cross-section of LC cell with parallel layers of molecules and wave vectors k, k_x and k_y

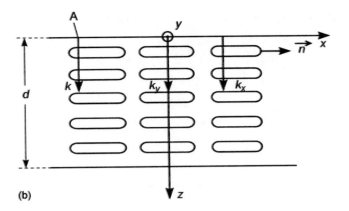

Figure 3.4 (continued)

with the Jones vector introduced in Equation (3.23)

$$J = \begin{pmatrix} J_\xi \\ J_\eta \end{pmatrix} = \begin{pmatrix} E_{\xi 0} \\ 0 \end{pmatrix}. \tag{3.31}$$

J represents a plane harmonic wave entering the cell at $z=0$ with wave vector k in Figure 3.4(b) perpendicular to A. The electric field E oscillates for all times t in a straight line; this is called a *linearly polarized wave*.

To obtain the components J_x and J_y of the Jones vector parallel and perpendicular to the director n of the parallel aligned LC molecules in the cell with thickness d in Figure 3.4(b), we rotate the ξ-η coordinates clockwise into the x-y -plane according to Equation 3.28 which provides

$$\begin{pmatrix} J_x \\ J_y \end{pmatrix} = R(-\alpha) \begin{pmatrix} E_{\xi 0} \\ 0 \end{pmatrix} = \begin{pmatrix} E_{\xi 0} \cos\alpha \\ E_{\xi 0} \sin\alpha \end{pmatrix}. \tag{3.32}$$

While propagating through the cell, the component J_x parallel to $\vec{n}$ experiences the refractive index $n_\parallel$ and the wave vector k_x perpendicular to A or parallel to the unit vector z_0 in Figure 3.4(b), whereas J_y sees the refractive index $n_\perp$ and the wave vector k_y parallel to z_0. The wave vectors are

$$\vec{k}_x = 2\pi \frac{n_\parallel}{\lambda} \vec{z}_0 \tag{3.33}$$

and

$$\vec{k}_y = 2\pi \frac{n_\perp}{\lambda} \vec{z}_0 = 2\pi \frac{n_\parallel}{\lambda} \vec{z}_0 - 2\pi \frac{\Delta n}{\lambda} \vec{z}_0, \tag{3.34}$$

where Equation (2.5) has been used.

The Jones vector in the plane in the distance $r = z \cdot z_0$ from the top plane in Figure 3.4(a) is

$$\begin{pmatrix} J_{zx} \\ J_{zy} \end{pmatrix} = \begin{pmatrix} J_x e^{-i\vec{k}_x \vec{r}} \\ J_y e^{-\vec{k}_y \vec{r}} \end{pmatrix} = \begin{pmatrix} J_x e^{-ik_x z} \\ J_y e^{-i(k_x z - 2\pi(\Delta n/\lambda)z)} \end{pmatrix}. \tag{3.35}$$

With

$$N_x = e^{-ik_x z} \tag{3.36}$$

and

$$N_y = e^{-i(k_x - 2\pi(\Delta n/\lambda)z)}, \tag{3.37}$$

we obtain the matrix equation by also inserting J_x and J_y from Equation (3.32):

$$\begin{pmatrix} J_{zx} \\ J_{zy} \end{pmatrix} = \begin{pmatrix} N_x & 0 \\ 0 & N_y \end{pmatrix} \begin{pmatrix} J_x \\ J_y \end{pmatrix} = \begin{pmatrix} N_x & 0 \\ 0 & N_y \end{pmatrix} R(-\alpha) \begin{pmatrix} E_{\zeta 0} \\ 0 \end{pmatrix}. \tag{3.38}$$

The evaluation of the result Equation (3.38) in the x'-y'-coordinates rotated from the x-y-plane by the angle γ requires the components $J_{zx'}$ and $J_{zy'}$, given as

$$\begin{pmatrix} J_{zx'} \\ J_{zy'} \end{pmatrix} = R(\gamma) \begin{pmatrix} J_{zx} \\ J_{zy} \end{pmatrix} = R(\gamma) \begin{pmatrix} N_x & 0 \\ 0 & N_y \end{pmatrix} R(-\alpha) \begin{pmatrix} E_{\zeta 0} \\ 0 \end{pmatrix}, \tag{3.39}$$

leading with Equations (3.36) and (3.37) to

$$J_{zx'} = E_{\zeta 0} \left(\cos\alpha\cos\gamma \, e^{-ik_x z} + \sin\alpha\gamma \, e^{-i(k_x z - 2\pi(\Delta nz/\lambda)} \right) \tag{3.40}$$

and

$$J_{zy'} = E_{\zeta 0} \left(-\cos\alpha\sin\gamma \, e^{-ik_x z} + \sin\alpha\cos\gamma \, e^{-i(k_x z - 2\pi(\Delta nz/\lambda)} \right). \tag{3.41}$$

From Equations (3.35) and (3.32), we derive the scalars according to Equation (3.10) for the components E_x and E_y in the distance z

$$E_x = E_{\zeta 0}\cos\alpha\cos(\omega t + \varphi - k_x z) \tag{3.42}$$

and

$$E_y = E_{\zeta 0}\sin\alpha\cos\left(\omega t + \varphi - k_x z + 2\pi\frac{\Delta n}{\lambda}z\right). \tag{3.43}$$

Equations (3.42) and (3.43) represent the electric field in the x-y-coordinates after having travelled the distance z through the Fréedericksz cell, whereas Equations (3.40) and (3.41) give the Jones vectors in the x'-y'-coordinates at distance z. These equations will be evaluated with respect to the amplitude modulation in the next two chapters, and the phase modulation in Section 3.2.4. In the remaining portion of this chapter, we investigate the polarization of the light while it travels through the Fréedericksz cell.

We start with Equations (3.42 and 3.43), and calculate the locus of the vector $\vec{E} = x_0\,E_x + y_0\,E_y$ in the x-y-plane as time evolves. To this aim, we eliminate $\tau = \omega t + \varphi$ and, after some algebraic steps presented in Appendix 3, reach the result (Born and Wolf, 1980)

$$\left(\frac{E_x}{A_x}\right)^2 + \left(\frac{E_y}{A_y}\right)^2 - 2\frac{E_x\,E_y}{A_x\,A_y}\cos\delta = \sin^2\delta \tag{3.44}$$

where

$$\delta = 2\pi\frac{\Delta n}{\lambda}\,z, \tag{3.45}$$

$$A_x = E_{\xi 0}\cos\alpha \tag{3.46}$$

and

$$A_y = E_{\xi 0}\sin\alpha. \tag{3.47}$$

This is a conic. From Equations (3.42) and (3.43), it can be seen that the conic must lie within the rectangular region bordered by the lines $x = \pm A_x$ and $y = \pm A_y$, as shown in Figure 3.5. Hence, the conic is an ellipse and light is called *elliptically polarized*. Equation (3.44) indicates that the principal axes of the ellipse are not parallel to the x- and y-axes; they

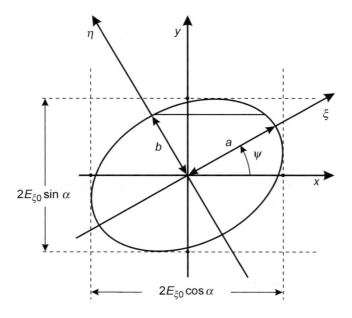

Figure 3.5 The ellipse as locus for the vector of the electric field

are parallel to the ξ- and η-axes in Figure 3.5, in which the equation for the ellipse is

$$\left(\frac{E_\xi}{a}\right)^2 + \left(\frac{E_\eta}{b}\right)^2 = 1. \tag{3.48}$$

The principal axes a and b are given by

$$a^2 = A_x^2 \cos^2 \Psi + A_y^2 \sin^2 \Psi + 2A_x A_y \cos\delta \cos\Psi \sin\Psi \tag{3.49}$$

$$b^2 = A_x^2 \sin^2 \Psi + A_y^2 \cos^2 \Psi - 2A_x A_y \cos\delta \cos\Psi \sin\Psi. \tag{3.50}$$

The angle Ψ for the rotation is determined by

$$\tan 2\Psi = \frac{2A_x A_y}{A_x^2 - A_y^2} \cos\delta. \tag{3.51}$$

A derivation for Equations (3.48) through (3.51) can be found in Born and Wolf (1980), and is repeated in Appendix 3.

The sense of revolution of the E-vector with increasing time is determined by Equations (3.42) and (3.43). If E_y is time-wise ahead of E_x, this happens for $0 < \delta < \pi$, i.e., for $\sin \delta > 0$ with δ in Equation (3.45), then the rotation of E is right-handed elliptically polarized, as shown in Figure 3.6; if E_y lags E_x, that happens for $\pi < \delta < 2\pi$ or $\sin \delta < 0$, the rotation is left-handed elliptically polarized (see Figure 3.6). The polarization dependent upon δ is depicted in Figure 3.7. The sense of revolution may be verified by plotting E_x and E_y with increasing time until the change of one of these components alters its sign.

Two special cases are the linear and the circular polarization. For

$$\delta = \nu\pi, \quad \nu = 0, \pm 1, \pm 2, \ldots, \tag{3.52}$$

Equation (3.44) degenerates into

$$\left(\frac{E_x}{A_x} - (-1)^\nu \frac{E_y}{A_y}\right)^2 = 0 \quad \text{or} \quad \frac{E_y}{E_x} = (-1)^\nu \frac{A_y}{A_x}. \tag{3.53}$$

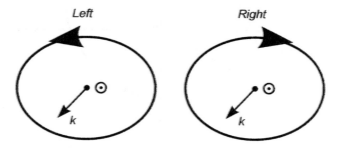

Figure 3.6 Right- and left-handed elliptically polarized light seen against the propagating wave with wave vector k. $\odot$ viewing against the arrow of k; $\otimes$ viewing in direction of the arrow

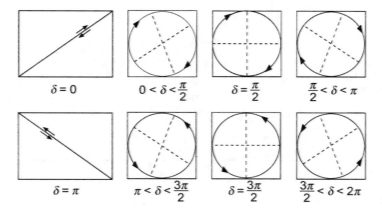

Figure 3.7 Elliptical, circular and linear polarization for different phase differences $\delta=2\pi(\Delta n/\lambda)z$ (Born and Wolf, 1980)

E_x and E_y lie on a straight line in Figure 3.7. This is the case of linearly polarized light. For $A_x=A_y$ and

$$\delta = \nu\frac{\pi}{2}, \quad \nu = \pm 1, \pm 3,\ldots, \tag{3.54}$$

Equation (3.44) provides the circle

$$E_x^2 + E_y^2 = A_x^2, \tag{3.55}$$

which represents circularly polarized light as also depicted in Figure 3.7.

Further characterizations of the polarization are given in Chapter 5. We are now ready to evaluate the results obtained so far for the Fréedericksz cell.

3.2.2   The transmissive Fréedericksz cell

In Figure 3.8 we continue the discussion of the transmissive cell begun with Figures 3.4(a) and 3.4(b). The incoming linearly polarized light enters at an angle α in Figure 3.8. The linear polarization occurs again according to Equation (3.52) at $\delta=2\pi(\Delta n/\lambda)z=\nu\pi$, $\nu=\pm 1,\pm 2,\ldots$, or for $\nu=1$ at the smallest z-value

$$z = \frac{1}{2}\frac{\lambda_0}{\Delta n}, \tag{3.56}$$

where λ_0 is the pertinent wavelength.

At the output of the cell for $z=d$, where d is the thickness of the cell, the retardation is

$$d\Delta n = \frac{\lambda_0}{2}. \tag{3.57}$$

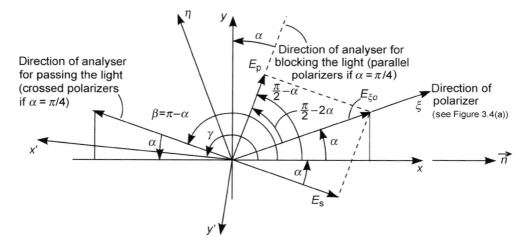

Figure 3.8 Angles of polarizer and analyser for the Fréedericksz cell

This retardation is associated with a change of phase by $\delta = \pi$ after the wave has propagated the distance d through the cell. The cell operates as a $\lambda/2$-plate. Obviously, the retardation is the phase shift measured in parts of λ_0.

For linear polarization with wavelength λ_0 corresponding to the angular frequency ω_0 of the electric field, the components in the Equations (3.42) and (3.43) at $z = d$ are

$$E_x = A_x \cos(\omega_0 t + \varphi - k_x d) \tag{3.58}$$

$$E_y = A_y \cos(\omega_0 t + \varphi - k_x d + \pi) = -A_y \cos(\omega_0 t + \varphi - k_x d). \tag{3.59}$$

Equations (3.58) and (3.59) reveal the angle β of the linearly polarized light at $z = d$ as

$$\frac{E_y}{E_x} = \tan\beta = -\frac{A_y}{A_x}. \tag{3.60}$$

Due to Equations (3.46) and (3.47), we obtain on the other hand for the light at $z = 0$

$$\frac{E_y}{E_x} = \frac{A_y}{A_x} = \tan\alpha. \tag{3.61}$$

Equations (3.60) and (3.61) indicate

$$\beta = \pi - \alpha. \tag{3.62}$$

as shown in Figure 3.8, where all important angles for a Fréedericksz cell are drawn.

The result for a wavelength λ_0 at the output $z = d$ of a cell without a voltage applied is linearly polarized light at an angle $\beta = \pi - \alpha$, where α is the angle of the incoming linearly polarized light. If we place the analyser in the direction $\beta = \pi - \alpha$ the light can pass representing the normally white mode. The analyser perpendicular to β that is at an angle $\pi/2 - \alpha$ in Figure 3.8 blocks the light representing the normally black mode. We will to investigate these two modes in greater detail.

We choose the angle γ for which the x'-y'-plane in Figure 3.8 is rotated from the x-y-plane as $\gamma = \beta = \pi - \alpha$. This provides, along with Equations (3.40) and (3.41),

$$J_{zx'} = E_{\xi 0}\left(-\cos^2\alpha\, e^{-ik_x z} + \sin^2\alpha\, e^{-ik_x z}\, e^{i2\pi(\Delta nz/\lambda)}\right) \tag{3.63}$$

$$J_{zy'} = E_{\xi 0}\left(-\cos\alpha\sin\alpha\, e^{-ik_x z} - \sin\alpha\cos\alpha\, e^{-ik_x z}\, e^{i2\pi(\Delta nz/\lambda)}\right) \tag{3.64}$$

or for the electrical field

$$E_{zx'} = E_{\xi 0}\left(-\cos^2\alpha\cos k_x z + \sin^2\alpha\cos\left(k_x z - 2\pi\frac{\Delta nz}{\lambda}\right)\right) \tag{3.65}$$

and

$$E_{zy'} = -E_{\xi 0}\cos\alpha\sin\alpha\left(\cos k_x z + \cos\left(k_x z - 2\pi\frac{\Delta nd}{\lambda}\right)\right). \tag{3.66}$$

For the wavelength $\lambda = \lambda_0$ in Equation (3.57), we obtain at $z = d$

$$J_{dx'} = -E_{\xi 0}\, e^{-ik_x d} \tag{3.67}$$

$$J_{dy'} = 0 \tag{3.68}$$

as expected, since we know already that at $z = d$ light with wavelength λ_0 is linearly polarized in the direction $\beta = \pi - \alpha$. For this case, the Jones vectors provide the components of the electrical field as

$$E_{dx'} = -E_{\xi 0}\cos(\omega t + \varphi - k_x d) \tag{3.69}$$

$$E_{dy'} = 0. \tag{3.70}$$

Now we place the analyser perpendicular to the angle $\beta = \pi - \alpha$ that is in the direction with angle $\gamma = \pi/2 - \alpha$ in Figure 3.8. For this case $J_{zx'}$ is identical to $-J_{zy'}$ in Equation (3.64) and (3.66) and $J_{zy'}$ is identical with $J_{zx'}$ in Equations (3.63) and (3.64). Hence, we investigate Equations (3.63) through (3.66) for both cases. The intensity[2] $I'_x = |J_{dx'}|^2$ for $z = d$ is, with Equation (3.63),

$$I_{x'} = E_{\xi 0}^2\left(\cos^4\alpha + \sin^4\alpha - 2\cos^2\alpha\sin^2\alpha\cos\frac{2\pi d\Delta n}{\lambda}\right) \tag{3.71}$$

---

[2]Intensity is $0.5|J_{dx'}|^2$, as factor 0.5 is not essential in this book, it is deleted.

or

$$I_{x'} = E_{\xi 0}^2 \left(1 - \sin^2 2\alpha \cos^2 \frac{\pi d \Delta n}{\lambda} \right). \tag{3.72}$$

For $I_{y'} = |J_{dy'}|^2$, Equation (3.64) yields

$$I_{y'} = E_{\xi 0}^2 \sin^2 \alpha \cos^2 \frac{\pi d \Delta n}{\lambda}. \tag{3.73}$$

In order to learn how to choose α, we now consider the case of a large enough voltage across the LC cell to fully orient the LC molecules apart from two thin layers on top of the orientation layer, due to $\Delta \varepsilon > 0$ in parallel to the electric field. The linearly polarized light coming in at angle α no longer experiences birefringence, as it is only exposed to the refractive index $n_\perp$. It reaches the plane $z = d$ with the phase shift $2\pi(n_\perp d / \lambda)$. Its component E_p passing an analyser with the angle $(\pi/2) - \alpha$ in Figure 3.8 is

$$
\begin{aligned}
E_p &= E_{\xi 0} \cos \left(\frac{\pi}{2} - 2\alpha \right) \cos \left(\omega t + \varphi - 2\pi \frac{n_\perp d}{\lambda} \right) \\
&= E_{\xi 0} \sin 2\alpha \cos \left(\omega t + \varphi - 2\pi \frac{n_\perp d}{\lambda} \right),
\end{aligned} \tag{3.74}
$$

whereas the component E_s passing an analyser with the angle $\pi - \alpha$ in Figure 3.8 is

$$E_s = E_{\xi 0} \cos 2\alpha \cos \left(\omega t + \varphi - 2\pi \frac{n_\perp d}{\lambda} \right). \tag{3.75}$$

The intensities belonging to E_p and E_s are

$$I_p = E_{\xi 0}^2 \sin^2 2\alpha \overline{\cos^2 \left(\omega t + \varphi - 2\pi \frac{n_\perp d}{\lambda} \right)} \tag{3.76}$$

and

$$I_s = E_{\xi 0}^2 \cos^2 2\alpha \overline{\cos^2 \left(\omega t + \varphi - 2\pi \frac{n_\perp d}{\lambda} \right)}. \tag{3.77}$$

The bar over the cos terms means the average over time needed for calculating the intensity. The maximum of I_p occurs for $\alpha = \pi/4$, which (according to Figure 3.8) places the polarizer and analyser in parallel. I_s assumes a maximum for $\alpha = 0$, for which again the polarizer and analyser pertaining to E_s are in parallel.

In Figures 3.9 and 3.10, the intensities $I_{y'}$ and $I_{x'}$ in Equations (3.73) and (3.72) are plotted versus $x = (\pi d \Delta n / \lambda)$ and $\lambda = (\pi d \Delta n / x)$. $I_{y'}$ in Figure 3.9 becomes zero at

$$x = \frac{\pi}{2} \tag{3.78}$$

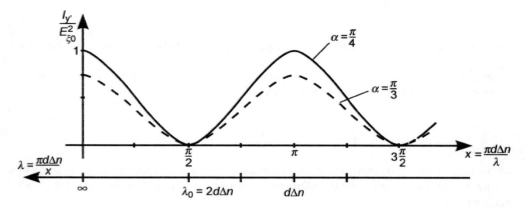

Figure 3.9 The intensity $I_{y'}$ of the Fréedericksz cell for two values of α in Equation (3.73)

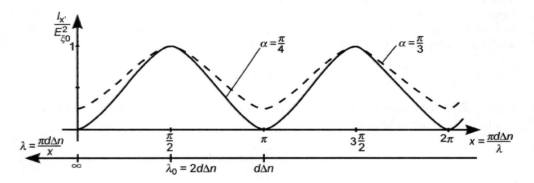

Figure 3.10 The intensity $I_{x'}$ of the Fréedericksz cell for two values of α in Equation (3.72)

or

$$\lambda = \lambda_0 = 2d\Delta n, \tag{3.79}$$

from which

$$d = \frac{\lambda_0}{2\Delta n} \tag{3.80}$$

follows. The function $\cos^2 x$ is lowered around $x = \pi/2$ by the multiplication with $\sin^2 2\alpha$. This is most welcome, as it enhances the black state which is imperfect by the suppression of only λ_0. This is demonstrated by two values for α in Figure 3.9. The intensity $I_{x'}$ in Equation (3.72) exhibits the same dependence on x as $I_{y'}$, and is plotted in Figure 3.10, demonstrating that at $x = \pi/2$

$$I_{x'} = E_{\xi 0}^2 \tag{3.81}$$

is the maximum intensity independent of α, which passes an analyser placed in the direction x'.

We are now ready to determine the contrast $C(\alpha)$ for the normally black and normally white cell as a function of α. C is defined as

$$C = \frac{L_{\text{max}}}{L_{\text{min}}}, \tag{3.82}$$

where L_{max} is the maximum luminance assumed to be proportional to the maximum intensity, whereas L_{min} stands for the minimum luminance assumed to be proportional to the minimum intensity.

We first investigate the normally white mode. In the field free state, the incoming linearly polarized light with angle α in Figure 3.11(a) again generates linear polarization for a wavelength λ_0 at the angle $\beta = \pi - \alpha$ in Equation (3.62) with the intensity $I_{x'}$ given in Equation (3.72) representing the white state. If a large enough field is applied, the light reaches the analyser linearly polarized in the direction α independent of λ. Hence, the analyser in Figure 3.11(a) allows the component

$$E_{\text{on}} = E_{\xi 0} \cos 2\alpha \neq f(\lambda) \tag{3.83}$$

to pass. This represents the black state. From Equations (3.72) and (3.83), we obtain the contrast in Equation (3.82) as

$$C = \sqrt{\frac{1 - \sin^2 2\alpha \cos \pi d \, \Delta n / \lambda}{\cos^2 2\alpha}}. \tag{3.84}$$

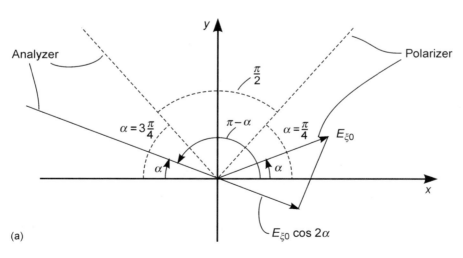

(a)

Figure 3.11 The angles of the electric field and the polarizers in a normally white Fréedericksz cell with linearly polarized light at the output $d = \lambda / 2 \Delta n$. (a) Crossed polarizers; (b) parallel polarizers

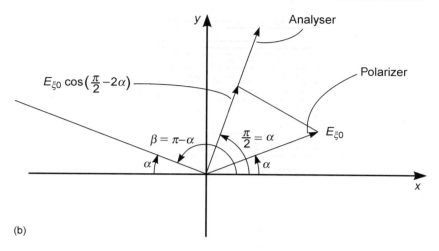

(b)

Figure 3.11 (continued)

The optimum contrast is reached for $\alpha = \pi/4$ for which $C \to \infty$ because the denominator in Equation (3.84) is zero for all wavelengths. Further, for $\alpha = \pi/4$ the numerator is maximum. The case $\alpha = \pi/4$ is shown with dotted lines in Figure 3.11(a). The analyser is perpendicular to the linear polarized light at the output if a field is applied, and hence provides blocking of light independent of λ.

The normally black mode is shown in Figure 3.11(b). The analyser is perpendicular to the angle $\beta = \pi - \alpha$, and allows the intensity $I_{y'}$ in Equation (3.73) to pass. This represents the black state. If a large enough field is applied, the light with the electrical field

$$E_{\text{on}} = E_{\xi 0} \cos\left(\frac{\pi}{2} - 2\alpha\right) = E_{\xi 0} \sin 2\alpha \tag{3.85}$$

independent of wavelength can pass the analyser according to Figure 3.11(b). This is the white state. The contrast is with Equations (3.85) and (3.73)

$$C(\lambda) = \frac{1}{|\cos \pi d \, \Delta n/\lambda|} \neq f(\alpha). \tag{3.86}$$

For the single wavelength λ_0 in Equation (3.79), C is infinite as λ_0 is blocked; this does not apply for other wavelengths in the light. Therefore, contrast in the normally black state is inferior to the contrast in the normally white state in Equation (3.84). An optimum C dependent on α does not exist. For $\alpha = \pi/4$ the normally black cell has two parallel polarizers. This configuration will be used for reflective cells.

3.2.3 The reflective Fréedericksz cell

A reflective cell is depicted in Figure 3.12(a) (Uchida, 1999; Lueder *et al.*, 1998c). The polarizer transmits linearly polarized light again at an angle of $\pi/4$ to the x-axis in Figure 3.12(b). The illumination is provided either by ambient light or by an external light source

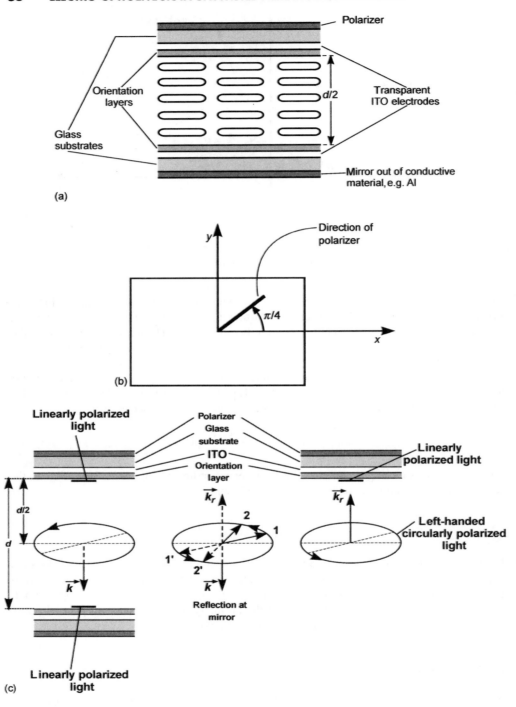

Figure 3.12 The reflective Fréedericksz cell. (a) Cross-section; (b) top view; (c) explanation of the operation of a reflective cell in the field-free state

above the polarizer. After having travelled through the cell in Figure 3.12(a) with half the thickness $d/2$ of a transmissive cell, the light is reflected at the mirror usually made of Al and finally exits through the same polarizer. Thus, the reflective cell saves one polarizer. The reflective cell can be designed according to the general principles, which will be outlined now (Lueder *et al.*, 1998). We first recall the operation of a Fréedericksz cell with parallel polarizers as depicted in the left column of Figure 3.12(c). We know that linearly polarized incoming light in the direction $\alpha=\pi/4$ to the x-axis is blocked at $z=d$ for wavelength λ_0. Due to Equation (3.80), the thickness is $d=\lambda_0/2\Delta n$. We determine at which value z the light is circularly polarized.

With Equation (3.45) this happens for the first time for $\delta=\pi/2$, reflecting in

$$z = \frac{\lambda_0}{4\Delta n} = \frac{d}{2}. \tag{3.87}$$

As $\sin\delta=\sin(\pi/2)>0$ the light is right-handed circularly polarized seen against $\vec{k}$, as indicated in the middle column of Figure 3.12(c). The highly conductive mirror in the middle column in Figure 3.12(c) reflects this light which is drawn with solid lines in two positions from 1 to 2 for increasing time. Since the mirror cannot sustain an electric field as the high conductance shortens the fields, the field vectors 1 and 2 have to be compensated by vectors $1'$ and $2'$ of the reflected light shown with dashed lines. The reflected light propagates in the direction of the wave vector $\vec{k}_r$, and it represents left-handed circularly polarized light seen against k_r. The upwards travelling reflected light in the right column in Figure 3.12(c) images the downward moving light in the left column, and is blocked in the polarizer as linearly polarized light, the same way as the downward travelling light is at its lower polarizer. The described imaging of the downward wave by the reflected wave can no longer take place if the mirror is not placed at $z=d/2$. Therefore, many reflective cells are constructed according to the principle discussed. If a voltage V is applied to the reflective cell, the LC molecules orient themselves in parallel to the electric field for $\Delta\varepsilon>0$. The linearly polarized light travels downward, is reflected, reaches the polarizer unchanged apart from a phase shift and passes the polarizer. Hence, the reflective cell exhibits the same electro-optical performance as the transmissive cell. The performance is given for the normally black Fréedericksz cell with parallel polarizers defined by $\alpha=\pi/4$ and Equations (3.73) and (3.85).

The normally white cell with crossed polarizers cannot be transformed into a reflective version as this version has only one polarizer able to realize only parallel polarizers. This fact, however, renders the reflective cell somewhat more economic as the added mirror is cheaper than the saved polarizer.

The surface of the mirror and the lower edge of the LC material in Figure 3.12(a) are supposed to be located at $z=d/2$, which is not exactly feasible because of the presence of the ITO and the orientation layers. As both layers are very thin, around 100 nm each, this does not show up in the performance of the cell.

3.2.4 The Fréedericksz cell as a phase-only modulator

So far we have treated the Fréedericksz cell as an amplitude modulator, as it is required for realizing grey shades in displays. The phase was of no importance. In coherent optical signal processing, the phase of the light wave is crucial; a voltage controlled phase shift without altering the amplitude is often required. A component with that performance belongs to the group of Spatial Light Modulators (SLMs). Another SLM is a pixellated optical multiplier in the form of an LCD placed behind a picture in Figure 3.13. Each area in front of the pixels of the multiplier is multiplied by the grey shade in those pixels. In other words, it is multiplied by a value 1 corresponding to a fully transparent pixel, a value 0 corresponding to a black pixel, and all values $\varepsilon(0, 1)$ corresponding to the grey shades. As this multiplication occurs with the speed of light and with all pixels operating in parallel, extremely high processing speeds are feasible with the SLMs of an electro-optical processor (Lu and Saleh, 1990).

The explanation of the SLM for phase-shifts starts with the most general Equations (3.40) and (3.41) of the Fréedericksz cell containing the arbitrary angle α of the polarizer at the input and the pertinent angle γ of the analyser (Figures 3.4(a) and 3.8).

The Equations (3.40) and (3.41) yield, for $\gamma=\alpha$ (that is, for parallel polarizers) the Jones vectors $J_{z\xi}$ and $J_{z\eta}$ measured in the ξ-η coordinates in Figure 3.4(a)

$$J_{z\xi} = \cos^2\alpha\,e^{-i2\pi(n_\| z/\lambda)} + \sin^2\alpha\,e^{-i(2\pi(n_\| z/\lambda)-2\pi(\Delta nz/\lambda))} \tag{3.88}$$

$$J_{z\eta} = -\cos\alpha\sin\alpha\,e^{-i2\pi(n_\| z/\lambda)} + \cos\alpha\sin\alpha\,e^{-i(2\pi(n_\| z/\lambda)-2\pi(\Delta nz/\lambda))} \tag{3.89}$$

The Jones vector $J_{z\xi}$ of the light passing through the analyser parallel to the polarizer is, for $\alpha=0$, π and $\alpha=\pi/2$ and for no voltage applied,

$$J_{z\xi/\alpha=0,\pi} = e^{-i2\pi(n_\| z/\lambda)} \tag{3.90}$$

and

$$J_{z\xi/\alpha=\pi/2} = e^{-i2\pi(z/\lambda)(n_\|-\Delta n)}. \tag{3.91}$$

The magnitude is constant, whereas the phase changes with the distance z from the input.

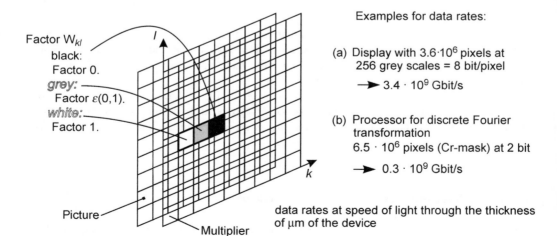

Factor W_{kl}
  black:
  Factor 0.
grey: ——
  Factor $\varepsilon(0,1)$.
white: ——
  Factor 1.

Picture

Multiplier

Examples for data rates:

(a) Display with $3.6\cdot10^6$ pixels at 256 grey scales = 8 bit/pixel
  → $3.4\cdot10^9$ Gbit/s

(b) Processor for discrete Fourier transformation
  $6.5\cdot10^6$ pixels (Cr-mask) at 2 bit
  → $0.3\cdot10^9$ Gbit/s

data rates at speed of light through the thickness of μm of the device

Figure 3.13 An LCD used as an SLM operating as a multiplier

$|J_{z\xi}|$ from Equation (3.88) is plotted in Figure 3.14 versus α and z. The constant magnitude 1 for $\alpha=0$, $(\pi/2)$ and π independent of z is visible as well as the maximum amplitude modulation for $\alpha=\pi/4$. However, we want arc $J_{z\xi}$ to change with the voltage V across the cell. To this aim, we consider the Fréedericksz cell in Figure 3.15, where a voltage has been applied to tilt the molecules by an angle φ. The linearly polarized light E_0 stemming from the polarizer with angle $\alpha=0$ to the x-axis has to meet the boundary condition at the transition from the polarizer into the cell. The tangential components have to be equal on both sides, which means they are E_0 in Figure 3.15. This indicates that the light wave has the wave vector $\vec{k}$ parallel to the z-axis. If no voltage V is applied, the LC-molecules are parallel to the surface and the x-axis, yielding

$$|k| = 2\pi \frac{n_{\|}}{\lambda}. \tag{3.92}$$

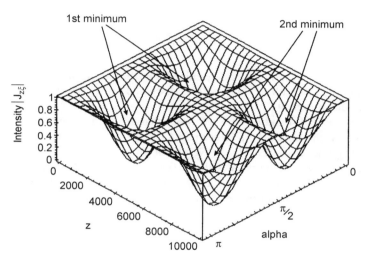

Figure 3.14 $|J_{z\xi}|$ in Equation (3.88) plotted versus α and z

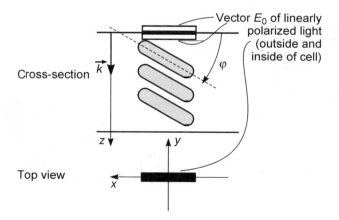

Figure 3.15 The linearly polarized light in parallel ($\alpha=0$) to the projection of the long axis of the LC molecules into the $x-y$ plane

With a large enough voltage V, all molecules are perpendicular to the surface or the x-axis, yielding

$$|k| = 2\pi \frac{n_\perp}{\lambda}. \tag{3.93}$$

From Equations (3.92) and (3.93), we detect the voltage induced change of the refraction index $n(V) \in \lfloor n_\perp, n_\parallel \rfloor$ with $n_\parallel$ for the lower voltage and $n_\perp$ for the higher voltage. Along with Equation (3.90), for a given wavelength λ_0 and for $z=d$, this provides

$$J_{z\xi}/\alpha{=}0,\pi = e^{-i(2\pi/\lambda_0)n(V)d} \tag{3.94}$$

with

$$arc\, J_{z\xi} = -\frac{2\pi}{\lambda_0}n(V)d. \tag{3.95}$$

How n changes with V cannot yet be determined. For that we need the propagation of light obliquely to the LC molecules, which will be discussed in Chapter 6. So far, $arc\, J_{z\xi}(V)$ is determined by measurements.

For $\lambda_0=0.5\,\mu$, $n_\perp=1.4$ and $n_\parallel=1.5$ we notice that $arc\, J_{z\xi}$ may change from $2.8 \cdot 2\pi \cdot d$ to $3.2\pi \cdot d$. Obviously, the larger d the larger is the change of $arc\, J_{z\xi}$, however, the necessary addressing voltage increases. For $d=3\mu$ we obtain $arc\, J_{z\xi} = \in [8.4 \cdot 2\pi, 9 \cdot 2\pi]$, a change of $0.6 \cdot 2\pi$. Figure 3.16 shows in curve 1 with $\alpha=0$ a phase-only SLM with a maximum phase shift of 0.75π. Due to Equation (3.88), the other curves with $\alpha \neq 0$ cannot possess a constant magnitude, as is also visible in Figure 3.14.

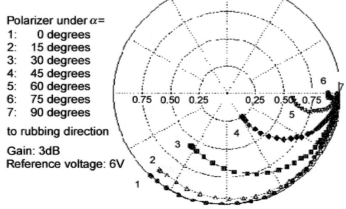

Figure 3.16 Measured phase-shift curves of a Fréedericksz cell

For $\alpha = \pi/2$ the linearly polarized light in Figure 3.15 always encounters the refractive index $n_\perp$ independent of the tilt, and hence is independent of V (curve 7 in Figure 3.16), which is of no use for a phase shifter.

3.2.5 The DAP cell or the vertically aligned cell

The cell operating with the deformation of aligned phases, called the DAP cell (Glueck, 1995), is the inverse of the Fréedericksz cell. In the field-free state the LC molecules are perpendicularly (or in other words, homeotropically) aligned to the surface of both substrates, as depicted in Figure 3.17. This cell is also called a Vertically Aligned (VA) LCD. In this situation, incoming linearly polarized light with a wave vector $\vec{k}$ in parallel to the z-axis in Figure 3.17 does not encounter birefringence, and arrives at the second substrate with an unchanged state of its polarization. If the analyser is parallel to the polarizer, the full light can pass representing the normally white state. If the analyser is crossed with the polarizer, the light is blocked at the output for all wavelengths and independent of d. This is the normally black state. The cell exhibits an extremely good black state, since the blocking is again independent of λ. Further, the molecules on the orientation layer are also, contrary to the Fréedericksz cell, vertically aligned. A low black value in the denominator of the contrast in Equation (3.82) is most beneficial for a high contrast. The main attraction of the DAP cell is this extremely high contrast, reaching values of more than 500 : 1.

If an electrical field is applied, the LC molecules orient themselves perpendicularly to the field as $\Delta\varepsilon < 0$. This alignment corresponds to the same alignment of the Fréedericksz cell

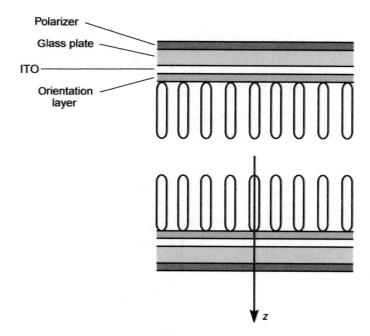

Figure 3.17 The DAP cell or Vertically Aligned (VA) cell in the field-free state

in the field-free state. Hence, all results in Equations (3.40) through (3.87) also apply to the DAP cell which is exposed to an electric field. The DAP cell is as well suited for phase-only modulators, as the pertinent Equations (3.90) and (3.95) also hold if a voltage V is applied. However, for the voltage-dependent refractive index $n(V)$, we obtain $n(V) \in \lfloor n_{\|}, n_{\perp} \rfloor$, but contrary to the Fréedericksz cell with $n_{\perp}$ for the lower voltage and $n_{\|}$ for the higher voltage. The homeotropic alignment of the molecules in the DAP cell requires special care. It is achieved by a spin-coated monomolecular silane-layer disolved in ethyl alcohol which is polymerized in the presence of humidity. The high polarity of silane thus generated anchors the polar LC molecules perpendicular to the surface. If a voltage is applied, all molecules are supposed to tilt in the same direction, since they have to end up all in parallel to each other and parallel to the plane of the substrates. This is realized by a small uniformly oriented pretilt of around 1° to 2° off the normal of the surface. A larger pretilt must be avoided, since it degrades the black state. The polymerized silane layer is uniformly rubbed with a carbon fibre brush to generate the grooves for the orientation of the molecules. As an alternative, this pretilted uniform orientation is produced with a very high manufacturing yield by a SiO_2 layer obliquely evaporated or sputtered under an angle of 2° off the normal. This alternative also achieves a very high contrast exceeding 500 : 1. The sputtering of this SiO_2 layer is explained in Figure 3.18. The DAP cell is, like a Fréedericksz cell, designed as a $\lambda/2$-plate with a retardation $\Delta nd = \lambda/2$, and hence $d = \lambda/2\Delta n$. For most commercially available LC materials exhibiting $\Delta n = 0.08$, this leads for $\lambda = 550$ nm to a cell thickness of $d = 3.4\,\mu$. The reflective version is a $\lambda/4$-plate with a thickness of $d = 1.7\,\mu$, which is often too thin for a high yield fabrication because small particles could easily cause shorts. The search for electro-optical effects with a larger cell thickness leads to the HAN cells and the Twisted-Nematic cells (TN-cells), which are covered in the next chapter and in Chapter 4.

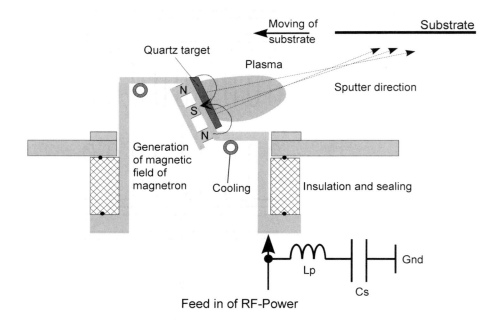

Figure 3.18 The sputtering of an SiO_2 orientation layer under an oblique angle of 70°

A reflective DAP cell with a thickness $d/2$ can be constructed in the same way as a Fréedericksz cell.

3.2.6 The HAN cell

The Hybrid Aligned Nematic cell (HAN cell) represents a mixture between the Fréedericksz cell and the DAP cell (Glueck, 1995). We investigate the reflective version in Figure 3.19(a), because among untwisted cells they have turned out to be more important. On the plate with the polarizer the LC molecules are in the field-free state in Figure 3.19(a) oriented all in the same direction parallel to the surface of the plate like in a Fréedericksz cell, whereas on the plate with the mirror they are homeotropically oriented. The linearly polarized light enters the cell at an angle $\alpha \neq 0$ to the x-axis, as in the transmissive and reflective Fréedericksz cells. The optical anisotropy Δn changes with z described by $\Delta n(z)$. At $z=0$ the light wave encounters the full anisotropy Δn, meaning that $\Delta n(0)=\Delta n$. This is only true for $\alpha \neq 0$. At $z=d$ the light encounters no anisotropy as the medium is isotropic with a refraction index $n_\perp$, reflecting in $\Delta n(d)=0$. Assuming a linear change of $\Delta n(z)$ we obtain $\Delta n(z)$ in Figure 3.19(b), leading to an effective retardation R of

$$R = \int_0^d \Delta n(z)dz = \frac{1}{2}\Delta nd. \tag{3.96}$$

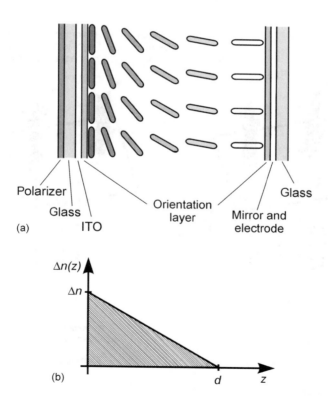

Polarizer · Glass · ITO · Orientation layer · Mirror and electrode · Glass

(a)

(b)

Figure 3.19 The reflective HAN cell. (a) Cross-section; (b) optical anisotropy $\Delta n(z)$

We know from Equations (3.45) and (3.54) that a retardation of $\lambda/4$ transforms linearly into circularly polarized light as desired at the mirror of a reflective cell. From $(1/2)\cdot\Delta nd=\lambda/4$ we obtain

$$d = \frac{\lambda}{2\Delta n}. \tag{3.97}$$

This is twice the thickness of the reflective Fréedericksz and DAP cells, resulting in a higher fabrication yield. This advantage of the HAN cell was brought about by lowering the effective bi-refringence. We shall encounter the same effect again with TN cells.

The reflection of the circularly polarized light in the field-free state is depicted in Figure 3.20(a) (Glueck, 1995). After the reflection the wave reaches the polarizer rotated by 90° and is blocked. If a field is applied in Figure 3.20(b), the molecules orient themselves due to $\Delta\varepsilon>0$ in parallel to the field, birefringence does not take place, and the reflected wave passes the polarizer. The cell is normally black.

For the homeotropic alignment of the molecules, an obliquely evaporated or sputtered SiO_2 orientation layer is again a good solution.

3.2.7 The π cell

So far we have not dealt with the time needed to switch an LC cell from the black state to the white state, or vice versa. Dynamics of a cell are based on mechanical properties of the

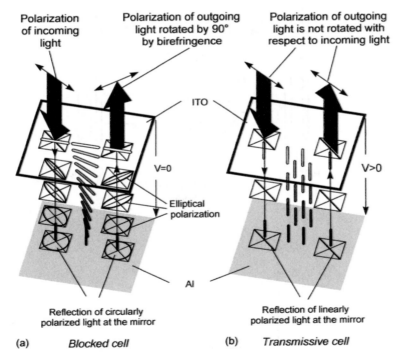

(a)   *Blocked cell*   (b)   *Transmissive cell*

Figure 3.20 The operation of a reflective HAN cell. (a) In the field-free state; (b) if a voltage is applied

LC material, and will be discussed in Section 3.2.8. Some information on switching speed can, however, be derived from the field of directors. That's how a very fast switching cell, the π cell, has been found (Bos and Koehler, 1984), as will be outlined later in this chapter.

The Fréedericksz cell and the TN cell exhibit a uniform alignment of the molecules with a pretilt α in the opposite direction on the substrates, as shown in Figure 3.21(a) for the on-stage of the cell. After the electric field has been switched off, the molecules relax to the off-state resulting in a flow of LC material and a tilting of the molecules. The molecules around the centre of the cell are exposed to a torque, causing a back-flow of the material and trying to rotate them through a large angle to the position in the off-state. This slows down the switching process. Figure 3.21(b) depicts the pretilt of the π cell pointing in the same direction on both substrates. In this case, the molecules in the centre of the cell experience almost no torque while they relax into the off-stage. The angles by which the relaxing molecules have to be tilted are smaller, resulting in a much higher switching speed.

The phenomenon of a 'back-flow' causes the 'optical bounce' in the electro-optical response to a rectangular voltage in Figure 3.22(a), whereas the response of the π cell in Figure 3.22(b) does not exhibit the prolonged relaxation time. In a TN cell switching off takes, as a rule, 3 to 4 ms, whereas the π cell requires only around 1 ms. However, after 1 ms the relaxation is not yet completely finished, which may not be noticeable at a high enough switching frequency. If the uncompleted relaxation is disturbing, a holding voltage of around 2V or a polymer stabilization (Vithana and Faris, 1997) arrests the relaxation.

As in the HAN cell, the optical anisotropy Δn in the π cell for $V_{LC}=0$ changes along the z-axis, which is denoted by $\Delta n(z)$. As a consequence, the optical retardation R from the input at $z=0$ to the output at $z=d$ is

$$R = \int_{z=0}^{d} \Delta n(z)dz, \qquad (3.98)$$

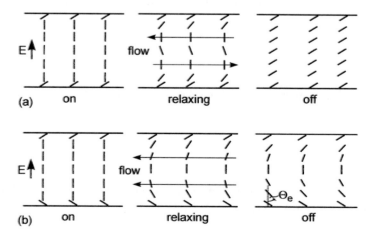

Figure 3.21 The pretilt angles and the relaxation to the off-state. (a) Of a TN cell; (b) of a π cell

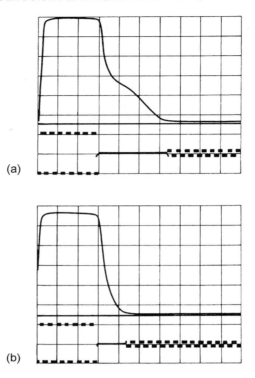

Figure 3.22 The electro-optical response to a square voltage pulse. (a) Of a TN cell with a prolonged relaxation; and (b) of a π cell with a fast relaxation

where $\Delta n(z)$ depends upon the angle Θ_e in Figure 3.21(b). We shall derive $\Delta n(\Theta_e)$ in Chapter 6, leading to Equation (6.29). For a π cell with crossed polarizers, the transmission T is obtained by a straightforward calculation with Jones vectors, yielding

$$T = \frac{1}{2}\sin^2\frac{\pi}{\lambda}R. \tag{3.99}$$

A normally white cell has the first transmission maximum at $(\pi/\lambda)R=(\pi/2)$, or

$$R = \frac{\lambda}{2}, \tag{3.100}$$

corresponding to a $\lambda/2$-plate. We introduce an effective anisotropy Δn_{eff} for the entire optical path in the cell, which is given by

$$\Delta n_{\mathrm{eff}}\,d = R, \tag{3.101}$$

providing, with Equation (3.100),

$$d = \frac{\lambda}{2} \frac{1}{\Delta n_{\text{eff}}}.$$ (3.102)

The term Δn_{eff} renders this result similar to Equation (3.97).

The wide viewing angle inherent of the π cell is discussed in Section 6.3.2.

3.2.8 Switching dynamics of untwisted nematic LCDs

We assume that the LC molecules are anchored on the surface at $z=0$ at an angle Θ_0 to the normal and at $z=d$ under the angle Θ_{d}, as shown in Figure 3.23. In the field-free state the field of directors is defined by the equilibrium state with minimum free energy. After applying an electric field in the form of a step function the voltage $V(t)$ across the cell has to exceed a threshold V_{th} before the molecules are able to rotate in order to assume the position imposed by the field. The threshold is caused by the intermolecular forces, which first have to be overcome by the forces of the field. The transition to the new voltage imposed field of directors is called the *Fréedericksz transition*. The dynamic of this transition (Degen, 1980; Priestley, Wojtowicz and Sheng, 1979) is governed by the interaction between the electric torques forcing the directors into positions parallel for $\Delta\varepsilon>0$ or perpendicular for $\Delta\varepsilon<0$ to the electric field and the mechanical torques trying to restore the field-free state. These torques are the only mechanical influences if the molecules do not undergo a translatory movement. A magnetic field is as a rule not applied in LC applications. The transient between the states of the director field is calculated by adding all free energies, and by taking the functional first derivative with respect to the angle Θ in Figure 3.23. The torques are related to splay, twist and bend with the elastic constants K_{11}, K_{22}, K_{33} and the rotational kinematic viscosity η, as well as to the dielectric torque dependent on $\Delta\varepsilon$. For the derivation of the results we refer to special publications (Labrunie and Robert, 1973; Saito and Yamamoto, 1978). In Saito and Yamamoto (1978), expressions for the rise time T_r and the decay time T_d for the reorientation of the LC molecules induced by a voltage step with amplitude V were derived. The results depend upon the tilt angles Θ_0 and Θ_d of the

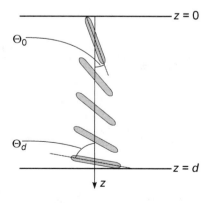

Figure 3.23 The anchoring of LC molecules at $z=0$ and $z=d$

molecules. T_r and T_d translate directly in the rise time and the decay time of the luminance as it changes directly with the director field. The results for general angles Θ_d and Θ_0 are:

$$
T_r = \frac{\eta d^2}{\pi^2 K_{11}} + \left|\left| \left\{ 1 + K\cos^2\left(\frac{\Theta_d + \Theta_0}{2}\right) + K\left(\frac{\Theta_d - \Theta_0}{\pi}\right)^2 \cos\left(\Theta_d + \Theta_0\right) - \frac{\Delta\varepsilon}{\pi K_{11}} g_b V^2 \right\} \right.\right.
$$

$$
+ \left\{ \left[1 + K\left(\frac{\pi}{\Theta_d - \Theta_0}\right)^2 \right] \sin\left(\Theta_d + \Theta_0\right) - \frac{\Delta\varepsilon}{\pi^2 K_{11}} g_a V^2 \right\}
$$

$$
\left. \left. \times \left[K\left(\frac{\Theta_d - \Theta_0}{\pi}\right)^2 \sin\left(\Theta_d + \Theta_0\right) + 2\frac{\Delta\varepsilon}{\pi^2 K_{11}} g_c V^2 \right] \right|\right|^{-1/2}
\tag{3.103}
$$

with

$$
K = \frac{K_{33} - K_{11}}{K_{11}},
$$

$$
g_a = \frac{1}{2R^2}\left[E - \frac{2}{R}(PG + QF) + \frac{3Q^2 G}{R^2} \right], \qquad g_b = \frac{1}{2R^2}\left(F - \frac{2QG}{R} \right), \qquad g_c = \frac{G}{2R^2}
$$

$$
P = \int_0^1 \left\{ \frac{\varepsilon\cos 2(\Delta\Theta Z_n + \Theta_0)}{[I + \varepsilon\cos^2\{\Delta\Theta Z_n + \Theta_0\}]^2} + \frac{\varepsilon^2 \sin^2 2(\Delta\Theta Z_n + \Theta_0)}{[1 + \varepsilon\cos^2(\Delta\Theta Z_n + \Theta_0)]^3} \right\} \sin^2(\pi Z_n) dZ_n
$$

$$
Q = -\int_0^1 \left\{ \frac{\varepsilon\sin(\Theta Z_n + \Theta_0)}{[I + \varepsilon\cos^2(\Delta\Theta + \Theta_0)]^2} \right\} \sin(\pi Z_n) dZ_n
$$

$$
R = \int_0^1 \frac{1}{I + \varepsilon\cos^2(\Delta\Theta Z_n + \Theta_0)} dZ_n, \qquad G = \frac{\sin(\Theta_d + \Theta_0)}{[I + \varepsilon\cos^2((\Theta_d + \Theta_0)/2)]^3}
$$

$$
F = \frac{2}{[1 + \varepsilon\cos^2(\Theta_d + \Theta_0)/2]^2}\left(\cos(\Theta_d + \Theta_0) + \frac{\varepsilon\sin^2(\Theta_2 + \Theta_0)}{I + \varepsilon\cos^2((\Theta_d + \Theta_0)/2)} \right)
$$

$$
\Delta\Theta = \Theta_d - \Theta_0, \qquad \varepsilon = \frac{\Delta\varepsilon}{\varepsilon_\perp} \quad \text{and} \quad Z_n = \frac{z}{d}.
$$

For $\Theta_d = \Theta_0 = \pi/2$ we obtain the rise time for the Fréedericksz cell as

$$T_r = \frac{\eta d^2}{\pi^2 K_{11}} \left(\frac{\Delta \varepsilon}{\pi^2 K_{11}} V^2 - 1 \right)^{-1} \quad \text{for} \quad V > V_{\text{th}} = \pi \sqrt{\frac{K_{11}}{\Delta \varepsilon}} \qquad (3.104)$$

and for $\Theta_d = \Theta_0 = 0$ the rise time of the DAP cell as

$$T_r = \frac{\eta d^2}{\pi^2 K_{33}} \left(\frac{|\Delta \varepsilon|}{\pi^2 K_{33}} V^2 - 1 \right)^{-1} \quad \text{for} \quad V > V_{\text{th}} = \pi \sqrt{\frac{K_{33}}{|\Delta \varepsilon|}}. \qquad (3.105)$$

These two results have already been published in Labrunie and Robert (1973).

The threshold voltage in both cases can be detected from the denominators of T_r in the Equations (3.104) and (3.105) as points where T_r becomes infinite. Obviously, T_r increases with the viscosity η and the square of the thickness d independent of Θ_d and Θ_0. Figure 3.24 shows the normalized rise time $T_{rn} = T_r/\eta d^2/\pi^2 K_{11}$ versus the normalized voltage $V_n = V/\pi\sqrt{K_{11}/\Delta\varepsilon}$ calculated from Equation (3.103) for a p-type nematic with $\Delta\varepsilon = 0.55$ and $K = 0.16$ for various angles Θ_d and Θ_0. The Fréedericksz cell (planar cell) with $\Theta_d = \Theta_0 = 90°$ exhibits a larger rise time than all of the other cells, including the HAN cell with $\Theta_0 = 0$ and $\Theta_d = \pi/2$. The pronounced decrease of T_{rn} at $V_n^2 > 1$, as shown in Equation (3.104), is also clearly visible in Figure 3.24(a). Figure 3.24(b) depicts the normalized rise time $T_{rn} = T_r/\pi\sqrt{K_{33}/|\Delta\varepsilon|}$ versus the normalized voltage $V_n = V/\pi\sqrt{K_{33}/|\Delta\varepsilon|}$, again calculated from

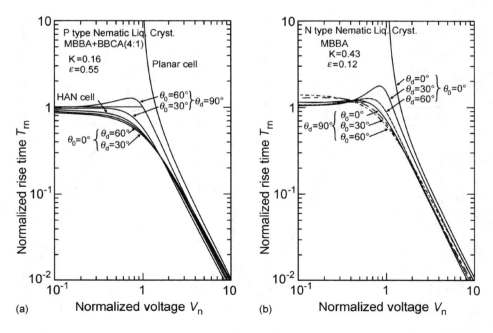

Figure 3.24 Normalized rise time T_{rn} versus normalized voltage V_n with various tilt angles θ_d and θ_0. (a) For p-type and (b) n-type nematic LCs

Equation (3.103), but this time for an n-type nematic LC with $\Delta\varepsilon = -0.12$ and $K=0.43$ for various angles Θ_d and Θ_0. In this case, the rise time of the DAP cell with $\Theta_d=\Theta_0=0$ exceeds the rise time of all other cells. Thus, in both cases, the Fréedericksz cell and the DAP cell are slower than all of the other cells with different combinations of pretilt angles. The decrease of T_{rm} with increasing V_n again takes place only for $V_n^2 > 1$.

Finally, Figure 3.25(a) and 3.25(b) depict T_{rm} versus V_n with $K=(K_{33}-K_\parallel)/K_\parallel$ as a parameter for a p-type and an n-type nematic LC. The Fréedericksz cell in Figure 3.25(a) and the DAP cell in Figure 3.25(b) are independent of K and slower than all the HAN cells with different values of K. The shorter rise time of the HAN cell over the other cells can phenomenologically be explained by the fact that half of the molecules are already rotated in the direction imposed by the field, horizontally for $\Delta\varepsilon < 0$ and vertically for $\Delta\varepsilon > 0$. The decay time T_d is derived in Saito and Yamamoto (1978) as

$$
T_d = \frac{\eta d^2}{\pi^2 K_{11}} \left| \left\{ 1 + K\cos^2\frac{\Theta_d + \Theta_0}{2} + K\left(\frac{\Theta_2 - \Theta_0}{\pi}\right)^2 \cdot \cos\left(\Theta_d + \Theta_0\right) \right\}^2 \right.
$$

$$
\left. + 2K^2\left(\frac{\Theta_d - \Theta_0}{\pi}\right)^2\left(1 + \left(\frac{\Theta_d - \Theta_0}{\pi}\right)^2\right)\sin^2(\Theta_d + \Theta_0) \right|^{-1/2},
$$

(3.106)

which is independent of the applied voltage V and of $\Delta\varepsilon$, and has the same factor outside the magnitude sign as T_r in Equation (3.103). For $\Theta_d=\Theta_0=\pi/2$ we obtain T_d of the

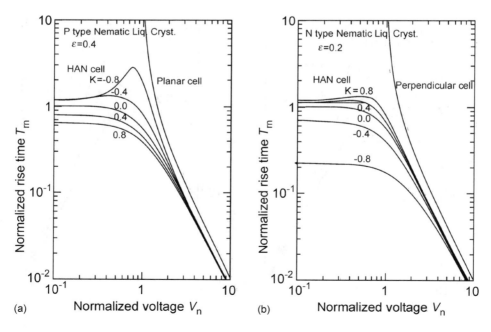

Figure 3.25 Normalized rise time T_{rm} versus normalized voltage V_n with the ratio K of elastic constants as parameter (a) for p-type and (b) n-type nematic LCs

Fréedericksz cell as

$$T_d = \frac{\eta d^2}{\pi^2 K_{11}},$$ (3.107)

and for $\Theta_d = \Theta_0 = 0$ T_d of the DAP cell as

$$T_d = \frac{\eta d^2}{\pi^2 K_{33}}$$ (3.108)

whereas the decay time for the HAN cell is obtained by putting $\Theta_d = \pi/2$ and $\Theta_0 = 0$, yielding

$$T_d = \frac{\eta d^2}{\pi^2 K_{11} \sqrt{(I + K/2)^2 + 5K^2/8}}.$$ (3.109)

A comparison between the HAN cell and the Fréedericksz cell which is valid for p-type nematic LCs reveals for the same cell-thickness

$$T_{dn} = \frac{T_{dHAN}}{T_{dFreed}} = \left(\sqrt{(1 + K/2)^2 + 5K^2/8} \right)^{-1}.$$ (3.110)

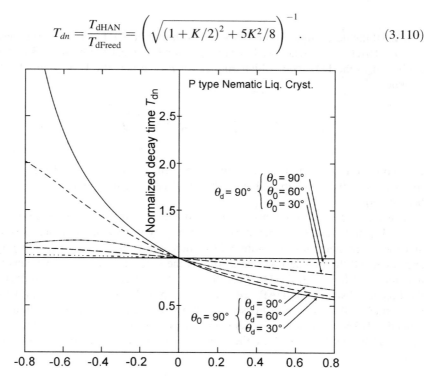

Figure 3.26 The ratio T_{dn} in Equation (3.110) versus K for a p-type nematic LC

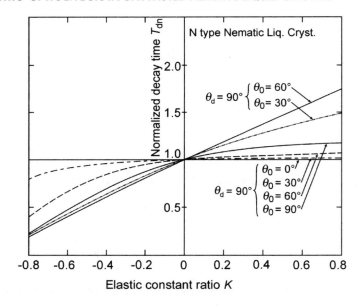

Figure 3.27 The ratio T_{dn} in Equation (3.111) versus K for a n-type nematic LC

For $K > 0$ the decay time of the HAN cell is shorter, and for $-1 < K < 0$ longer than that of the Fréedericksz cell, whereas they are equal for $K = 0$ reached by $K_{11} = K_{33}$. Comparing the HAN cell to the DAP cell, which applies for n-type nematic LCs, yields for the same cell thickness

$$T_{dn} = \frac{T_{dHAN}}{T_{dFreed}} = \left(\sqrt{(1 + K/2)^2) + 5K^2/8} \right)^{-1} (K + 1). \qquad (3.111)$$

In contrast to the Fréedericksz cell, the decay time of the HAN cell for $K > 0$ is longer, and for $-1 < K < 0$ shorter than that of the DAP cell. Again, for $K = 0$ the two decay times become equal.

The ratios T_{dn} in Equations (3.110) and (3.111) are plotted in Figures 3.26 and 3.27 versus K with Θ_d and Θ_0 as parameters. For the p-type nematic in Figure 3.26, T_{dn} decreases, and for n-type nematics in Figure 3.27 it increases with increasing K. For p-type nematics and for $K > 0$ the Fréedericksz cell exhibits the longest decay time, whereas for the n-type nematics and for $K > 0$ the DAP cell has the shortest decay time. The reflective version of the Fréedericksz and the DAP cell with cell gap $d/2$ have rise times and decay times four times smaller than their transmissive counterparts because of the proportionality to d^2. The reflective HAN cell requires the same thickness d as the transmissive Fréedericksz and DAP cells, and hence does not share the enhancement of switching speed of the other reflective cells.

4

Electro-optic Effects in Twisted Nematic Liquid Crystals

4.1 The Propagation of Polarized Light in Twisted Nematic Liquid Crystal Cells

The Twisted Nematic cell (TN cell) is the most widely commercially used LC cell. It was proposed by Schadt and Helfrich (1971), and is therefore also termed the Schadt–Helfrich cell. The theoretical investigation is based on Jones' vectors. Solutions for the light exiting a TN cell were given by Yeh (1998), Yeh and Gu (1999), Grinberg and Jacobson (1976) and Rosenbluth *et al.* (1998). The derivation relies on rotating back the coordinate system to the original coordinates in twisted media and on the Chebychev identity of matrices (Bodewig, 1959). On the other hand, the derivation of the results presented here rotates the coordinates with the twist of the layers. The further calculation is based on Specht (2000).

The planar wave with wave vector $\vec{k}$ at the input of the cell propagates along the z-axis in Figure 4.1. We investigate the propagation through the cell without considering the light reflected at the LC molecules or absorbed in the cell. The incoming light is linearly polarized in the ξ-direction at an angle α to the x-axis with the electrical field $E_{\xi 0}$ giving the Jones vector

$$J_1 = \begin{pmatrix} J_x \\ J_y \end{pmatrix} = R(-\alpha) \begin{pmatrix} E_{\xi 0} \\ 0 \end{pmatrix} \tag{4.1}$$

in the x-y-coordinates. The LC molecules in the x-y-plane are all anchored in the rubbing grooves parallel to the x-direction. The directors of all molecules in the cell are parallel to the x-y plane and form a helix with the z-direction as the axis, and with the linear

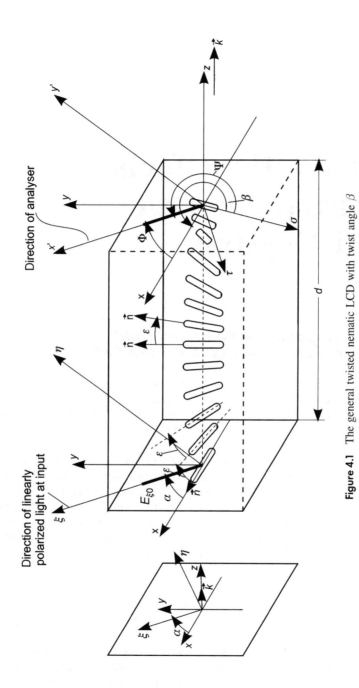

Figure 4.1   The general twisted nematic LCD with twist angle β

twist angle

$$\varphi = \alpha_0 z, \tag{4.2}$$

a pitch p given by

$$\alpha_0 p = 2\pi, \tag{4.3}$$

and a twist angle

$$\beta = \alpha_0 d \tag{4.4}$$

at $z=d$. For calculations, the helix is cut into slices parallel to the x-y plane. In each slice, all molecules are assumed to be parallel. The slices are rotated from the previous slice by the angle ε. The angle ε corresponds to a thickness d_ε with

$$\frac{d_\varepsilon}{\varepsilon} = \frac{p}{2\pi}$$

or

$$d_\varepsilon = \frac{p}{2\pi}\varepsilon \tag{4.5}$$

for each slice. The twist angle at $z=d$ is from Equation (4.4) with α_0 in Equation (4.3)

$$\beta = \frac{2\pi d}{p}, \tag{4.6}$$

whereas the number of slices in the cell is with Equation (4.5)

$$s = \frac{d}{d_\varepsilon} = \frac{2\pi d}{p\varepsilon}. \tag{4.7}$$

The Jones vector J_1 at the input is translated into the vector O_1 at the output of the first slice with thickness d_ε. Its component J_{1x} is parallel to the x-axis and the component J_{1y} is parallel to the y-axis, and hence (as already known from the Fréedericksz cell), without the need to rotate the input vector, we obtain

$$O_1 = \begin{pmatrix} J_{1x} \\ J_{1y} \end{pmatrix} = \begin{pmatrix} e^{-ik_x d_\varepsilon} & 0 \\ 0 & e^{-ik_y d_\varepsilon} \end{pmatrix} \begin{pmatrix} J_x \\ J_y \end{pmatrix}$$
$$= e^{-i2\pi(\bar{n}/\lambda)d_\varepsilon} \begin{pmatrix} e^{-i\pi(\Delta n/\lambda)d_\varepsilon} & 0 \\ 0 & e^{i\pi(\Delta n/\lambda)d_\varepsilon} \end{pmatrix} \begin{pmatrix} J_x \\ J_y \end{pmatrix} \tag{4.8}$$

or

$$O_1 = T(\varepsilon)\begin{pmatrix} J_x \\ J_y \end{pmatrix} \tag{4.9}$$

with

$$\bar{n} = \frac{n_\parallel + n_\perp}{2},$$

where $\vec{k}_x$ and $\vec{k}_y$ in Equations (3.33) and (3.34) were used.

The further propagation of the light through the $s-1$ remaining rotated slices is depicted in Figure 4.2 with the rotation matrices

$$R_\nu(\varepsilon) = \begin{pmatrix} \cos\varepsilon & \sin\varepsilon \\ -\sin\varepsilon & \cos\varepsilon \end{pmatrix} \nu = 2, 3 \ldots s - 1 \tag{4.10}$$

and the transmission matrices

$$T_\nu(\varepsilon) = T(\varepsilon) = e^{-i2\pi(\bar{n}/\lambda)d_\varepsilon} \begin{pmatrix} e^{-i\pi(\Delta n/\lambda)d_\varepsilon} & 0 \\ 0 & e^{i\pi(\Delta n/\lambda)d_\varepsilon} \end{pmatrix} \nu = 2, 3 \ldots s. \tag{4.11}$$

The Jones vector O_s at the output at $z=d$ measured in the coordinates σ and τ with the angle β to the x-axis in Figure 4.1 is, with Equations (4.1) and (4.9)

$$O_s = \begin{pmatrix} O_{s\sigma} \\ O_{s\tau} \end{pmatrix} = (T(\varepsilon)R(\varepsilon))^{s-1} \cdot T(\varepsilon)R(-\alpha) \begin{pmatrix} E_{\xi 0} \\ 0 \end{pmatrix} \tag{4.12}$$

or with Equations (4.10), (4.11), d_ε in Equation (4.5), s in Equation (4.7) and

$$a = \frac{\Delta n}{2\lambda} p \tag{4.13}$$

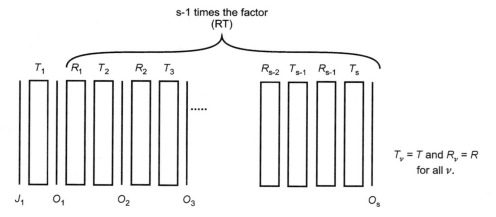

Figure 4.2 The propagation of light from the Jones vector J_1 at the input to the Jones vector O_s at the output through the transmission matrices T_ν and the rotation matrices R_ν

$$\begin{pmatrix} O_{s\sigma} \\ O_{s\tau} \end{pmatrix} = \left[e^{-i(\bar{n}/\lambda)p\varepsilon} \begin{pmatrix} e^{-ia\varepsilon} & 0 \\ 0 & e^{ia\varepsilon} \end{pmatrix} \begin{pmatrix} \cos\varepsilon & \sin\varepsilon \\ -\sin\varepsilon & \cos\varepsilon \end{pmatrix} \right]^{(2\pi d/p\varepsilon)}$$
$$R^{-1}(\varepsilon)T^{-1}(\varepsilon)R(-\alpha)\begin{pmatrix} E_{\xi 0} \\ 0 \end{pmatrix} \tag{4.14}$$

We want to determine the lim for $\varepsilon \to 0$ of this expression, providing an infinitesimally small thickness of the slices, and hence the exact solution. From Equation (4.14), we obtain

$$\begin{pmatrix} O_{s\sigma} \\ O_{s\tau} \end{pmatrix} = e^{-i2\pi(\bar{n}d/\lambda)} \lim_{\varepsilon \to 0}\left[\begin{pmatrix} e^{-ia\varepsilon} & 0 \\ 0 & e^{ia\varepsilon} \end{pmatrix} \begin{pmatrix} \cos\varepsilon & \sin\varepsilon \\ -\sin\varepsilon & \cos\varepsilon \end{pmatrix} \right]^{(2\pi d/p\varepsilon)}$$
$$\lim_{\varepsilon \to 0}\begin{pmatrix} \cos\varepsilon & -\sin\varepsilon \\ \sin\varepsilon & \cos\varepsilon \end{pmatrix} R(-\alpha)\begin{pmatrix} E_{\xi 0} \\ 0 \end{pmatrix}. \tag{4.15}$$

For an easier calculation of the lim for $\varepsilon \to 0$, we transform

$$T(\varepsilon)R(\varepsilon) = \begin{pmatrix} e^{-ia\varepsilon} & 0 \\ 0 & e^{ia\varepsilon} \end{pmatrix} \begin{pmatrix} \cos\varepsilon & \sin\varepsilon \\ -\sin\varepsilon & \cos\varepsilon \end{pmatrix} = \begin{pmatrix} e^{-ia\varepsilon}\cos\varepsilon & e^{-ia\varepsilon}\sin\varepsilon \\ -e^{ia\varepsilon}\sin\varepsilon & e^{ia\varepsilon}\cos\varepsilon \end{pmatrix} \tag{4.16}$$

into a diagonal matrix

$$D = M^{-1}T(\varepsilon)R(\varepsilon)M, \tag{4.17}$$

where D contains the eigenvalues, and the columns of M are the eigenvectors of $T(\varepsilon)R(\varepsilon)$ (Specht, 2000). The eigenvalues $\xi_{1,2}$ are obtained by $|T(\varepsilon)R(\varepsilon) - \xi I| = 0$, where I is the unity matrix, as

$$\xi_{1,2} = \cos a\varepsilon \cdot \cos\varepsilon \pm i\sqrt{1 - \cos^2 a\varepsilon \cos^2\varepsilon} \tag{4.18}$$

providing

$$D = \begin{pmatrix} \xi_1 & 0 \\ 0 & \xi_2 \end{pmatrix}. \tag{4.19}$$

Since $|\xi_1| = |\xi_2| = 1$, the eigenvalues can be rewritten as

$$\xi_1 = e^{i\varphi} = e^{i\arctan\sqrt{1 - \cos^2 a\varepsilon \cos^2\varepsilon}/\cos a\varepsilon \cos\varepsilon} \tag{4.20}$$

and

$$\xi_2 = e^{-i\varphi} = e^{-i\arctan\sqrt{1 - \cos^2 a\varepsilon \cos^2\varepsilon}/\cos a\varepsilon \cos\varepsilon} \tag{4.21}$$

with

$$\tan\varphi = \frac{\mathrm{Im}\xi_1}{\mathrm{Re}\xi_1}. \tag{4.22}$$

On the other hand, ξ_1 and ξ_2 in Equation (4.19) describe the transmission of a wave through a slice with the thickness d_e in Equation (4.5). Therefore, with the wave vector $\vec{k'}$ of this slice, we obtain

$$\xi_1 = e^{ik'd_e} = e^{i \arctan \sqrt{1 - \cos^2 \alpha \varepsilon \cos^2 \varepsilon} / \cos \alpha \varepsilon \cos \varepsilon} \tag{4.23}$$

and

$$\xi_2 = e^{-ik'd_e} = e^{-i \arctan \sqrt{1 - \cos^2 \alpha \varepsilon \cos^2 \varepsilon} / \cos \alpha \varepsilon \cos \varepsilon} \tag{4.24}$$

with

$$k' = \frac{\arctan\left(\sqrt{1 - \cos^2 \alpha \varepsilon \cos \varepsilon} / \cos \alpha \varepsilon \cos \varepsilon\right)}{p \varepsilon / 2\pi}. \tag{4.25}$$

This finally leads to

$$D = \begin{pmatrix} e^{-ik'd_e} & 0 \\ 0 & e^{ik'd_e} \end{pmatrix} \tag{4.26}$$

with k' in Equation (4.25) and d_e in Equation (4.5).

The eigenvectors $V=(V_1, V_2)$ with the components $V_{x1,2}$ and $V_{y1,2}$ are calculated from Equation (4.16) by solving

$$(T(\varepsilon)R(\varepsilon) - \xi_i I)V_i = 0 \quad i = 1, 2$$

as

$$\begin{pmatrix} V_{x1,2} \\ V_{y1,2} \end{pmatrix} = \begin{pmatrix} e^{-ia} \sin \varepsilon \cdot r_{1,2} \\ (\xi_{1,2} - e^{-ia\varepsilon} \cos \varepsilon)r_{1,2} \end{pmatrix} \tag{4.27}$$

with the two arbitrary constants r_1 and r_2. The transformation matrix M is, from Equation (4.27)

$$M = \begin{pmatrix} V_{x1} & V_{x2} \\ V_{y1} & V_{y2} \end{pmatrix} = \begin{pmatrix} r_1 e^{-ia\varepsilon} \sin \varepsilon & r_2 e^{-ia\varepsilon} \sin \varepsilon \\ r_1(\xi_1 - e^{-ia\varepsilon} \cos \varepsilon) & r_2(\xi_2 - e^{-ia\varepsilon} \cos \varepsilon) \end{pmatrix}, \tag{4.28}$$

with the magnitudes of the eigenvectors $\|V_1\|$ and $\|V_2\|$ given by

$$\|V_1\|^2 = r_1^2[\sin^2 \varepsilon + (\xi_1 - e^{-ia\varepsilon} \cos \varepsilon)(\bar{\xi}_1 - e^{ia\varepsilon} \cos \varepsilon)]$$

and

$$\|V_2\|^2 = r_2^2[\sin^2 \varepsilon + (\xi_2 - e^{-ia\varepsilon} \cos \varepsilon)(\bar{\xi}_2 - e^{ia\varepsilon} \cos \varepsilon)],$$

or with ξ_1 and ξ_2 in Equation (4.18),

$$\|V_1\|^2 = r_1^2 2(1 - \cos\varepsilon\cos(a\varepsilon + \varphi)) \tag{4.29}$$

and

$$\|V_2\|^2 = r_2^2 2(1 - \cos\varepsilon\cos(a\varepsilon - \varphi)) \tag{4.30}$$

with φ from Equation (4.22).

We continue with the normalized matrices M:

$$M = \begin{pmatrix} \dfrac{r_1 e^{-ia\varepsilon}\sin\varepsilon}{\|V_1\|^2} & \dfrac{r_2 e^{-ia\varepsilon}\sin\varepsilon}{\|V_2\|^2} \\[2ex] \dfrac{r_1(e^{i\varphi} - e^{-ia\varepsilon}\cos\varepsilon)}{\|V_1\|^2} & \dfrac{r_2(e^{-i\varphi} - e^{-ia\varepsilon}\cos\varepsilon)}{\|V_2\|^2} \end{pmatrix}. \tag{4.31}$$

Based on Equation (4.17), we can represent $T(\varepsilon)R(\varepsilon)$ with the known matrices D in Equation (4.26) and M in Equation (4.31) as

$$T(\varepsilon)R(\varepsilon) = MDM^{-1}. \tag{4.32}$$

$[T(\varepsilon)R(\varepsilon)]^{2\pi d/p\varepsilon}$ in Equations (4.14) assumes the form

$$T(\varepsilon)R(\varepsilon)^{2\pi d/p\varepsilon} = MDM^{-1}MDM^{-1}\ldots MDM^{-1} = MD^{2\pi d/p\varepsilon}M^{-1}. \tag{4.33}$$

The evaluation of $D^{2\pi d/p\varepsilon}$, with d_ε in Equation (4.5) and D in Equation (4.26), provides

$$\lim_{\varepsilon\to 0} D^{2\pi d/p\varepsilon} = \begin{pmatrix} \lim_{\varepsilon\to 0} e^{-ik'd} & 0 \\ 0 & \lim_{\varepsilon\to 0} e^{ik'd} \end{pmatrix}. \tag{4.34}$$

k' in Equation (4.25) leads to

$$\lim_{\varepsilon\to 0} k' = \frac{2\pi}{p}\lim_{\varepsilon\to 0}\frac{\arctan\left(\sqrt{1 - \cos^2 a\varepsilon\cos^2\varepsilon}\,/\cos a\varepsilon\cos\varepsilon\right)}{\varepsilon}.$$

As both the numerator and denominator tend to zero for $\varepsilon\to 0$, the application of Hospital's rule is needed, yielding

$$\lim_{\varepsilon\to 0} k' = \frac{2\pi}{p}\lim_{\varepsilon\to 0}\frac{a\sin a\varepsilon\cos\varepsilon + \sin\varepsilon\cos a\varepsilon}{\sqrt{1 - \cos^2 a\varepsilon\cos^2\varepsilon}}. \tag{4.35}$$

The repetition of Hospital's rule provides

$$\lim_{\varepsilon\to 0} k' = \lim_{\varepsilon\to 0}\frac{(1 + a^2)\cos a\varepsilon\cos\varepsilon - 2a\sin\varepsilon\sin a\varepsilon}{\left((a\sin a\varepsilon\cos\varepsilon + \sin\varepsilon\cos a\varepsilon)/\sqrt{1 - \cos^2 a\varepsilon\cos^2\varepsilon}\right)}\cdot\frac{2\pi}{p}.$$

As the denominator of the last equation is identical with $\lim_{\varepsilon \to 0} k'/(2\pi/p)$ in Equation (4.35), we obtain

$$\lim_{\varepsilon \to 0} k' = \frac{2\pi}{p} \sqrt{1 + a^2} \tag{4.36}$$

ensuring

$$\lim_{\varepsilon \to 0} D^{2\pi d/p\varepsilon} = \begin{pmatrix} e^{-i(2\pi d/p)\sqrt{1+a^2}} & 0 \\ 0 & e^{i(2\pi d/p)\sqrt{1+a^2}} \end{pmatrix}. \tag{4.37}$$

The limit of M in Equation (4.31) is calculated in the following steps with $\varphi = \arctan \sqrt{1 - \cos^2 \alpha\varepsilon \cos\varepsilon}/\cos a\varepsilon \cos\varepsilon$: as the elements m_{ik} in the unnormalized matrix M in Equation (4.28) and the magnitudes in Equations (4.29) and (4.30) tend to zero for $\varepsilon \to 0$, we choose $r_1 = r_2 = 1/\varepsilon$ and evaluate the limits of the numerator and denominator in Equation (4.31) separately according to Hospital's rule. By doing so, we obtain for the numerators

$$m_{11} = \lim_{\varepsilon \to 0} \frac{e^{-ia\varepsilon} \sin\varepsilon}{\varepsilon} = 1 = m_{12} \tag{4.38}$$

and

$$m_{21} = \lim_{\varepsilon \to 0} \frac{e^{i \arctan \sqrt{1-\cos^2 \alpha\varepsilon \cos^2 \varepsilon}/\cos a\varepsilon \cos \varepsilon} - e^{-ia\varepsilon} \cos\varepsilon}{\varepsilon}.$$

With similar calculations as performed for D, we obtain the limit as

$$m_{21} = -ia + i\sqrt{1 + a^2} \tag{4.39}$$

and in the same way, also

$$m_{22} = -ia - i\sqrt{1 + a^2}. \tag{4.40}$$

For the magnitudes the evaluation of the limit value is performed at the square of the magnitudes in Equations (4.29) and (4.30), again leading with similar calculations as for D to

$$\lim_{\varepsilon \to 0} \|V_1\|^2 = 2\left((1 + a^2) - a\sqrt{1 + a^2}\right) \tag{4.41}$$

and

$$\lim_{\varepsilon \to 0} \|V_2\|^2 = 2\left((1 + \alpha^2) + \alpha\sqrt{1 + \alpha^2}\right). \tag{4.42}$$

Hence, the normalized M is, for $\varepsilon \to 0$,

$$M = \begin{pmatrix} \dfrac{1}{\sqrt{2\big((1+a^2) - a\sqrt{1+a^2}\big)}} & \dfrac{1}{\sqrt{2\big((1+a^2) + a\sqrt{1+a^2}\big)}} \\[2ex] \dfrac{-i\big(a - \sqrt{1+a^2}\big)}{\sqrt{2\big((1+a^2) - a\sqrt{1+a^2}\big)}} & \dfrac{-i\big(a + \sqrt{1+a^2}\big)}{\sqrt{2\big((1+a^2) + a\sqrt{1+a^2}\big)}} \end{pmatrix}, \tag{4.43}$$

from which, as M is a unitary matrix, we obtain

$$M^{-1} = \overline{M^{\mathrm{T}}} = \begin{pmatrix} \dfrac{1}{\sqrt{2\big((1+a)^2 - a\sqrt{1+a^2}\big)}} & \dfrac{i\big(a - \sqrt{1+a^2}\big)}{\sqrt{2\big((1+a^2) - a\sqrt{1+a^2}\big)}} \\[2ex] \dfrac{1}{\sqrt{2\big((1+a)^2 + a\sqrt{1+a^2}\big)}} & \dfrac{i\big(a + \sqrt{1+a^2}\big)}{\sqrt{2\big((1+a)^2 + a\sqrt{1+a^2}\big)}} \end{pmatrix}. \tag{4.44}$$

Inserting Equations (4.29), (4.30), (4.43) and (4.44) into Equation (4.28) provides, for $\varepsilon \to 0$,

$$(T(\varepsilon)R(\varepsilon))^{\mathrm{s}} = \begin{pmatrix} \dfrac{1}{\|V_1\|} & \dfrac{1}{\|V_2\|} \\[2ex] \dfrac{-i\big(a-\sqrt{1+a^2}\big)}{\|V_1\|} & \dfrac{-i\big(a+\sqrt{1+a^2}\big)}{\|V_2\|} \end{pmatrix} \begin{pmatrix} e^{-ik'd} & 0 \\ 0 & e^{ik'd} \end{pmatrix} \begin{pmatrix} \dfrac{1}{\|V_1\|} & \dfrac{i\big(a-\sqrt{1+a^2}\big)}{\|V_1\|} \\[2ex] \dfrac{1}{\|V_2\|} & \dfrac{i\big(a+\sqrt{1+a^2}\big)}{\|V_2\|} \end{pmatrix}$$

$$= \begin{pmatrix} \dfrac{1}{-i\big(a - \sqrt{1+a^2}\big)} & \dfrac{1}{-i\big(a + \sqrt{1+a^2}\big)} \end{pmatrix} \begin{pmatrix} \dfrac{e^{-ik'd}}{\|V_1\|^2} & 0 \\[1ex] 0 & \dfrac{e^{ik'd}}{\|V_2\|^2} \end{pmatrix} \begin{pmatrix} 1 & i\big(a - \sqrt{1-a^2}\big) \\ 1 & i\big(a + \sqrt{1-a^2}\big) \end{pmatrix}, \tag{4.45}$$

with k' in Equation (4.36). Performing the multiplications in Equation (4.45) results in

$$(T(\varepsilon)R(\varepsilon))^{\mathrm{s}} = \begin{pmatrix} \cos\gamma + i\beta_0 si\gamma & \beta si\gamma \\ -\beta si\gamma & \cos\gamma - i\beta_0 si\gamma \end{pmatrix}, \tag{4.46}$$

with

$$\gamma = \frac{2\pi d}{p}\sqrt{1+a^2}, \tag{4.47}$$

$$a = \frac{p\Delta n}{2\lambda}, \tag{4.48}$$

$$\beta = \frac{2\pi}{p}d, \tag{4.49}$$

and

$$\beta_0 = \pi \frac{d\Delta n}{\lambda} = a\beta. \tag{4.50}$$

The final result is, with Equation (4.15),

$$\begin{pmatrix} O_{s\sigma} \\ O_{s\tau} \end{pmatrix} = e^{-i2\pi(\bar{n}d/\lambda)} \begin{pmatrix} \cos\gamma + i\beta_0 si\gamma & \beta si\gamma \\ -\beta si\gamma & \cos\gamma - i\beta_0 si\gamma \end{pmatrix} R(-\alpha) \begin{pmatrix} E_{\xi 0} \\ 0 \end{pmatrix} \tag{4.51}$$

given in the σ-τ-coordinates in Figure 4.1, which are rotated by the twist angle -β from the x-y-coordinates. An evaluation of the results in the x'-y'-coordinates in Figure 4.1 requires a rotation by the angle ψ, resulting in

$$\begin{pmatrix} O_{sx'} \\ O_{sy'} \end{pmatrix} = R(-\psi) e^{-i2\pi(\bar{n}d/\lambda)} \begin{pmatrix} \cos\gamma + i\beta_0 si\gamma & \beta si\gamma \\ -\beta si\gamma & \cos\gamma - i\beta_0 si\gamma \end{pmatrix} R(-\alpha) \begin{pmatrix} E_{\xi 0} \\ 0 \end{pmatrix}. \tag{4.52}$$

This result will be discussed for three special cases, namely the Twisted Nematic cell (TN cell) with $\alpha=0$ and $\beta=\pi/2$, the Supertwist TN cell (STN-cell) with $\alpha=0$ and $\beta>\pi/2$, and the Mixed mode TN cell (MTN-cell) with $\alpha\neq0$ and $\beta\neq\pi/2$.

The derivation of Equation (4.52) also contains a proof of the Chebychev identity. The full length of the derivation was presented in a detailed manner, as it leads to a core result for LCDs. Further, the considerations involved are useful for solving a variety of special display problems.

4.2 The Various Types of TN Cells

4.2.1 The regular TN cell

This cell is the most widely used active matrix LCD (Schadt and Helfrich, 1971; Yeh, 1988), and represents a special case of the general result in Equation (4.51). The linearly polarized light is incident at an angle

$$\alpha = 0; \tag{4.53}$$

in Figure 4.1 that is parallel to the director of the LC molecules on top of the orientation layer. The twist angle is

$$\beta = \pi/2, \tag{4.54}$$

leading to a pitch in Equation (4.49) of $p=4d$. Hence, at $z=d$ the helix has reached a quarter of a turn. From these values follow a in Equation (4.48) as

$$a = 2\frac{d\Delta n}{\lambda} \tag{4.55}$$

and γ in Equation (4.47) as

$$\gamma = \frac{\pi}{2}\sqrt{1 + \left(\frac{2d\Delta n}{\lambda}\right)^2} = \frac{\pi}{2}\sqrt{1 + a^2}. \tag{4.56}$$

In Equation (4.51), the σ-axis is rotated by $\beta = \pi/2$ from the x-axis, whereas the τ-axis is parallel to the x-axis. The Jones vectors in the σ-τ axes assume, after insertion of Equations (4.54), (4.55) and (4.56) into Equation (4.51), the form

$$O_{s\sigma} = e^{-i2\pi(\bar{n}/\lambda)}\left(\cos\frac{\pi}{2}\sqrt{1 + \left(\frac{2d\Delta n}{\lambda}\right)^2} + i\frac{\pi d\Delta n}{\lambda}si\left(\frac{\pi}{2}\sqrt{1 + \left(\frac{2d\Delta n}{\lambda}\right)^2}\right)\right)E_{\xi 0} \tag{4.57}$$

and

$$O_{s\tau} = -e^{-i2\pi(\bar{n}/\lambda)}\frac{\pi}{2}si\left(\frac{\pi}{2}\sqrt{1 + \left(\frac{2d\Delta n}{\lambda}\right)^2}\right)E_{\xi 0}. \tag{4.58}$$

For parallel polarizers the analyser lies in the τ-axis. The intensity I_τ passing this analyser is (Gooch and Tarry, 1974)

$$I_\tau = (O_{s\tau\text{eff}})^2 = \frac{1}{2}|O_{s\tau}|^2 = \frac{1}{2}\frac{\sin^2\pi/2\sqrt{1 + (2d\Delta n/\lambda)^2}}{1 + (2d\Delta n/\lambda)^2}E_{\xi 0}^2. \tag{4.59}$$

The reduced transmission $T = I_\tau/E_{\xi 0}^2$ is plotted in Figure 4.3 versus $a = 2d\Delta n/\lambda$. Zeros occur for $\sqrt{1 + a^2} = 2, 4, 6, 8, \ldots$, and hence for $a = \sqrt{3}, \sqrt{15}, \sqrt{35}, \sqrt{63}$. In the first minimum we obtain, with $a = \sqrt{3}$ from Equation (4.55), the optical retardation

$$d\Delta n = \frac{\sqrt{3}}{2}\lambda \tag{4.60}$$

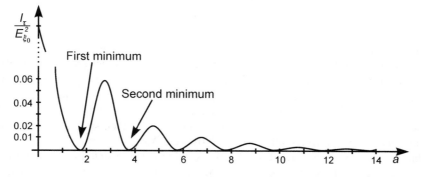

Figure 4.3 The intensity of light passing through a non-addressed TN-LCD with twist angle $\beta = \pi/2$ with $a = 2d\Delta n/\lambda$ according to Equation (4.59)

and the thickness d for a given $\lambda=\lambda_0$ as

$$d = \frac{\sqrt{3}\,\lambda_0}{2\,\Delta n}. \tag{4.61}$$

The d value is larger by a factor $\sqrt{3}$ than the corresponding one for the Fréederickzs or the DAP cell in Equation (3.80). The thicker cell can be manufactured with a larger yield, as the risk of generating shorts by dust particles in the cell is reduced. Obviously, the effective retardation in a twisted cell is smaller than in an untwisted one, resulting in a larger thickness. A similar effect is already known from the HAN cell.

According to Equation (4.61), the blocking of light with $\lambda_0=550$ nm requires with $\Delta n=0.15$ a thickness $d=3.1\,\mu$ when working at the first minimum.

Extinction of the light in Figure 4.3 occurs only for one wavelength λ. Neighbouring wavelengths can pass, which lights up the black state with a bluish-yellowish tint. Hence, this normally black mode provides a poor black state. It will be enhanced by compensation foils introduced in Chapter 6.

In the addressed state with the maximum voltage around 6V all LC molecules have aligned for $\Delta\varepsilon>0$ in parallel to the electric field. The incoming linearly polarized light experiences no bi-refringence and arrives unchanged at the parallel analyser, which can be passed. The transmission T of light for this normally black cell is depicted versus the reduced voltage across the cell in Figure 2.13.

In the operation with crossed polarizers leading to the normally white mode, the analyser lies in the σ-direction in Figure 4.1. The pertinent Jones vector $O_{s\sigma}$ for the field free state is given in Equation (4.57), from which the intensity I_σ passing the analyser follows as

$$I_\sigma = (O_{s\sigma\,\mathrm{eff}})^2 = \frac{1}{2}|O_{s\sigma}|^2 = \frac{1}{2}\left(\cos^2\gamma + \left(\frac{\pi d\Delta n}{\lambda}\right)^2 si^2\gamma\right)E_{\xi0}^2 \tag{4.62}$$

with γ in Equation (4.56). This leads to

$$I_\sigma/E_{\xi0}^2 = \frac{1}{2}\frac{\cos^2(\pi/2)\sqrt{1+a^2}+a^2}{1+a^2} \tag{4.63}$$

with a in Equation (4.55).

This reduced intensity is plotted versus a in Figure 4.4. The values are considerably larger than in the normally black mode in Figure 4.3. If the cell is fully addressed, the LC molecules orient themselves parallel to the field, and hence the incoming light, linearly polarized along the x-axis, reaches the crossed analyser unchanged, and is hence blocked independent of the wavelength. This generates an excellent black state if a field is applied. Therefore, the normally white cell is the preferred LCD.

The black state is independent of the thickness d. In the white state in Equation (4.63) and in Figure 4.4, the first maximum of the intensity lies in the vicinity of the first maximum of the cos term occurring at $\sqrt{1+a^2}=2$, and hence $a=\sqrt{3}$, leading again with a in Equation (4.55) to $d=\sqrt{3}\lambda_0/2\Delta n$, as for the normally black mode. Further maxima occur around

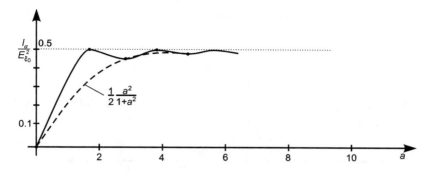

Figure 4.4 The intensity of light passing through a non-addressed normally white TN-LCD with twist angle $\beta=\pi/2$ and with $a=2d\Delta n/\lambda$ according to Equation (4.63)

$\sqrt{1+a^2} = 2, 4, 6, \ldots$, or $a = \sqrt{3}, \sqrt{15}, \sqrt{35}, \ldots$, as for the minima in the normally black mode. The maxima are of equal height $1/2$.

4.2.2 The supertwisted nematic LC cell (STN-LCD)

Linear polarized light is fed into the STN cell (Scheffer and Nehring, 1998; Nehring and Scheffer, 1990) again with $\alpha=0$ in Figure 4.1. The twist angle β at $z=d$ is, however, $\beta>\pi/2$, with values typically between π and $3(\pi/2)$. With the twist angle β in Equation (4.49) and a in Equation (4.48), we rewrite γ in Equation (4.47) with p from Equation (4.49) as

$$\gamma = \beta\sqrt{1+a^2} \quad \text{with } a = \pi d\Delta n/\beta\lambda. \tag{4.64}$$

With γ in Equation (4.64), and β_0 in Equation (4.50), Equation (4.52) governing the field free state provides for the coordinates x' and y' with the angle ψ in Figure 4.1, and again, drawn in Figure 4.5

$$\begin{pmatrix} O_{sx'} \\ O_{sy'} \end{pmatrix} = e^{-i2\pi(\bar{n}d/\lambda)} \begin{pmatrix} \cos\psi - \sin\psi \\ \sin\psi \cos\psi \end{pmatrix} \begin{pmatrix} \cos\gamma+i\beta_0 si\gamma & \beta si\gamma \\ -\beta si\gamma & \cos\gamma-i\beta_0 si\gamma \end{pmatrix} \begin{pmatrix} \cos\alpha & -\sin\alpha \\ \sin\alpha & \cos\alpha \end{pmatrix} \begin{pmatrix} E_{\xi 0} \\ 0 \end{pmatrix} \tag{4.65}$$

leading to

$$O_{sx'} = e^{-i2\pi(\bar{n}d/\lambda)} E_{\xi 0}\{\cos\alpha[\cos\psi(\cos\gamma + i\beta_0 si\gamma + \beta\sin\psi si\gamma] \\ + \sin\alpha[\beta\cos\psi si\gamma - \sin\psi(\cos\gamma - i\beta_0 si\gamma)]\} \tag{4.66}$$

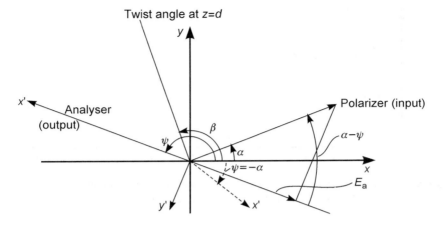

Figure 4.5   Angles and coordinates for an STN display

and

$$O_{sy'} = e^{-i2\pi(\bar{n}d/\lambda)} E_{\xi0}\{\cos\alpha[\sin\psi(\cos\gamma + i\beta_0 si\gamma) - \beta\cos\psi si\gamma]$$
$$+ \sin\alpha[\beta\sin\psi si\gamma + \cos\psi(\cos\gamma - i\beta_0 si\gamma)]\}. \qquad (4.67)$$

For

$$\gamma = \pi, \quad 2\pi, \quad 3\pi, \ldots, \qquad (4.68)$$

Equations (4.66) and (4.67) yield

$$O_{sx'} = e^{-i2\pi(\bar{n}d/\lambda)} E_{\xi0}\{(-1)^{\nu}\cos\alpha\cos\psi - (-1)^{\nu}\sin\alpha\sin\psi\}$$
$$= e^{-i2\pi(\bar{n}d/\lambda)} E_{\xi0}(-1)^{\nu}\cos(\alpha+\psi), \quad \nu = 1, 2, 3, \ldots, \qquad (4.69)$$

and

$$O_{sy'} = 0. \qquad (4.70)$$

$O_{sx'}$ reaches its maximum

$$O_{sx'} = e^{-i2\pi(\bar{n}d/\lambda)}(-1)^{\nu}E_{\xi0} \qquad (4.71)$$

for

$$\psi = -\alpha. \qquad (4.72)$$

If the axis x' of the analyser is at an angle $\psi = -\alpha$ to the x-axis, the maximum intensity $I_{sx'} = E_{\xi0}^2$ passes the cell. For energy reasons the intensity in the y'-direction must be zero, which is met in Equation (4.70). The direction x' for maximum intensity is shown by dashed lines in Figure 4.5. The condition in Equation (4.68) for maximum intensity yields for a from Equation (4.64)

$$a = \sqrt{\frac{(\nu\pi)^2}{\beta^2} - 1} \tag{4.73}$$

with the first realizable value for $\nu=2$. For $\beta=3(\pi/2)$, we obtain the values

$$a = \sqrt{7}/3, \quad \sqrt{3}, \quad \sqrt{55}/3,\ldots, \quad \text{for} \quad \nu = 2,\ 3,\ 4,\ldots, \tag{4.74}$$

From Equation (4.64), we calculate the thickness

$$d = \frac{a\beta\lambda}{\pi\Delta n}. \tag{4.75}$$

For $\lambda=550\,\text{nm}$, $\beta=3(\pi/2)$, $a = \sqrt{7}/3$ and $\Delta n=0.05$ the thickness is $d=14.5\,\mu\text{m}$. As the thickness in Equation (4.75) is proportional to β, STN displays operate with thicker cells than regular TN displays.

The intensity passing the analyser at an angle ψ describes the normally white state, with the maximum for $\psi=-\alpha$. If a high enough field is applied, the linear polarized light reaches the analyser at the angle α, resulting (according to Figure 4.5) in a component

$$|E_a| = E_{\xi0}|\cos(\alpha - \psi + \pi)| = E_{\xi0}|\cos(\alpha - \psi)|. \tag{4.76}$$

For $\alpha=\pi/4$, and hence $\psi=-\alpha=-\pi/4$, we obtain $E_a=0$ independent of λ. This normally white state exhibits an excellent black state, and works with crossed polarizers.

For the optimum $\psi=-\alpha$ and the normally black state, the analyser has to be placed, due to Equation (4.70), in the y'-direction, which is for $\psi=-\alpha=-\pi/4$ parallel to the polarizer. In this case, however, $O_{sy'}$ holds only for one wavelength in Equation (4.75). The white state after a field has been applied fully passes the analyser.

As the normally white state exhibits a black independent of wavelength, it is the preferred mode of operation.

If the large thickness of an STN cell is decreased, the transmission falls below optimum and some luminance is sacrificed, but a wider viewing angle is obtained. The reason for this will be explained in the chapter on compensation foils later.

The larger twist angle β in supertwist nematic LCDs has a pronounced effect on the transmitted luminance versus voltage curve in Figure 4.6 by rendering the transition from the white state to the black state much steeper. As will be explained in Chapter 12, this enhanced steepness is required for addressing an STN cell with a larger number of lines without losing too much contrast. The increase in steepness with increased β is now explained phenomenologically. In the transition from, lets say, the white state to the black state, the LC molecules have to be tilted by a torque stemming from the applied electrical field. They finally end up parallel to the field. A smaller torque is needed if the molecules exhibit a larger twist angle from layer to layer as the restoring force by the vertically neighbouring molecules becomes weaker. Hence, a smaller voltage is required to achieve the tilt angles.

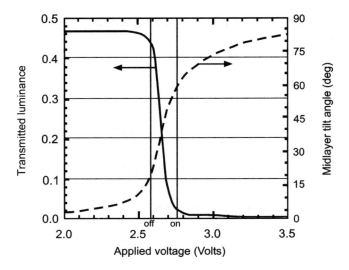

Figure 4.6 Transmitted luminance and midlayer tilt versus the voltage across an STN cell with a twist of $\beta=240°$, an off-voltage of 2.58 V and an on-voltage of 2.75 V for addressing 240 lines (Scheffer and Nehring, 1998)

A calculation of this effect is based on fluid mechanics and liquid crystal continuum equations (Degen, 1980), where the mechanical parameters K_{11}, K_{22} and K_{33}, the dielectric constants $\varepsilon_\perp$ and $\varepsilon_\parallel$, the pretilt angle at the orientation layers and, of course, the twist angle β and d/p play a role. The calculations are similar to the electro-optical investigations of TN cells in Section 4.1, because propagation matrices based on the mechanical properties are established for a sequence of twisted layers, and are finally multiplied.

Figure 2.12 shows that the tilt of the molecules is larger in the midlayer due to the restoring forces of the molecules anchored on top of the orientation layer. Figure 4.7 depicts the midlayer tilt versus the voltage V_{LC} across the cell with twist angles β as a parameter. The larger is β, the greater is the slope of the curves. For $\beta=3\pi/2(270°)$ the curve rises perpendicularly in the centre portion. For β greater than 270° the curves become double-valued, causing bistability and hysteresis. Therefore, the twist must not exceed 270°. STN cells use twists between 180° and 270°, where 240° is often encountered as the electro-distortional curves are steep, but sufficiently removed from bistability.

To sustain these large twists, chiral compounds have to be added. The chiral dopant imparts an intrisic twist, given by d/p, to the helical structure. On the other hand, the twist angle is also imposed by the angular difference Φ_T of the rubbing directions. A matching of the two constraints requires $d/p=\Phi_T/2\pi$. In practical STN cells $d/p<\Phi_T/2\pi$ is chosen which compresses the helical structure in order to avoid the phenomenon of stripes. These stripes are generated if the condition that the local optic axis changes orientation only along the spatial coordinate perpendicular to the layer is not met. (Nehring and Scheffer, 1990). They result in scattering of light, rendering the display unacceptable.

In Figure 4.8, the influence of the pretilt on the electro-distortional curve is depicted. The larger the pretilt, the smaller the voltage needed to tilt the molecules. This is understandable, as the molecules with larger pretilt are already rotated in the right direction.

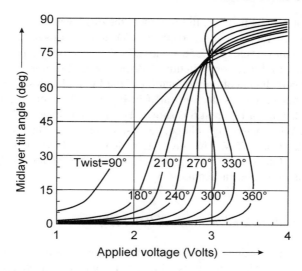

Figure 4.7 Midlayer tilt versus voltage of an STN cell with twist angle β as a parameter (Scheffer and Nehring, 1998)

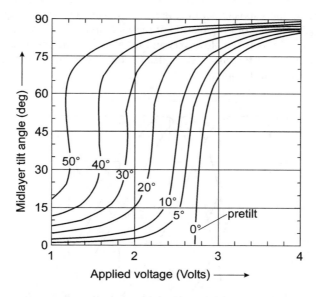

Figure 4.8 The influence of the pretilt angle on the electro-distortional curve of the midlayer in an STN cell (Scheffer and Nehring, 1998)

Finally Figure 4.6 shows both the midlayer tilt and the transmitted luminance versus V_{LC} for a twist of 240°. The transition from white to black is considerably steeper than for the regular 90° TN cell in Figure 2.13. It exhibits an extended linear range leading to a good grey scale operation. It is interesting to note that a sufficient black state occurs even when the midlayer has not yet reached 90°. In Figure 4.6, the display assumes a desired grey shade if voltages in between the on-voltage (fully black) and the off-voltage (fully white) are applied. The off and on voltages are 2.58 V and 2.75 V, respectively representing a comparatively small voltage difference for switching a display with 240 lines.

The voltage induced change of the orientation of the LC molecules also affects the colour appearance of an image. This is demonstrated in Figure 4.9, where the luminance vs. the wavelength of the display in Figure 4.6 is shown. For an off-voltage of 2.58 V, the colouring is greenish-yellow. With increasing voltage, the display becomes bluer, ending up with dark blue at the on-voltage of 2.75 V. This mode is called the yellow mode, which is used in inexpensive displays. Other colours are generated with larger values for $d\Delta n$. More on this colour generation is presented in Section 4.3.

4.2.3 The mixed mode twisted nematic cell (MTN cell)

In this mode the polarizer is at an angle $\alpha \neq 0$ to the *x*-axis while the director exhibits a twist β, in most cases $\beta = \pi/2$. The twist starts with the LC molecules anchored parallel to the *x*-axis. This situation gives rise to two modes of propagation of light from which the term *mixed mode* is derived.

The first mode is based on linearly polarized light feeding in both a wave with a field component parallel to the *x*-axis experiencing $n_\parallel$ and a wave with a *y*-component experiencing $n_\perp$. As $n_\parallel > n_\perp$, the speed of propagation of the *x*-component is smaller than the speed of the *y*-component; the *x*-axis is the slow axis and the *y*-axis is the fast one. The effect at the output is based on this bi-refringence. Examples for this mode are all the untwisted nematic cells that were described in Chapter 3.

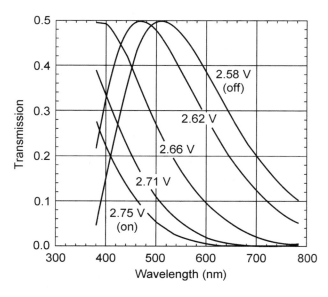

Figure 4.9 Transmitted spectrum of a 240° STN display with the addressing voltage as a parameter (Scheffer and Nehring, 1998)

The second mode stems from the twist of the molecules given in Equation (4.51) for $\alpha=0$, yielding

$$\begin{pmatrix} O_{s\sigma} \\ O_{s\tau} \end{pmatrix} = e^{-i2\pi(\bar{n}d/\lambda)} \begin{pmatrix} (\cos\gamma + i\beta_0 si\gamma)E_{\xi 0} \\ -\beta si\gamma E_{\xi 0} \end{pmatrix} \tag{4.77}$$

with

$$\gamma = \sqrt{\beta^2 + (\pi d\Delta n/\lambda)^2} \tag{4.78}$$

derived from Equation (4.64).

For $\beta \ll \pi d\Delta n/\lambda$ Equation (4.77) can be approximated by

$$\begin{pmatrix} O_{s\sigma} \\ O_{s\tau} \end{pmatrix} = e^{-i2\pi(\bar{n}d/\lambda)} \begin{pmatrix} (\cos\pi d\Delta n/\lambda + i\sin\pi d\Delta n/\lambda)E_{\xi 0} \\ O \end{pmatrix}$$
$$= e^{-i2\pi(\bar{n}d/\lambda)} \begin{pmatrix} e^{i\pi d\Delta n/\lambda}E_{\xi 0} \\ O \end{pmatrix} \tag{4.79}$$

where β_0 from Equation (4.50) was used. This result indicates that the component in the σ-axis represents the entire propagating wave. This axis rotates with the twist of the LC molecules we introduced during the derivation of the general result in Equation (4.51). Hence, the wave is guided by the twist. This waveguiding effect, also called the *adiabatic following*, is the second mode of propagation.

The derivation of the transmission in the MTN mode starts with the general Equations (4.66) and (4.67) for the Jones vectors in the x'-y'-coordinates in Figure 4.1. We know from Equation (4.72) that maximum transmission takes place for $\psi = -\alpha$, and the analyser in the ψ-direction. We now pursue this case, and obtain from Equations (4.66) and (4.67) for $\psi = -\alpha$

$$O_{sx'} = e^{-i2\pi(\bar{n}d/\lambda)} E_{\xi 0}(\cos\gamma + i\beta_0 si\gamma \cos 2\alpha) \tag{4.80}$$

$$O_{sy'} = e^{-i2\pi(\bar{n}d/\lambda)} E_{\xi 0}(-\beta si\gamma - i\beta_0 si\gamma \sin 2\alpha). \tag{4.81}$$

The polarizers are crossed with a 90° angle for $\alpha=\pi/4$, which as is shown in our discussion of Equation (4.76), yields a normally white state with an excellent black state. For $\alpha=\pi/4$, Equations (4.80) and (4.81) yield

$$O_{sx'} = e^{-i2\pi(\bar{n}d/\lambda)} E_{\xi 0}\cos\gamma \tag{4.82}$$

and

$$O_{sy'} = -e^{-i2\pi(\bar{n}d/\lambda)} E_{\xi 0} si\gamma(\beta + i\beta_0). \tag{4.83}$$

The intensity $I_{sx'}$ of the light passing through the polarizer in the x' direction in the normally white state with $\alpha = -\psi = \pi/4$ is

$$I_{sx'} = \cos^2 \gamma E_{\xi0}^2 \qquad (4.84)$$

with

$$\gamma = \beta\sqrt{1+a^2}, \quad a = \pi d\Delta n/\lambda.$$

$I_{sx'}$ is plotted in Figure 4.10 versus a for $\beta = \pi/2$; zeros are at $a = 0, \sqrt{8}, \sqrt{24}, \ldots$, whereas maxima occur at $a = \sqrt{3}, \sqrt{15}, \sqrt{35}, \ldots$. The maximum at $a = \sqrt{3}$ leads to $d = \sqrt{3}\lambda/\pi\Delta n$. The fully on state offers an excellent black independent of λ. The normally black state with parallel polarizers possesses a black state for $O_{sy'} = 0$, that is, for $\gamma(\lambda) = \pi$. More on MTNs is presented in the following two chapters.

4.2.4 Reflective TN cells

The basic structure of a reflective cell is depicted in Figure 3.12(a). After the light has passed the polarizer placed at an angle α to the x-axis, it is reflected at the rear mirror. Contrary to the untwisted case, we place the mirror at any distance z, and not only at $z=d/2$. Then the transmission from the input to the mirror is given by the Jones vectors $O_{s\sigma}$ and $O_{s\tau}$ in Equation (4.51) as

$$\begin{pmatrix} O_{s\sigma} \\ O_{s\tau} \end{pmatrix} = TR(-\alpha) \begin{pmatrix} E_{\xi0} \\ 0 \end{pmatrix} = T \begin{pmatrix} E_{\xi0}\cos\alpha \\ E_{\xi0}\sin\alpha \end{pmatrix} \qquad (4.85)$$

with the transition matrix

$$T = \begin{pmatrix} \cos\gamma + i\beta_0 si\gamma & \beta si\gamma \\ -\beta si\gamma & \cos\gamma - i\beta_0 si\gamma \end{pmatrix}. \qquad (4.86)$$

The σ-τ-coordinates are shown in Figure 4.1 with the twist angle β from the x-axis. The components $O_{s\sigma}$ and $O_{s\tau}$ are reflected at a mirror placed at $z=d$ in Figure 4.1. If the mirror is

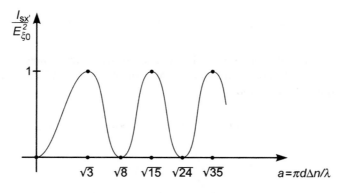

Figure 4.10 The reduced intensity of a mixed mode TN display

metallic it cannot sustain an electric field. For this reason, an incident elliptically polarized wave in Figure 4.11 generates the electric field shown with dashed lines in Figure 4.11. This field is reflected and propagates in the z-direction towards the input. The dashed field has a phase shift from the incident field. This phase shift is omitted in the future calculation as it is of no importance for the intensity to be calculated. Therefore, neglecting the phase shift caused by the reflection, we assume the vector with the components $O_{s\sigma}$ and $O_{s\tau}$ at $z \neq 0$ as the input for a wave travelling in the opposite direction towards the input. The pertinent transmission matrix T_r is, according to the principle of reciprocity in physics (Yeh and Gu, 1999),

$$T_r = T',\tag{4.87}$$

where T' stands for the transpose of T. With T' we obtain

$$\begin{pmatrix} O_{sx} \\ O_{sy} \end{pmatrix} = T' \begin{pmatrix} O_{s\sigma} \\ O_{s\tau} \end{pmatrix},\tag{4.88}$$

where O_{sx} and O_{sy} are the reflected Jones vectors at the input in the x-y-coordinates. To obtain the component passing through the polarizer, we have to rotate the coordinates by α, resulting in

$$\begin{pmatrix} O_{s\xi} \\ O_{s\eta} \end{pmatrix} = R(\alpha) \begin{pmatrix} O_{sx} \\ O_{sy} \end{pmatrix}.\tag{4.89}$$

$O_{s\xi}$ and $O_{s\eta}$ are the components in the ξ-η-coordinates in Figure 4.1.
   Insertion of Equations (4.85) and (4.88) into Equation (4.89) yields

$$\begin{pmatrix} O_{s\xi} \\ O_{s\eta} \end{pmatrix} = R(\alpha)T'T \begin{pmatrix} E_{\xi 0} \cos\alpha \\ E_{\xi 0} \sin\alpha \end{pmatrix}.\tag{4.90}$$

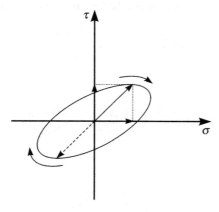

Figure 4.11 Incident (——) and reflected (- - -) elliptically polarized light at the metallic mirror of a reflective cell

With T from Equation (4.86), we obtain

$$T'T = \begin{pmatrix} \underbrace{(\cos\gamma + i\beta_0 si\gamma)^2 + \beta^2 si^2\gamma}_{t_{11}} & \underbrace{i2\beta\beta_0 si^2\gamma}_{t_{12}} \\ \underbrace{i2\beta\beta_0 si^2\gamma}_{t_{12}} & \underbrace{(\cos\gamma - i\beta_0 si\gamma)^2 + \beta_0^2 si^2\gamma}_{t_{11}} \end{pmatrix}, \tag{4.91}$$

and with Equation (4.90)

$$O_{s\xi} = (\mathrm{Re}t_{11} + t_{12}\sin 2\alpha + i\mathrm{Im}t_{11}\cos 2\alpha)E_{\xi 0}$$

$$O_{s\eta} = (-i\mathrm{Im}t_{11}\sin 2\alpha + t_{12}\cos 2\alpha)E_{\xi 0},$$

where the abbreviations in Equation (4.91) have been used. The intensity I_ξ passing through the polarizer is

$$I_\xi/E_{\xi 0}^2 = |O_{s\xi}|^2/E_0^2 = \mathrm{Re}t_{11}^2 + (2\beta\beta_0 si^2\gamma\sin 2\alpha + \mathrm{Im}t_{11}\cos 2\alpha)^2$$

or

$$I_\xi/E_{\xi 0}^2 = \cos^2\gamma + (\beta^2 - \beta_0^2)si^2\gamma + (2\beta\beta_0 si^2\gamma\sin 2\alpha + 2\beta_0\cos\gamma si\gamma\cos 2\alpha)^2. \tag{4.92}$$

The intensity perpendicular to the polarizer is

$$I_\eta/E_{\xi 0}^2 = |O_{s\eta}|^2/E_{\xi 0}^2 = (-\mathrm{Im}t_{11}\sin 2\alpha + 2\beta\beta_0 si^2\gamma)^2$$

or

$$I_\eta/E_{\xi 0}^2 = (-2\beta_0\cos\gamma si\gamma\sin\alpha + 2\beta\beta_0 si^2\gamma)^2. \tag{4.93}$$

An maximum for I_ξ dependent on α is reached by $d\,I_\xi/d\,\alpha = 0$. This leads to

$$\tan 2\alpha = \frac{\beta}{\gamma}\tan\gamma = \frac{\tan\gamma}{\sqrt{1 + a^2}} \tag{4.94}$$

and for

$$\tan 2\alpha = -\frac{\gamma}{\beta}\frac{1}{\tan\gamma} = -\frac{\sqrt{1 + a^2}}{\tan\gamma} \tag{4.95}$$

with

$$a = \frac{p\Delta n}{2\lambda} = \frac{\pi d\Delta n}{\beta\lambda} \tag{4.96}$$

with γ from Equations (4.47) and (4.49) and a from Equations (4.48) and (4.49).

For 2α in Equation (4.95), the term in parentheses in Equation (4.92) vanishes, hence this provides the minimum intensity

$$\min I_\xi / E_\xi^2 = (\mathrm{Re} t_{11})^2 = \cos^2\gamma + (\beta^2 - \beta_0^2)si^2\gamma, \tag{4.97}$$

whereas 2α in Equation (4.94) yields the maximum intensity

$$\max I_\xi / E_\xi^2 = 1, \tag{4.98}$$

the largest intensity possible.

The minimum intensity in Equation (4.97) becomes zero for

$$\tan\gamma = \frac{\gamma}{\sqrt{\beta^2 - \beta_0^2}} = \pm\sqrt{\frac{a^2 + 1}{a^2 - 1}}. \tag{4.99}$$

For this value Equation (4.95) provides the pertinent angle α of the polarizer as

$$\tan 2\alpha = \mp\sqrt{a^2 - 1} \tag{4.100}$$

which has a solution for $a \geq 1$.

From the results for the field-free reflective cell reached so far, we derive the conditions for a normally black cell. We recall that in a reflective cell in Figure 3.12(c), the polarizer and the analyser are identically oriented at an angle α to the x-axis.

The minimum intensity I_ξ / E_ξ^2 in Equation (4.97) with the angle of the polarizer in Equation (4.100) provides the black state. It can be zero for $\tan\gamma$ in Equation (4.99), that is for

$$\tan\beta\sqrt{1 + a^2} = \pm\sqrt{\frac{a^2 + 1}{a^2 - 1}} \tag{4.101}$$

providing

$$\tan 2\alpha = \mp\sqrt{a^2 - 1}. \tag{4.102}$$

Equation (4.101) must be solved numerically. A table of solutions can be found in Yeh and Gu (1999).

Two examples are:

β	a	$\Delta nd/\lambda$	α
$\pi/4$	1.16092	0.29023	$-15.2637°$
$\pi/2$	3.32399	1.66200	$36.2459°$

Once a in Equation (4.101) has been found, $\Delta nd/\lambda$ and α can be calculated from Equations (4.96) and (4.102). The black state is achieved only for a given λ, which determines d in Equation (4.96); $\lambda = 550\,\mathrm{nm}$ and $\Delta n = 0.05$ yield $d = 3.19\,\mu\mathrm{m}$ in the first case and $d = 18.26\,\mu\mathrm{m}$ in the second case in the table above.

The solutions for normally black reflective cells with $\alpha \neq 0$ discussed so far belong to the mixed mode TN cells introduced in Chapter 4. For $\alpha = 0$ Equation (4.81) requires $a = 1$, and from Equation (4.99) $\tan \gamma = \infty$ or $\gamma = \pi/2 = \beta\sqrt{1 + a^2} = \sqrt{2}\beta$ leading to a twist angle $\beta = \pi/2\sqrt{2}$. The regular TN cell with $\beta = \pi/2$ does not permit a solution for a reflective cell with $I_\xi = 0$ at a given wavelength λ. Therefore, mixed mode TN cells are preferred for reflective TN displays.

As normally white reflective TN cells are seldom used due to their poorer performance, they are not dealt with here. A description can be found in Yeh and Gu (1999).

4.3 Electronically Controlled Birefringence for the Generation of Colour

We consider the mixed mode TN cell for the angle of the polarizer $\alpha = \pi/4$ with the intensity

$$I_{sx'} = \cos^2 \beta \sqrt{1 + \pi d \Delta n/\lambda}$$

in Equation (4.84). The maxima of $I_{sx'}$ for $a = \pi d \Delta n/\lambda = \sqrt{3}, \sqrt{15}, \sqrt{35}, \ldots$, are plotted in Figure 4.10. A plot of $I_{sx'}$ versus $\lambda = \pi d \Delta n/a$ with maxima for $a = \sqrt{3}, \sqrt{15}, \sqrt{35}, \ldots$, is shown in Figure 4.12. By changing Δn we are able to shift the wavelength λ of the maxima, or in other words, we can shift the colour of the transmitted intensity. The birefringence $\Delta n = n_\parallel - n_\perp = n_e - n_0$ can be changed by applying a voltage V_{LC} which gradually tilts the LC molecules in Figure 4.13 for $\Delta \varepsilon > 0$ from the homogeneous alignment in the field-off state to the homeotropic alignment. This effect is called Electronically Controlled Birefringence (ECB). Normally incident light in Figure 4.13 always experiences the same index $n_\perp$, whereas $n_\parallel$ changes with the tilt angle $\Theta_e(V_{LC})$, which depends upon V_{LC}. In Chapter 6, $n_\parallel(\Theta_e)$ will be derived as

$$n_e(\Theta_e) = \frac{n_0 n_e}{\sqrt{n_0^2 \sin^2 \Theta_e + n_e^2 \cos^2 \Theta_e}}. \tag{4.103}$$

The dependence of Θ_e on V_{LC} can be determined from the mechanical properties of the LC material, which is not discussed further here. The result is a voltage dependent birefringence

$$\Delta n(V_{LC}) = n_e(\Theta_e(V_{LC})) - n_0. \tag{4.104}$$

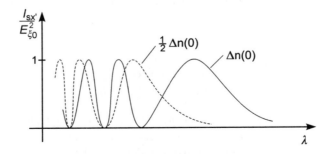

Figure 4.12 The reduced intensity of a mixed mode TN display versus $\lambda = \pi d \Delta n/a$ with Δn (V_{LC})

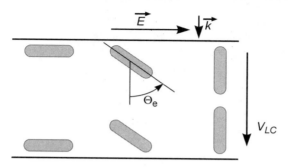

Figure 4.13 The tilt of the LC molecules under the influence of a voltage V_{LC} for $\Delta\varepsilon > 0$

Figure 4.12 shows the intensity $I_{sx'}/E_{\xi 0}^2$ for the field free state with $\Delta n(0)$ in Figure 4.10 and for a $\Delta n(V_{LC}) = (1/2)\Delta n(0)$ with dashed lines as an example. Figure 4.9 shows measured transmission curves of an STN display versus λ, with V_{LC} as a parameter.

The generation of colour by ECB replaces the colour filter by applying an individual voltage to each pixel corresponding to the desired colour. The achievable colours are not fully saturated because the spectral curves are not narrow enough, and neighbouring colours overlap. The reflective cells are constructed according to Section 4.2.4.

5

Descriptions
of Polarization

In Section 3.2, elliptically, circularly and linearly polarized light and its right- and left-handedness were introduced. Since polarized light is an essential feature of many liquid crystal cells, we have to familiarize ourselves with two more means for the characterization of polarization.

5.1 The Characterizations of Polarization

Obviously, the state of polarization is defined by three numbers, such as A_x, A_y and δ in Equations (3.44) through (3.55). The first equivalent characterization based on those three parameters are the Stokes parameters (Born and Wolf, 1980)

$$S_0 = A_x^2 + A_y^2, \tag{5.1}$$

$$S_1 = A_x^2 - A_y^2 \tag{5.2}$$

and

$$S_2 = 2A_x A_y \cos \delta. \tag{5.3}$$

The fourth parameter

$$S_3 = 2A_x A_y \sin \delta \tag{5.4}$$

is dependent on the first three parameters as

$$S_0^2 = S_1^2 + S_2^2 + S_3^2, \tag{5.5}$$

but is used later on for the geometrical interpretation of the Stokes parameters by the Poincaré sphere.

A second set of equivalent parameters (Born and Wolf, 1980) is derived in the Appendix 3, and is based on the principal axes a and b of the ellipse in Figure 3.5, its angle ψ to the x-axis, and on its ellipticity $e = \pm b/a = \tan \Omega$. The relations to A_x, A_y and δ are, according to Appendix 3, Equations (27), (28) and (29):

$$a^2 + b^2 = A_x^2 + A_y^2, \tag{5.6}$$

$$\tan 2\psi = (\tan 2\gamma_0) \cos \delta, \tag{5.7}$$

$$\sin 2\Omega = \sin 2\gamma_0 \sin \delta, \tag{5.8}$$

with

$$\tan \gamma_0 = A_y/A_x. \tag{5.9}$$

The relations between the Stokes parameters and the parameters $a^2 + b^2$ and Ω are also as derived in Appendix 3. They are

$$S_0 = a^2 + b^2 \tag{5.10}$$

$$S_1 = S_0 \cos 2\Omega \cos 2\psi \tag{5.11}$$

$$S_2 = S_0 \cos 2\Omega \sin 2\psi \tag{5.12}$$

$$S_3 = S_0 \sin 2\Omega. \tag{5.13}$$

Equations (5.10) through (5.13) allow geometric interpretation by the Poincaré sphere in Figure 5.1. The radius of the sphere is $S_0 = a^2 + b^2 = A_x^2 + A_y^2$, which stands for the intensity of the light. A point P on the sphere is given by the two angles 2ψ and 2Ω in Figure 5.1, resulting in the Cartesian coordinates S_1, S_2 and S_3 of point P as given in Equations (5.11) through (5.13).

Each point P on the sphere represents a given state of polarization, which is defined by Equations (5.1) through (5.4), or (5.10) through (5.13). From Equation (25) in Appendix 3, we know that $2\Omega\varepsilon[-\pi/2, 0)$ stands for left-handed and $2\Omega\varepsilon(0,\pi/2]$ for right-handed polarization; $2\Omega\varepsilon(0,\pi/2]$ yields $S_3 > 0$ in Equation (5.13). Hence, the points above the equatorial

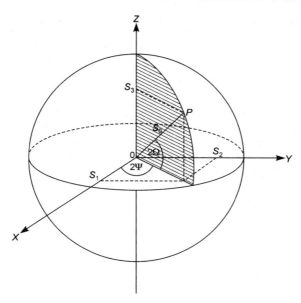

Figure 5.1 The Poincaré sphere as a description of polarization

plane represent right-handed polarization; $2\Omega\varepsilon[-\pi/2, 0)$ indicates left-handed polarization, and yields $S_3 < 0$. Therefore, the points below the equatorial plane represent left-handed polarization. For linearly polarized light, Equation (3.52) provides $\delta = m\pi, m = 0, 1, 2, \ldots$, yielding $S_3 = 0$ in Equation (5.4). Hence, all linearly polarized states lie on the equator. Finally, for circularly polarized light, defined by $A_x = A_y$ and $\delta = \pi/2$ or $-\pi/2$ in Equation (3.54), Equations (5.2) and (5.3) provide S_1, $S_2 = 0$. Thus, circularly polarized light is represented by the north pole for right-handed polarization and by the south pole for left-handed polarization.

Poincaré's sphere provides a useful visualization of the various states of polarization. Its method of defining the state of polarization is listed in Table 5.1 (Born and Wolf, 1980). A further very compact representation of polarization (Yeh, 1988) is based on complex numbers associated with a ratio χ of components of the Jones vector, the Jones vector itself and a ratio χ_c of right-handed and left-handed circularly polarized light.

We consider the components

$$J_x = A_x e^{i\delta_x} \tag{5.14}$$

and

$$J_y = A_y e^{i\delta_y} \tag{5.15}$$

of a Jones vector, where A_x and A_y are real numbers, and

$$\delta_x = k_x z - \varphi \tag{5.16}$$

Table 5.1 Various representations of the state of polarization

Polarization	Poincaré sphere	χ	Jones vector $\begin{pmatrix} J_x \\ J_y \end{pmatrix}$	χ_c
Linear	Equator	x-axis	real, $\delta=0$	$\lvert\chi_c\rvert - 1$
Arbitrary angle γ_0 to x-axis	Equator	$\tan\gamma_0$	$\dfrac{1}{\sqrt{2}}\begin{pmatrix} \cos\gamma_0 \\ \sin\gamma_0 \end{pmatrix}$	$e^{-2i\gamma_0}$
$\gamma_0=0$ in x-direction	Equator	0	$\dfrac{1}{\sqrt{2}}\begin{pmatrix} 1 \\ 0 \end{pmatrix}$	$+1$
$\gamma_0=\pi/2$ in x-direction	Equator	∞	$\dfrac{1}{\sqrt{2}}\begin{pmatrix} 0 \\ 1 \end{pmatrix}$	-1
$\gamma_0 = \dfrac{\pi}{4}$	Equator	1	$\dfrac{1}{\sqrt{2}}\begin{pmatrix} 1 \\ 1 \end{pmatrix}$	$-i$
$\gamma_0 = 3\dfrac{\pi}{4}$	Equator	-1	$\dfrac{1}{\sqrt{2}}\begin{pmatrix} 1 \\ -1 \end{pmatrix}$	i
Right-handed circular	North pole	$+i$	$\dfrac{1}{\sqrt{2}}\begin{pmatrix} 1 \\ i \end{pmatrix}$	∞
Left-handed circular	South pole	$-i$	$\dfrac{1}{\sqrt{2}}\begin{pmatrix} 1 \\ -i \end{pmatrix}$	0
Right-handed elliptical	Above equator exept pole	Upper half plane exept i	Upper half plane exept $\dfrac{1}{\sqrt{2}}\begin{pmatrix} 1 \\ i \end{pmatrix}$	Lower half plane exept $\lvert\chi_c\rvert=1$ with $\gamma_0 \in (0, -\pi/2)$ see Equation (4.70)
Left-handed elliptical	Below equator exept pole	Lower half plane exept $-i$	Upper half plane exept $\dfrac{1}{\sqrt{2}}\begin{pmatrix} 1 \\ -i \end{pmatrix}$	Upper half plane exept $\lvert\chi_c\rvert=1$ with $\gamma_0 \in (0, \pi/2)$ see Equation (4.70)

and

$$\delta_y = k_y z - \varphi \tag{5.17}$$

stand for the phase angles. The equation

$$\chi = \frac{J_y}{J_x} = \frac{A_y}{A_y} e^{i(\delta_y - \delta_x)} = \tan\gamma_0 e^{i\delta} \tag{5.18}$$

describes the polarization, and is also listed in Table 5.1, however, with only the two parameters δ and $A_y/A_x = \tan\gamma_0$ given in Equation (5.18); a third parameter such as the intensity $A_x^2 + A_y^2$ is missing, but is not essential for the state of polarization. In the complex χ-plane in Figure 5.2 (Azzam and Bashara, 1972) with Re χ and Im χ as coordinates $\delta\varepsilon(0,\pi)$, that is, the upper half-plane, corresponding to $\delta > 0$, represents right-handed and hence the lower half-plane left-handed polarization; $\delta=0,\pi$, that is, the x-axis, contains the linear polarization as depicted in Figure 5.2. For $\delta=0,\pi$, the real-valued number

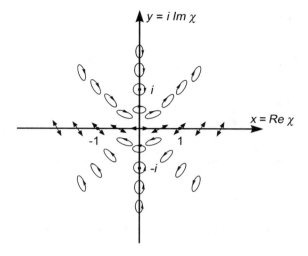

Figure 5.2   The complex plane for χ and the pertinent polarizations

$\chi = \pm \tan \gamma_0$ in Equation (5.9) stands for the angle γ_0 of the Jones vector; $\chi = 0$ means $A_y = 0$, and hence the electric field oscillates parallel to the x-axis, whereas at $\chi = \pm 1$ the oscillation occurs at an angle of $\pi/4$ and $3\pi/4$ to the x-axis, as is shown together with more examples in Figure 5.2. For $\delta = \pm \pi/2$ and $A_x = A_y$ representing circular polarization, we obtain $\chi = \pm i$. For $\delta \neq 0$, $\pm \pi/2$, π we obtain a point in the χ-plane with an elliptical polarization where the angle ψ of the principal axis is given with δ and γ_0 by Equations (5.7) and (5.18).

The characterization of the polarization state can also be based on the Jones vector (Yeh, 1988)

$$J = \begin{pmatrix} J_x \\ J_y \end{pmatrix} = \begin{pmatrix} A_x \ e^{i\delta_x} \\ A_y \ e^{i\delta_x} \end{pmatrix}.$$

$$(5.19)$$

J is represented in the cartesian coordinates in Figure 5.3 with the unity vectors x_0 and y_0. The components J_x and J_y in Equation (5.19) are complex. J will be normalized by satisfying the condition $\bar{J}J = 1$. We translate the results obtained for χ defined by Equation (5.18) into the Jones vectors J in Equation (5.19). From Equations (5.19) and (5.18), for the normalized form, we obtain

$$J_x = \cos \gamma_0 \qquad (5.20)$$

and

$$J_y = \sin \gamma_0 e^{i\delta} \qquad (5.21)$$

or

$$J = \begin{pmatrix} \cos \gamma_0 \\ \sin \gamma_0 e^{i\delta} \end{pmatrix}. \qquad (5.22)$$

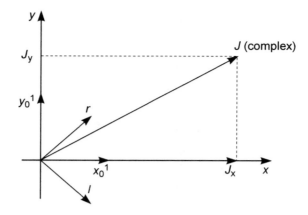

Figure 5.3 Jones vector J with its complex components J_x and J_y and the unity vectors r and l for circular polarization

As we already know, $\delta=0,\pi$ yields linear polarization and real χ-values, and hence also real Jones vectors J_x and J_y in Equations (5.20) and (5.21).

For $\gamma_0=0$ the vector oscillates in the direction of the x-axis. For phase $\delta=\pm\pi/2, A_x=A_y$, and hence $\tan\gamma_0=1$ and $\chi=\pm i$ stands for right-handed and left-handed circular polarization, and translates with Equation (5.22) into the Jones vector for right-handed circularly polarized light

$$J_r = r = \frac{1}{\sqrt{2}}\begin{pmatrix} 1 \\ i \end{pmatrix} \tag{5.23}$$

and for left-handed circularly polarized light into

$$J_l = l = \frac{1}{\sqrt{2}}\begin{pmatrix} 1 \\ -i \end{pmatrix}. \tag{5.24}$$

These unity vectors are shown in Figure 5.3. Obviously, r and l are orthogonal, as also demonstrated by $\bar{r}\cdot l = 0$. All other points with $\delta\neq0$, $\pm\pi/2$ are elliptically polarized for $\delta>0$ right-handed in the upper and for $\delta<0$ left-handed in the lower half plane in Figure 5.3.

Any light wave the E-vector of which is a continuous function of time and space can be composed from two orthogonal components lying in the x_0 and the y_0 directions. If an analyser is placed parallel to x_0, then the linearly polarized p-wave of the light passes, whereas the linearly polarized s-wave perpendicular to it is absorbed. Similarly, the orthogonal directions r and l may be chosen as the basis of an orthogonal coordinate-system, called *circular coordinates*.

We establish the laws for the transformation from cartesian into circular coordinates. From Figure 5.3, it follows that

$$r = \frac{1}{\sqrt{2}}(x_0 + iy_0). \tag{5.25}$$

$$l = \frac{1}{\sqrt{2}}(x_0 - iy_0). \tag{5.26}$$

Solving for x_0 and y_0 yields the transformation from circular to cartesian coordinates

$$x_0 = \frac{1}{\sqrt{2}}(r + l) \tag{5.27}$$

and

$$y_0 = \frac{-i}{\sqrt{2}}(r - l). \tag{5.28}$$

The vector E can be described in cartesian coordinates as

$$E = x_0 E_x + y_0 E_y,$$

where the components E_x and E_y are real. Replacing E by the Jones vector J with the complex components J_x and J_y, still in cartesian coordinates, we obtain

$$J = x_0 J_x + y_0 J_y. \tag{5.29}$$

In circular coordinates, J has the form

$$J = r J_r + l J_l. \tag{5.30}$$

Inserting Equations (5.27) and (5.28) into Equation (5.29) yields

$$J = \frac{1}{\sqrt{2}}(r + l)J_x - \frac{i}{\sqrt{2}}(r - l)J_y = \frac{1}{\sqrt{2}}r(J_x - iJ_y) + \frac{1}{\sqrt{2}}l(J_x + iJ_x),$$

from which, with Equation (5.30), follows

$$J_r = \frac{1}{\sqrt{2}}(J_x - iJ_y) \tag{5.31}$$

and

$$J_l = \frac{1}{\sqrt{2}}(J_x + iJ_y). \tag{5.32}$$

The matrix equation for this result is with the vector J_{ci} for the circular coordinates

$$J_{ci} = \begin{pmatrix} J_r \\ J_l \end{pmatrix} = \frac{1}{\sqrt{2}} \begin{pmatrix} 1 & -i \\ 1 & +i \end{pmatrix} \begin{pmatrix} J_x \\ J_y \end{pmatrix}, \tag{5.33}$$

or with the transformation matrix Ci from cartesian to circular coordinates

$$Ci = \frac{1}{\sqrt{2}} \begin{pmatrix} 1 & -i \\ 1 & +i \end{pmatrix}$$

(5.34)

and

$$Ci^{-1} = \frac{1}{\sqrt{2}} \begin{pmatrix} 1 & 1 \\ i & -i \end{pmatrix},$$

(5.35)

$$J_{ci} = \begin{pmatrix} J_r \\ J_l \end{pmatrix} = Ci \begin{pmatrix} J_x \\ J_y \end{pmatrix},$$

(5.36)

or

$$Ci^{-1}J_{ci} = \begin{pmatrix} J_x \\ J_y \end{pmatrix} = J.$$

(5.37)

The transmission of a light wave through a medium with transmission matrix T is, in cartesian coordinates according to Equation (4.9),

$$J_2 = TJ_1,$$

(5.38)

where J_1 is the Jones vector at the input and J_2 at the output of the medium. With Equations (5.36) and (5.37), Equation (5.38) is translated into the Jones vectors J_{2ci} and J_{1ci} in circular coordinates

$$CiJ_2 = J_{2ci} = CiTCi^{-1}J_{1ci},$$

(5.39)

where

$$T_{ci} = CiTCi^{-1}$$

(5.40)

is the transmission matrix into circular coordinates. Cartesian and circular coordinates are especially well suited for the investigation of polarized light propagating through isotropic and anisotropic optical systems, as the polarized wave can always be described by orthogonal linearly polarized components in cartesian coordinates, or by orthogonal right-handed and left-handed polarized components in circular coordinates.

For circular coordinates, the polarization can also be characterized by the ratio (Yeh, 1988)

$$\chi_c = \frac{J_r}{J_l}$$

(5.41)

similar to χ in Equation (5.18). The properties of χ_c are derived from Equations (5.31) and (5.32), and the properties of the Jones vector in Table 5.1. The results are also listed in Table

5.1. For linear polarization, the Jones vectors J_x and J_y are real, leading from Equations (5.31) and (5.32) to

$$\chi_c = \frac{J_r}{J_l} = \frac{J_x - iJ_y}{J_x + iJ_y} = \frac{\overline{J_x + iJ_y}}{J_x + iJ_y}, \tag{5.42}$$

and hence,

$$|\chi_c| = 1. \tag{5.43}$$

For the remaining cases the Jones vectors J_x and J_y in the second last column of Table 5.1 and Equations (5.41), (5.31) and (5.32) yield the χ_c-values given in the last column of the table.

A general Jones vector for elliptical polarization is given in cartesian coordinates by Equation (5.22), leading to $\chi = J_y/J_x = e^{i\delta} \tan \gamma_0$, which is already known, and to

$$\chi_c = \frac{J_x - iJ_y}{J_x + iJ_y} = \frac{\cos \gamma_0 - ie^{i\delta} \sin \gamma_0}{\cos \gamma_0 + ie^{i\delta} \sin \gamma_0}$$

or

$$\chi_c = e^{-i2 \arctan \frac{\cos \delta \tan \gamma_0}{1 + \sin \delta \tan \gamma_0}}. \tag{5.44}$$

In the next chapter, a differential equation for χ in Equation (5.18) will be investigated.

5.2 A Differential Equation for the Propagation of Polarized Light through Anisotropic Media

So far we have described the propagation of polarized light through an anisotropic medium by Jones matrices. An alternative consists in solving a differential equation for the polarization χ, according to Azzam and Bashara (1972a,b). As in all of our investigations, reflections at the LC layers are not included. This approach will reveal additional insight into the change of the polarization if linearly, elliptically or circularly polarized light propagates parallel to the optical axis z in Figure 4.1.

We first have to derive some equations on the polarization $\chi = J_y/J_x$ and the transmission matrix T of light represented by the Jones vectors J_1 at the input and J_2 at the output of an anisotropic medium, with the thickness z and with both of its two plane boundaries parallel to the x-y-plane in Figure 4.1. The pertinent matrix equation is

$$J_2 = \begin{pmatrix} J_{2x} \\ J_{2y} \end{pmatrix} = T(z)J_1 = \begin{pmatrix} t_{11} & t_{12} \\ t_{21} & t_{22} \end{pmatrix} \begin{pmatrix} J_{1x} \\ J_{2y} \end{pmatrix}. \tag{5.45}$$

The transmission matrix for a slice of thickness Δz located at z is

$$T(z, \Delta z) = T(z + \Delta z)T^{-1}(z). \tag{5.46}$$

A matrix N is defined with the unity matrix I as

$$N(z) = \lim_{\Delta z \to 0} \frac{T(z, \Delta z) - I}{\Delta z} = \lim_{\Delta z \to 0} \frac{T(z + \Delta z)T^{-1}(z) - I}{\Delta z} \tag{5.47}$$

or

$$N(z) = \lim_{\Delta z \to 0} \frac{T(z + \Delta z) - T(z)}{\Delta z} T^{-1}(z) = \frac{dT(z)}{dz} T^{-1}(z), \tag{5.48}$$

and hence

$$N(z)T(z) = \frac{dT(z)}{dz}. \tag{5.49}$$

Inserting Equations (5.49) and (5.45) into the derivative of Equation (5.45) provides

$$\frac{dJ_2}{dz} = \frac{dT(z)}{dz} J_1 = N(z)T(z)J_1 = N(z)J_2. \tag{5.50}$$

Expanding Equation (5.50) yields, for any Jones vector $J = \begin{pmatrix} J_x \\ J_y \end{pmatrix}$ dependent on z with

$$N(z) = \begin{pmatrix} n_{11} & n_{12} \\ n_{21} & n_{22} \end{pmatrix}, \tag{5.51}$$

$$\frac{dJ_x}{dz} = n_{11}J_x + n_{12}J_y \tag{5.52}$$

$$\frac{dJ_y}{dz} = n_{21}J_x + n_{22}J_y.$$

The polarization $\chi = J_y/J_x$ exhibits the first derivative

$$\frac{d\chi}{dz} = \frac{1}{J_x} \frac{dJ_y}{dz} - \frac{J_y}{J_x^2} \frac{dJ_x}{dz}. \tag{5.53}$$

Substituting Equation (5.52) and χ into Equation (5.53) provides

$$\frac{d\chi}{dz} = n_{12}\chi^2 + (n_{22} - n_{11})\chi + n_{21}. \tag{5.54}$$

This is an ordinary, nonlinear, first order differential equation for the polarization $\chi(z)$.

The derivation of Equation (5.54) was based on the transmission matrix T in x-y-coordinates. It is, however, the same for a transmission matrix in any coordinate system, e.g. also in circular coordinates and the pertinent Jones vector with its two components. Furthermore, the polarization χ can be based on the components of linearly or of circularly polarized light.

Equation (5.54) is now applied to the investigation of the propagation of light in the anisotropic medium in a TN cell presented in Chapter 4 (Section 4.1). The twist angle is

$$\varphi = \alpha_0 z. \tag{5.55}$$

We determine $T(z, \Delta z)$ in the first part of Equation (5.47) by considering a slice cut out of the LC cell, as in Chapter 4. The slice with thickness Δz located at z and with the rotation angle from the previous slice $\Delta\varphi = \alpha_0\Delta z$ exhibits the transmission matrix

$$
T(z, \Delta z) = \begin{pmatrix} e^{-ik_x\Delta z} & 0 \\ 0 & e^{-ik_y\Delta z} \end{pmatrix} \begin{pmatrix} \cos\alpha_0\Delta z & \sin\alpha_0\Delta z \\ -\sin\alpha_0\Delta z & \cos\alpha_0\Delta z \end{pmatrix}
$$

$$
= \begin{pmatrix} \cos\alpha_0\Delta z\,e^{-ik_x\Delta z} & \sin\alpha_0\Delta z\,e^{-ik_x\Delta z} \\ -\sin\alpha_0\Delta z\,e^{-ik_y\Delta z} & \cos\alpha_0\Delta z\,e^{-ik_y\Delta z} \end{pmatrix}.
$$

This approach only considers the rotation from the previous slice, and not the rotation from $z = 0$ to the slice under investigation at $z \neq 0$. Therefore, the coordinate system is automatically rotated with the sequence of slices. This differs from Grinberg and Jacobson (1976), where the result after each slice is rotated back to the original coordinates at $z = 0$. The approach presented exhibits less computational complexity. However, it should be borne in mind that the final result is automatically formulated in the cartesian coordinates rotated by $\varphi_d = \alpha_0 d$ in Equation (5.55).

According to Equation (5.47) we consider

$$
N(z) = \lim_{\Delta z \to 0} \begin{pmatrix} \dfrac{\cos\alpha_0\Delta z\,e^{-ik_x\Delta z} - 1}{\Delta z} & \dfrac{\sin\alpha_0\Delta z\,e^{ik_x\Delta z}}{\Delta z} \\ \dfrac{-\sin\alpha_0\Delta z\,e^{-ik_\Delta z}}{\Delta z} & \dfrac{\cos\alpha_0\Delta z\,e^{-ik_y\Delta z} - 1}{\Delta z} \end{pmatrix}
$$

from which we obtain, with Hopital's rule,

$$
N(z) = \begin{pmatrix} -ik_x & \alpha_0 \\ -\alpha_0 & -ik_y \end{pmatrix}. \tag{5.56}
$$

$N(z)$ is independent of z as $T(z, \Delta z)$ is the same for any location z. As the twist of the T cell matches the rotation of circular polarization, we continue the investigation in circular coordinates with the transmission matrix $T(z, \Delta z)$ transformed into these coordinates by Equation (5.40) with Ci and Ci^{-1} in Equations (5.34) and (5.35). The derivation of $N(z)$ in Equation (5.56) shows that we can also apply the transformation to $N(z)$, yielding $N_{ci}(z)$ for circular coordinates as

$$
N_{ci}(z) = \frac{1}{\sqrt{2}}\begin{pmatrix} 1 & -i \\ 1 & i \end{pmatrix}\begin{pmatrix} -ik_x & \alpha_0 \\ -\alpha_0 & -ik_y \end{pmatrix}\frac{1}{\sqrt{2}}\begin{pmatrix} 1 & 1 \\ i & -i \end{pmatrix} = \begin{pmatrix} -i\dfrac{k_x+k_y}{2}+i\alpha_0 & -i\dfrac{k_x-k_y}{2} \\ -i\dfrac{k_x-k_y}{2} & -i\dfrac{k_x+k_y}{2}-i\alpha_0 \end{pmatrix},
$$

$$\tag{5.57}$$

from which follows

$$n_{12} = n_{21} = -i\frac{k_x - k_y}{2} = -ig_0, \tag{5.58}$$

$$n_{22} - n_{11} = -i2\alpha_0, \tag{5.59}$$

and the differential equation for circular polarization χ_c,

$$\frac{d\chi_c}{dz} = ig_0\chi_c^2 - i2\alpha_0\chi_c - ig_0. \tag{5.60}$$

This nonlinear differential equation with constant coefficient is Riccati's differential equation, and is separable into

$$\frac{d\chi_c}{ig_0\chi_c^2 - i2\alpha_0\chi_c - ig_0} = dz,$$

providing the general solution with the integration constant C

$$\frac{1}{\rho}\arctan\frac{ig_0\chi_c - i\alpha_0}{\rho} = z + C \tag{5.61}$$

with

$$\rho = \sqrt{g_0^2 + \alpha_0^2}. \tag{5.62}$$

Solving for χ_c provides

$$\chi_c = \frac{\rho \tan \rho(z + C) + i\alpha_0}{ig_0}. \tag{5.63}$$

The initial state of polarization at $z=0$ is $\chi_c = \chi_0$, for which Equation (5.63) is satisfied with

$$\tan \rho C = \frac{1}{\rho}(ig_0\chi_0 - i\alpha_0). \tag{5.64}$$

Expanding the tan in Equation (5.63) and inserting Equation (5.64) finally establishes the solution

$$\chi_c = \frac{(\rho - i\alpha_0 \tan \rho z)\chi_0 - ig_0 \tan \rho z}{-ig_0 \tan \rho z \cdot \chi_0 + \rho + i\alpha_0 \tan \rho z}. \tag{5.65}$$

5.3 Special Cases for Propagation of Light

5.3.1 Incidence of linearly polarized light

The special incident state at $z=0$ to be investigated is linear polarization with an angle γ_0 to the x-axis given according to Table 5.1 by

$$\chi_0 = e^{-i2\gamma_0} = \cos 2\gamma_0 - i \sin 2\gamma_0, \tag{5.66}$$

the insertion of which into Equation (5.65) yields

$$\chi_c = \frac{\rho \cos 2\gamma_0 - \alpha_0 \tan \rho z \sin 2\gamma_0 - i(g_0 \tan \rho z + \alpha_0 \tan \rho z \cos 2\gamma_0 + \rho \sin 2\gamma_0)}{\rho - g_0 \tan \rho z \sin 2\gamma_0 + i(\alpha_0 \tan \rho z - g_0 \tan \rho z \cos 2\gamma_0)}. \tag{5.67}$$

Linear polarization at $z=d$ is reached for $|\chi_c| = 1$ (see Table 5.1). Calculating the square of the magnitudes in the numerator and the denominator of Equations (5.67) gives, for $|\chi_c| = 1$, the condition

$$\tan \rho z(\alpha_0 \tan \rho z \cos 2\gamma_0 + \rho \sin 2\gamma_0) = 0,$$

which is met for

$$\tan \rho z = 0, \text{ giving } \rho z = \nu\pi, \nu = 0, 1, 2, \ldots, \tag{5.68}$$

and for

$$\tan \rho z = -\frac{\rho}{\alpha_0} \tan 2\gamma_0. \tag{5.69}$$

We now treat a TN cell with a twist of $\beta=\pi/2$ at $z=d$. For this Equation (5.55) yields

$$\alpha_0 = \frac{\pi}{2d}. \tag{5.70}$$

With ρ in Equation (5.62), α_0 in Equation (5.70), g_0 in Equation (5.58) and k_x and k_y in Equations (3.33) and (3.34), the first solution given in Equation (5.68) results in

$$z = d = \frac{\sqrt{3}}{2} \frac{\lambda_0}{\Delta n}. \tag{5.71}$$

This result represents linearly polarized light with the wavelength λ_0 at an angle $\beta=\pi/2$ at $z=d$ independent of γ_0. Hence, especially for $\gamma_0=0$, Equation (5.69) leads to the same solution. A crossed analyser with angle $\pi/2$ to the polarizer allows this light to pass (normally white mode), whereas a parallel analyser blocks it (normally black mode). For twist angles $\beta > \pi/2$ at $z=d$, Equation (5.55) gives a different α_0. This is the case for supertwist LCDs, discussed in Chapter 4 (Section 4.2.2).

Additional solutions are found for $\gamma_0 \neq 0$. Equation (5.69) yields after insertion of ρ and α_0 from Equations (5.62) and (5.70) for $z=d$

$$\tan^2 \rho d = \frac{g_0^2 + (\beta/d)^2}{(\beta/d)^2} \tan^2 2\gamma_0, \tag{5.72}$$

where β stands for the twist angle at $z=d$. This is a nonlinear Equation for d which can be solved numerically.

Inserting Equation (5.69) into Equation (5.65) for $\chi_0 = e^{-i2\gamma_0}$ provides the polarization χ_c for the second occurrence of a linear polarization at $z=d$, as

$$\chi_c = \frac{(\rho + i\rho \tan 2\gamma_0)e^{-i2\gamma_0} + i(g_0/\alpha_0)\rho \tan 2\gamma_0}{ig_0(\rho/\alpha_0) \tan 2\gamma_0 e^{-i2\gamma_0} + \rho - i\rho \tan 2\gamma_0}$$

$$= \frac{(\cos 2\gamma_0 + i \sin 2\gamma_0)e^{-i2\gamma_0} + i(g_0/\alpha_0) \sin 2\gamma_0}{i(g_0/\alpha_0) \sin 2\gamma_0 e^{-i2\gamma_0} + \cos 2\gamma_0 - i \sin 2\gamma_0}.$$

With $\cos 2\gamma_0 + i \sin 2\gamma_0 = e^{i2\gamma_0}$, this simplifies to

$$\chi_c = e^{i2\gamma_0}. \tag{5.73}$$

This indicates linear polarization with a negative angle $-\gamma_0$ to the x-axis.

As mentioned before, the coordinate system has automatically been rotated at $z=d$ by the angle $\beta = \alpha_0 d$, leading to an additional factor for the phase shift of $e^{-i2\alpha_0 d}$ in the polarization χ_c. This results in a total polarization

$$\chi_{c\,total} = e^{-i2(\alpha_0 d - \gamma_0)}. \tag{5.74}$$

We now calculate under which conditions χ_c in Equation (5.67) represents circularly polarized light. Right-handed circular polarization occurs for $\chi_c = \infty$, leading to

$$\tan \rho z = \frac{\rho}{g_0 \sin 2\gamma_0} \tag{5.75}$$

and

$$\cos 2\gamma_0 = \frac{\alpha_0}{g_0}. \tag{5.76}$$

The light is left-handed polarized for $\chi_c = 0$, yielding

$$\tan \rho z = \frac{\rho}{\alpha_0} \cot an 2\gamma_0 \tag{5.77}$$

and

$$\cos 2\gamma_0 = -\frac{\alpha_0}{g_0}. \tag{5.78}$$

The tan ρz in Equation (5.75) has two solutions, one for

$$\rho z = \arctan\frac{\rho}{g_0}\frac{1}{\sin 2\gamma_0},\tag{5.79}$$

and one for

$$\rho z = \pi + \arctan\frac{\rho}{g_0}\frac{1}{\sin 2\gamma_0}.\tag{5.80}$$

The same difference of π applies to the solution of Equation (5.77).

5.3.2 Incident light is circularly polarized

The result in Equation (5.65) is able to handle general input polarizations other than the linear polarization treated so far. For right-handed circular polarization characterized by $\chi_0 = \infty$ in Table 5.1, we obtain, from Equation (5.65),

$$\chi_c = \frac{\rho - i\alpha_0 \tan \rho z}{-ig_0 \tan \rho z},\tag{5.81}$$

which is again right-handed circularly polarized if $\chi_c = \infty$. This occurs for

$$\tan \rho z = 0 \quad \text{that is, for} \quad \rho z = \nu\pi, \quad \nu = 1,2,3\ldots.$$

This is the same solution as in Equation (5.71) with

$$z = \frac{\sqrt{3}\lambda_0}{2\Delta n}.\tag{5.82}$$

Hence, both the linear and circular polarization at the input $z=0$ of a TN cell repeats itself for the wavelength λ_0 with the same value z in Equation (5.82).

Left-handed polarization at location z would require $\chi_c = 0$, which is impossible for Equation (5.81).

Linear polarization occurs for $|\chi_c| = 1$ in Equation (5.81), which leads to

$$\tan \rho z = \frac{\rho}{\sqrt{g_0^2 - \alpha_0^2}}\tag{5.83}$$

or

$$z = \frac{1}{\rho}\arctan\frac{\rho}{\sqrt{g_0^2 - \alpha_0^2}}.\tag{5.84}$$

At this point, linear polarization occurs as a response to right-handed polarized light at the input.

On the other hand, left-handed circular polarization at $z=0$ is given by $\chi_0=0$ (see Table 5.1), resulting, with Equation (5.65), in

$$\chi_c = \frac{-ig_0 \tan \rho z}{\rho + i\alpha_0 \tan \rho z} \qquad (5.85)$$

Left-handed circular polarization again occurs for $\chi_c=0$ that is for $\tan \rho z=0$ with $\rho z=\nu\pi, \nu=1,2,\ldots$ and z in Equation (5.83). Also in this case, right-handed polarization defined by $\chi_c=\infty$ is not possible in Equation (5.85). Hence, a TN cell fed at $z=0$ with right-handed (resp. left-handed) circularly polarized light only shows right-handed (resp. left-handed) light at $z>0$ inside the cell. Equation (5.85) leads to a linearly polarized solution for z in Equation (5.84).

6

Propagation of Light with an Arbitrary Incident Angle through Anisotropic Media

6.1 Basic Equations for the Propagation of Light

In Chapters 3 and 4 we treated the case of incident light with a wave vector k parallel to the z-axis in Figure 4.1. This direction was for both the untwisted and twisted case in the field-free state perpendicular to the director $\vec{n}$ on the alignment layer. The more complex case is an arbitrary angle of incidence of light with respect to the director, as shown in Figure 6.1. The medium is anisotropic, meaning that the relative dielectric constant ε_{ij} and the refractive indices n_{ij} depend upon the direction, e.g. measured in orthogonal coordinates x, y and z. We assume, however, that the medium is homogeneous in layers with parallel plane borders. It is possible to find principal coordinates x', y' and z' in Figure 6.1 in which the tensor ε of the dielectric constants exhibits only zeros off the main diagonal given by

$$\varepsilon = \begin{pmatrix} \varepsilon_{x'} & 0 & 0 \\ 0 & \varepsilon_{y'} & 0 \\ 0 & 0 & \varepsilon_{z'} \end{pmatrix} = \begin{pmatrix} n_{x'}^2 & 0 & 0 \\ 0 & n_{y'}^2 & 0 \\ 0 & 0 & n_{z'}^2 \end{pmatrix}. \tag{6.1}$$

The relation between the permittivity and the refractive index was introduced in Equation (2.6).

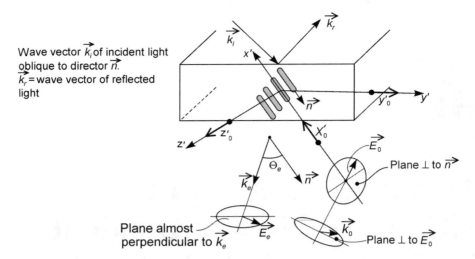

Wave vector $\vec{k}_i$ of incident light oblique to director $\vec{n}$.
$\vec{k}_r$ = wave vector of reflected light

Figure 6.1 The direction of $\vec{k}_0$ and $\vec{E}_0$ of the ordinary beam and of $\vec{k}_e$ and $\vec{E}_e$ of the extraordinary beam in a uniaxial LC medium

For the case of birefringent liquid crystals in Figure 6.1, the principal x'-axis is parallel to $\vec{n}$, and hence we obtain

$$\varepsilon_{x'} = \varepsilon_{\parallel}; \tag{6.2}$$

$$\varepsilon_{y'} = \varepsilon_{z'} = \varepsilon_{\perp}; \tag{6.3}$$

$$n_{x'} = n_{\parallel} = n_e; \tag{6.4}$$

$$n_{y'} = n_{z'} = n_{\perp} = n_0, \tag{6.5}$$

where the refractive index for the ordinary beam is n_0 and for the extraordinary beam n_e according to Equations (2.3) and (2.4). Two values of the dielectric constants and of the refractive indices are respectively equal, which defines a uniaxial system. Birefringent liquid crystals are uniaxial. If all three values in the pairs of Equations (6.2) (6.3) and (6.4) (6.5) are different, the medium is biaxial. The medium is isotropic if all three values are equal.

The problem to be solved is outlined in Figure 6.1. The given incident light with wave vector k_i gives rise to reflected light with wave vector k_r and, as we shall see, to the refracted ordinary beam with k_0 and the extraordinary beam with k_e. In this chapter, we shall determine the nature of the ordinary and extraordinary beams which, according to Maxwell's equations, can propagate through the LC cell. The derivation is based on Yeh (1988) and Zeile (2001).

The Jones vectors of the electric and magnetic planar and harmonic field are

$$\vec{E} = Ee^{i(\omega t - \vec{k}\vec{r})} \tag{6.6}$$

$$\vec{H} = He^{i(\omega t - \vec{k}\vec{r})} \tag{6.7}$$

with the components in the x', y' and z'-coordinates $E_\nu e^{i(\omega t - \vec{k}_\nu \vec{r}_\nu)}$ and $H_\nu e^{i(\omega t - \vec{k}_\nu \vec{r}_\nu)}$, $\nu = x'$, y' and z'. We start with Maxwell's equations

$$rot\,\vec{E} = \nabla x \vec{E} = -\frac{\partial \mu \vec{H}}{\partial t} \tag{6.8}$$

and

$$rot\,\vec{H} = \nabla x \vec{H} = \frac{\partial \varepsilon \vec{E}}{\partial t} = \frac{\partial \vec{D}}{\partial t}, \tag{6.9}$$

where $\vec{D} = \varepsilon \vec{E}$ stands for the electrical displacement and $\nabla x \vec{E}$ for

$$\nabla x \vec{E} = \begin{pmatrix} x'_0 & y'_0 & z'_0 \\ \dfrac{\partial}{\partial x'} & \dfrac{\partial}{\partial y'} & \dfrac{\partial}{\partial z'} \\ E_{x'} & E_{y'} & E_{z'} \end{pmatrix}. \tag{6.10}$$

x'_0, y'_0 and z'_0 are the unity vectors in the x', y' and z' directions in Figure 6.1.

Substitution for $\vec{E}$ and $\vec{H}$ from Equations (6.6) and (6.7), respectively, into Equations (6.8) and (6.9) yields

$$\vec{k} x \vec{E} = \omega \mu \vec{H} \tag{6.11}$$

and

$$\vec{k} x \vec{H} = -\omega \varepsilon \vec{E} = -\omega \vec{D}. \tag{6.12}$$

It is interesting to note from Equation (6.12) that the wave vector $\vec{k}$ is always perpendicular to $\vec{D}$. Only if ε is a scalar is it also perpendicular to $\vec{E}$. In the general case of birefringence, ε is a tensor, and hence k is in general no longer perpendicular to $\vec{E}$.

Inserting $\vec{H}$ in Equation (6.11) into Equation (6.12) provides

$$\vec{k} x (\vec{k} x \vec{E}) + \omega^2 \mu\, \varepsilon\, \vec{E} = 0. \tag{6.13}$$

Calculating the principal axes x', y' and z' in Figure 6.1 with ε in Equation (6.1) translates Equation (6.13) into

$$\begin{pmatrix} \omega^2 \mu \varepsilon_{x'} - k_{y'}^2 - k_{z'}^2 & k_{x'}k_{y'} & k_{x'}k_{z'} \\ k_{x'}k_{y'} & \omega^2 \mu \varepsilon_{y'} - k_{x'}^2 - k_{z'}^2 & k_{y'}k_{z'} \\ k_{x'}k_{z'} & k_{y'}k_{z'} & \omega^2 \mu \varepsilon_{z'} - k_{x'}^2 - k_{y'}^2 \end{pmatrix} \begin{pmatrix} E_{x'} \\ E_{y'} \\ E_{z'} \end{pmatrix} = 0. \tag{6.14}$$

This equation exhibits only nontrivial solutions if the determinant vanishes, resulting, with Equations (6.1), (6.4) and (6.5), with the wave vector in vacuum $k_{vac}=(\omega/c)$ from Equation (3.9) and c in Equation (3.3), in

$$
\begin{vmatrix}
k_{vac}^2 n_e^2 - k_{y'}^2 - k_{z'}^2 & k_{x'}k_{y'} & k_{x'}k_{z'} \\
k_{x'}k_{y'} & k_{vac}^2 n_0^2 - k_{x'}^2 - k_{z'}^2 & k_{y'}k_{z'} \\
k_{x'}k_{z'} & k_{y'}k_{z'} & k_{vac}^2 n_0^2 - k_{x'}^2 - k_{y'}^2
\end{vmatrix} = 0. \qquad (6.15)
$$

A longer but straightforward calculation finally leads to

$$
\left(\frac{k^2}{n_0^2} - k_{vac}^2 \right) \left(\frac{k_{y'}^2 + k_{z'}^2}{n_e^2} + \frac{k_{x'}^2}{n_0^2} - k_{vac}^2 \right) = 0, \qquad (6.16)
$$

where

$$
k^2 = k_{x'}^2 + k_{y'}^2 + k_{z'}^2. \qquad (6.17)
$$

The first solution of Equation (6.16) is

$$
k^2 = n_0^2 k_{vac}^2 = k_0^2 = k_{0x'}^2 + k_{0y'}^2 + k_{0z'}^2. \qquad (6.18)
$$

This is the square of the wave vector k_0 of the ordinary beam, as it depends only upon $n_0 = n_\perp$. Due to Equation (6.18), k_0 lies on a sphere with radius $n_0\, k_{vac}$, as shown in the cross-section in Figure 6.2(a).

The second solution of Equation (6.16) provides the extraordinary beam

$$
k_e = k_{ex'}x_0' + k_{ey'}y_0' + k_{ez'}z_0' \qquad (6.19)
$$

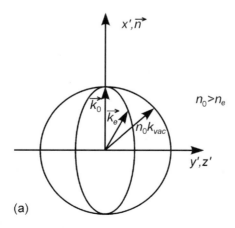

(a)

Figure 6.2   Cross-sections through the sphere of k_0 and the ellipsoid of revolution of k_e for (a) $n_0>n_e$, (b) $n_0<n_e$

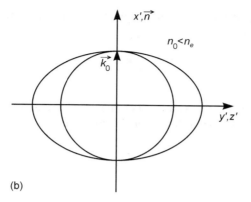

Figure 6.2   (continued)

as

$$\frac{k_{ey'}^2 + k_{ez'}^2}{n_e^2} + \frac{k_{ex'}^2}{n_0^2} = k_{vac}^2. \tag{6.20}$$

This represents an ellipsoid of revolution, depicted for $n_0 > n_e$ in Figure 6.2(a) and for $n_0 < n_e$ in Figure 6.2(b).

The vector $\vec{E}_0$ belonging to $\vec{k}_0$ is obtained by inserting Equation (6.18) into Equation (6.14), resulting with Equations (6.1), (6.6) and (3.9) in

$$
\begin{array}{rcl}
\left(k_{vac}^2\left(n_e^2 - n_0^2\right) + k_{0x'}^2\right)E_{x'} & + & k_{0x'}k_{0y'}E_{y'} & + & k_{0x'}k_{0z'}E_{z'} & = & 0 \\
k_{0x'}k_{0y'}E_{x'} & + & k_{0y'}^2 E_{y'} & + & k_{0y'}k_{0z'}E_{z'} & = & 0. \\
k_{0x'}k_{0z'}E_{x'} & + & k_{0y'}k_{0z'}E_{y'} & + & k_{0z'}^2 E_{z'} & = & 0
\end{array} \tag{6.21}
$$

Due to the singular coefficient matrix as demonstrated in Equation (6.15), the solution to Equations (6.21) depends upon an arbitrary constant C_0, and is

$$\vec{E}_0 = C_0(k_{0z}\,y_{0'} - k_{0y}z_{0'}). \tag{6.22}$$

Obviously, $\vec{E}_0$ lies in the $y_{0'} - z_{0'}$-plane, and hence $\vec{E}_0 \perp x_{0'}$ or $\vec{E}_0 \perp \vec{n}$ holds. In Figure 6.1, only one vector $\vec{E}_0$ of the possible vectors in the plane perpendicular to $\vec{n}$ is drawn. As a consequence of its direction, $\vec{E}_0$ only experiences the scalar permittivity $\varepsilon_\perp$, from which it follows that $\vec{k}_0 \perp \vec{E}_0$, as depicted in Figure 6.1. This $\vec{k}_0$ is also drawn in Figure 6.1. The unity vector O_0 in the direction of $\vec{E}_0 \perp \vec{k}_0$ and $\vec{E}_0 \perp \vec{n}$ is given by

$$O_0 = \frac{\vec{k}_0 \times \vec{n}}{|\vec{k}_0 \times \vec{n}|}. \tag{6.23}$$

The dielectric displacement $\vec{D}_0 = \varepsilon_\perp \vec{E}_0$ is parallel to $\vec{E}_0$.

The wave vector k_e of the extraordinary beam is derived from the second solution in Equation (6.16) defined by

$$\frac{k_{ey}^2 + k_{ez}^2}{n_e^2} + \frac{k_{ex}^2}{n_0^2} - k_{vac}^2 = 0. \tag{6.24}$$

For the determination of k_e, we assume a (so far) unknown angle Θ_e between $\vec{n}$ and $\vec{k}_e$, as shown in Figure 6.1. From this follows, for the components of k_e,

$$k_{ex'} = k_e \cos \Theta_e \tag{6.25}$$

and

$$k_{ey'}^2 + k_{ez'}^2 = k_e^2 \sin^2 \Theta_e, \tag{6.26}$$

and hence for Equation (6.24),

$$k_e^2 \left(\frac{\sin^2 \Theta_e}{n_e^2} + \frac{\cos^2 \Theta_e}{n_0^2} \right) = k_{vac}^2. \tag{6.27}$$

We introduce an effective refractive index

$$n_f = \frac{k_e}{k_{vac}} \tag{6.28}$$

and insert it into Equation (6.27), leading to

$$\frac{\sin^2 \Theta_e}{n_e^2} + \frac{\cos^2 \Theta_e}{n_0^2} = \frac{1}{n_f^2} \tag{6.29}$$

or

$$\left(\frac{n_f \sin \Theta_e}{n_e} \right)^2 + \left(\frac{n_f \cos \Theta_e}{n_0} \right)^2 = 1 \tag{6.30}$$

or

$$n_f = \frac{n_0 n_e}{\sqrt{n_0^2 \sin^2 \Theta_e + n_e^2 \cos^2 \Theta_e}}. \tag{6.31}$$

The ellipsoid of revolution in Equation (6.30) with the axis $n_f \sin \Theta_e$ and $n_f \cos \Theta_e$ is depicted in Figure 6.3(a).

This effective index of refraction will greatly simplify investigations of anisotropic media by treating the propagation of light with an angle Θ to $\vec{n}$ with a refractive index n_f in Equation (6.31), and with

$$k_e = n_f k_{vac}, \tag{6.32}$$

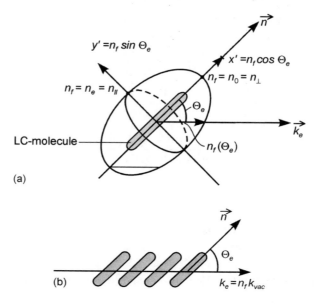

$y' = n_f \sin \Theta_e$

$x' = n_f \cos \Theta_e$

$n_f = n_o = n_\perp$

$n_f = n_e = n_\parallel$

Θ_e

LC-molecule

$\vec{n}$

$\vec{k}_e$

$n_f(\Theta_e)$

(a)

$\vec{n}$

Θ_e

(b) $k_e = n_f k_{vac}$

Figure 6.3   (a) The ellipsoid of revolution representing n_f in Equation (6.30); (b) propagation of the extraordinary beam obliquely through LC molecules

as depicted in Figure 6.3(b). The light with the wave vector k_0 provides the second refractive index $n_0 = n_\perp$.

$\vec{E}_e$ is determined by inserting Equation (6.24) into Equation (6.14) with the singular coefficient matrix in the form of Equation (6.15). The solution is

$$E_{ey'} = -\frac{k_{ex'}k_{ey'}}{k_{ey'}^2 + k_{ez'}^2}\left(\frac{n_e}{n_0}\right)^2 E_{ex'}, \tag{6.33}$$

$$E_{ez'} = -\frac{k_{ex'}k_{ez'}}{k_{ey'}^2 + k_{ez'}^2}\left(\frac{n_e}{n_0}\right)^2 E_{ex'}, \tag{6.34}$$

where $E_{ex'}$ is a free constant, as the determinant in Equation (6.15) is zero.

Another form of this solution is obtained by starting with

$$n_f = \frac{k_e}{k_{vac}} = \left(\frac{k_{ex'}^2 + k_{ey'}^2 + k_{ez'}^z}{((k_{ey'}^2 + k_{ez'}^2)/n_e^2) + (k_{ex'}^2/n_0^2)}\right)^{1/2}, \tag{6.35}$$

where k_{vac}^2 in Equation (6.24) was used.

The equation

$$\frac{n_f^2 - n_0^2}{n_0^2 n_e^2} = \frac{n_f^2}{n_0^2 n_e^2} - \frac{1}{n_e^2} = \frac{k_{ex'}^2 + k_{ey'}^2 + k_{ez'}^2}{n_0^2 n_e^2 \left(((k_{ey'}^2 + k_{ez'}^2)/n_e^2) + (k_{ex'}^2/n_0^2)\right)} - \frac{1}{n_e^2}$$

where Equation (6.35) was used, yields

$$\frac{n_f^2 - n_0^2}{n_0^2} = \frac{(n_e^2 - n_0^2)(k_{ey'}^2 + k_{ez'}^2)}{n_0^2(k_{ey'}^2 + k_{ez'}^2) + n_e^2 k_{ex'}^2} + \frac{1}{n_e^2}. \tag{6.36}$$

Similarly, we obtain

$$\frac{n_f^2 - n_e^2}{n_0^2} = \frac{(n_0^2 - n_e^2)k_{ex'}^2}{n_0^2(k_{ey'}^2 + k_{ez'}^2) + n_e^2 k_{ex'}^2} + \frac{1}{n_0^2}, \tag{6.37}$$

from which

$$\frac{n_f^2 - n_e^2}{n_f^2 - n_0^2} = -\left(\frac{n_e}{n_0}\right)^2 \frac{k_{ex'}^2}{k_{ey'}^2 + k_{ez'}^2} \tag{6.38}$$

follows. With Equation (6.38), Equations (6.33) and (6.34) assume the form

$$E_{ey'} = \frac{k_{ey'}}{k_{ex'}} \frac{n_f^2 - n_e^2}{n_f^2 - n_0^2} E_{ex'} \tag{6.39}$$

and

$$E_{ez'} = \frac{k_{ez'}}{k_{ex'}} \frac{n_f^2 - n_e^2}{n_f^2 - n_0^2} E_{ex'}. \tag{6.40}$$

$E_{ex'}$ can be chosen as a free constant. For reasons of symmetry, we choose

$$E_{ex'} = C_e \frac{k_{ex'}}{n_f^2 - n_e^2}, \tag{6.41}$$

with the arbitrary constant C_e yielding

$$\vec{E}_e = C_e \left(\frac{k_{ex'}}{n_f^2 - n_e^2} x_{0'} + \frac{k_{ey'}}{n_f^2 - n_0^2} y_{0'} + \frac{k_{ez'}}{n_f^2 - n_0^2} z_{0'} \right). \tag{6.42}$$

The pertinent electric displacement is $\vec{D}_e = \varepsilon \vec{E}_e = d_{e0} \vec{D}_e$, where ε is the tensor in Equation (6.1) and d_{e0} is the unity vector in the direction of $\vec{D}_e$. According to Equation (6.12), $\vec{D}_e \vec{k}_e = 0$ always holds. On the other hand, for $\vec{E}_e \vec{k}_e$ we obtain, with Equations (6.39) and (6.40), and the components of k_e,

$$\vec{E}_e \vec{k}_e = E_{ex'} k_{ex'} \left(1 - \frac{n_e^2}{n_0^2} \right) \neq 0, \tag{6.43}$$

indicating that $\vec{E}_e$ is not perpendicular to $\vec{k}_e$. For conventional LC materials, $|1 - (n_e^2/n_0^2)|$ lies between 0.1 and 0.2. Therefore, the difference in the directions of $\vec{k}_e$ and $\vec{E}_e$ is relatively small. The plane for $\vec{E}_e$ almost perpendicular to $\vec{k}_e$ is shown in Figure 6.1.

We determine the unity vector e_0 in

$$\vec{E}_e = e_0 E_e. \tag{6.44}$$

In Yeh (1988), the approximation $e_0 \approx d_{e0}$ is used, whereas the exact solution is derived from $\vec{E}_e = \varepsilon^{-1} \vec{D}_e$ for (Zeile, 2001)

$$e_0 = \frac{\varepsilon^{-1} \vec{D}_e}{|\varepsilon^{-1} \vec{D}_e|}. \tag{6.45}$$

The total solution consisting of the ordinary and extraordinary beam is

$$E_{\text{tot}} = \vec{E}_0 e^{-\vec{k}_0 \vec{r}} + \vec{E}_e e^{-\vec{k}_e \vec{r}}. \tag{6.46}$$

For the direction of $\vec{E}_0$ and $\vec{E}_e$, only the planes in which these vectors lie can be obtained. The specific direction in these planes is obtained by investigating the propagation of the incident light with wave vector $\vec{k}_i$ in the LC medium. This task will not be treated here; solutions can be found in Yeh (1988). We shall only briefly outline the steps taken for solving the problem. First, we have to satisfy the boundary conditions for the tangential components for the electrical field of the incident and the reflected light for the pertinent wave vectors outside the LC medium and for the field and the wave vectors of the ordinary and extraordinary beams inside the medium. Then, the known solutions for $\vec{k}_0$, $\vec{E}_0$, $\vec{k}_e$ and $\vec{E}_e$ with the free constants C_0 and C_e provide special solutions for $\vec{E}_0$ and $\vec{E}_e$ in their planes in Figure 6.1. The inclusion of the reflected beam leads to 2×2 matrices instead of the column matrices for the field used so far (Berreman, 1972).

6.2 Enhancement of the Performance of LC Cells

6.2.1 The degradation of picture quality

So far we have investigated the optical properties of an LC cell viewed perpendicularly to the glass plates. An oblique viewing direction is given by the azimuth angle ϕ and the off-axis angle Θ_e in Figure 6.4. As a rule, the larger Θ_e becomes, the more the optical performance is degraded. There is a loss of contrast and a grey level inversion, especially in the lower and upper vertical directions, as well as a change in colour for larger angles ϕ. A typical conoscopic image of an LCD, in which the contrast is plotted versus the two angles in Figure 6.4, is shown in Figure 6.5 (Haas, 1999). Characteristic curves for a display indicate where the contrast decreases below $10:1$, and where a grey shade inversion takes place. The meaning of a grey level inversion is explained in Figure 6.6, where the lower level of luminance $g8$ becomes brighter than the next brighter level $g7$ (Haas, 1999). A feature degrading with increasing Θ_e is the fairly equal spacing of the grey scales with equal steps in voltage V_{LC}. Thus, grey shade performance is dependent upon Θ_e and is obviously poor on the right side of Figure 6.6. Measures against degraded contrast and the associated colour

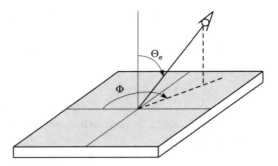

Figure 6.4 Viewing direction defined by azimuth angle Φ and off-axis angle Θ_e

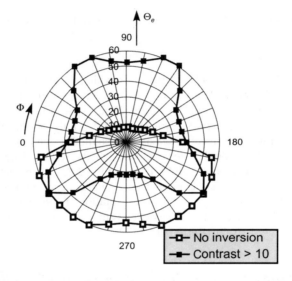

Figure 6.5 Curves of isocontrast and of limit for grey shade inversion of an LCD

changes for an increased viewing angle are compensation foils with negative birefringence, adding an inverse cell with negative or positive birefringence and the in-plane switching mode (IPS); the grey scale inversion is avoided by multi-domain pixels, IPS and by foils with positive birefringence (Haas, 1999; Yeh and Gu, 1999).

6.2.2 Optical compensation foils for the enhancement of picture quality

The enhancement of contrast

Crucial for the enhancement of contrast over a wide viewing angle is the improvement of the black state. In vertically aligned LCDs such as the DAP cell with crossed polarizers in the field-off state, or the TN cell with crossed polarizers in the field-on state, there is an

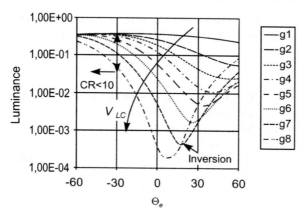

Figure 6.6 Grey level inversion versus Θ_e in the picture of an LCD

excellent black state independent of λ and d only when viewed in the vertical direction, for which the optical retardation is zero. When viewed under an oblique angle Θ_e, as depicted in Figure 6.7, the retardation is

$$\Delta n_f d_0 = (n_f(\Theta_e) - n_0)d_0 \neq 0, \tag{6.47}$$

with n_f in Equation (6.31) and d_0 standing for the effective cell thickness. The index ellipsoid in Figure 6.3(a) with $\vec{n}$ as the axis of revolution is drawn for one LC molecule in Figure 6.7, exhibiting a positive birefringence $\Delta n = n_e - n_0 > 0$. This is true for nematic LC materials.

We obtain a cigar-shaped index ellipsoid. Because of the non-vanishing retardation in Equation (6.47), the observer does not see a perfect black state at an angle Θ_e. Some light leaks through spoiling the black state. This leakage can be suppressed by adding a negative retardation to $\Delta n_f d_0$ in Equation (6.47) (Eblen *et al.*, 1994). This is achieved by a foil with a

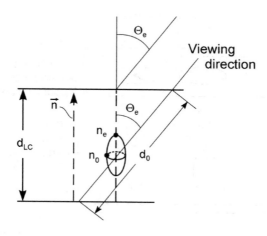

Figure 6.7 Optical retardation for oblique viewing angle $\Theta_e \neq 0$

negative birefringence due to the indexes $n'_e < n'_0$, as shown in Figure 6.8. The optical retardation along the viewing path in Figure 6.8 is required to be

$$\Delta n_f d_0 + \Delta n'_f\, d'_0 \approx 0, \qquad (6.48)$$

with $\Delta n'_f = n'_f(\Theta) - n'_0 < 0$ for the negative birefringence foil in Figure 6.8 with a pancake-shaped index ellipsoid. This shape may be realized by discotic liquid crystals, as mentioned in Chapter 2, from which the term *discotic foils* is derived.

A homogeneous and uniaxial birefringent plate with the optical axis perpendicular to the surface of the plate is called a c-plate. In Figure 6.8, the liquid crystal is a positive and the discotic film a negative c-plate.

The compensation in Equation (6.48) works from $\Theta_e = 0$ up to a fairly large angle around $\Theta_e = 60°$. This is demonstrated by the $10:1$ isocontrast curves in Figure 6.9(a) for an ECB cell with and without a discotic compensation film. The range without grey inversion has also been widened, as depicted in Figure 6.9(b). Left and right viewing symmetry is improved by placing one discotic film of half the thickness on each side of the LCD cell, as depicted schematically in Figure 6.10. The improved symmetry of contrast and limit of grey shade inversion is demonstrated in Figure 6.11 for the ECB cell of Figures 6.9(a) and (b). Discotic films do not add retardation in the vertical viewing direction as is desired for a good black state in this direction.

Compensation foils for LC molecules with different optical axes

The compensation discussed so far was based on vertically aligned LC molecules in Figure 6.7. In TN cells, however, the molecules close to the two orientation layers stay aligned with directors parallel to the surface of the glass plates and perpendicular to each other. Also, the optical retardation of these layers has to be compensated by negative birefringence, with discotic layers the directors of which are also parallel to the surface and perpendicular to

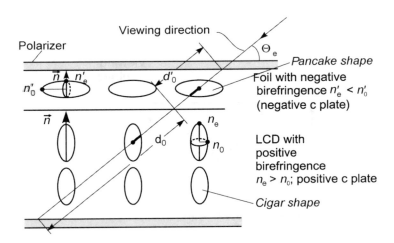

Figure 6.8   Compensation of retardation with a negative c plate

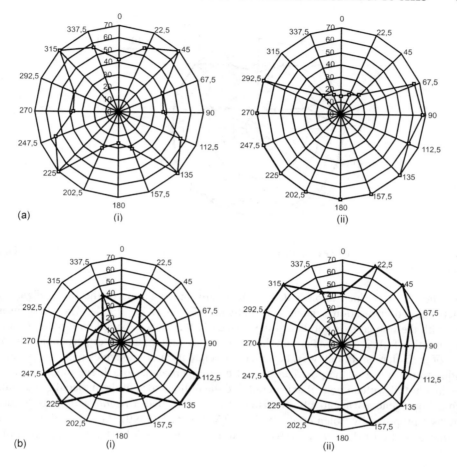

Figure 6.9 (a) A 10:1 isocontrast curve for an ECB cell with (ii) and without (i) a discotic compensation film (Haas, 1999); (b) the limit curves of grey inversion in an ECB cell with (ii) and without (i) a discotic film

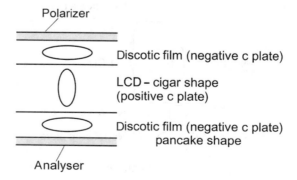

Figure 6.10   Discotic films of half thickness on either side of LCD

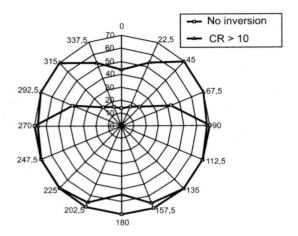

Figure 6.11   Symmetrical curves for 10:1 isocontrast and for limit of grey inversion of cell in Figures 6.9(a) and 6.9(b)

each other. This compensation is done individually on each side of the LCD. Figure 6.12 depicts this compensation scheme schematically (Eblen *et al.*, 1994). The compensation layer with the optical axis parallel to the surface of the layer is called an *a-plate*. In all the compensation foils introduced so far, the directors are either parallel or perpendicular to the plane of the polarizer. In that way, however, the retardation of molecules with an optical axis oblique to the plane of the polarizer cannot be compensated. To achieve this, a negatively birefringent foil with an optical axis parallel to the oblique axis of the molecules is required. Such a homogeneous negatively birefringent plate is called a *negative o-plate*. Solutions have been proposed by Eblen *et al.* (1997) and Witte *et al.* (1999). An alternative is the film introduced by Fuji (Mori *et al.*, 1997). According to Figure 6.13, the configuration of the

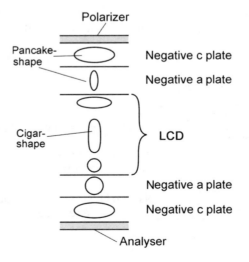

Figure 6.12   Compensation of the retardations of the midlayer and surface layer molecules

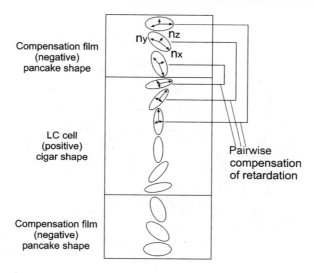

Compensation film
(negative)
pancake shape

ny nz
nx

LC cell
(positive)
cigar shape

Pairwise
compensation
of retardation

Compensation film
(negative)
pancake shape

Figure 6.13 The compensation of retardations of an LCD by discotic films with the same direction of optical axes

molecules of an LCD in the black state is repeated with parallel optical axes by the molecules of a negatively birefringent discotic film, for the known reasons of symmetry, with half the thickness on either side of the LCD. The molecules compensate each other pair-wise, as indicated in Figure 6.13. Had we the same number of molecules in the two discotic films and in the LCD, we would obtain a perfect compensation. As a rule, the number of molecules in the discotic film is smaller, but it still produces very good compensation. The vertical viewing direction inherently still exhibits a very good black state. Losses in the films render the display a little darker. The result is demonstrated by the isocontrast curve and the limit for grey scale inversion shown in Figure 6.14.

The black state of TN cells viewed perpendicular to the glass plates can be improved again by a pair-wise compensation but this time with a positively birefringent compensation film (Scheffer and Nehring, 1998). This is shown in Figure 6.15, in which a TN-LCD is followed by a second non-addressed TN-LCD cell. This set-up is mainly used for STN cells, and is called a Double STN Cell (DSTN). The uppermost molecule in the second cell is rotated by 90° with respect to the lowermost molecule in the first cell. Furthermore, the twists in the two cells are opposite. Otherwise, the configuration is the same. The operation is explained by recalling from Equations (2.10), (2.11) and (2.12) that the speed of propagation of light is larger if the E-field encounters a smaller index of refraction. For $n_\parallel > n_\perp$ we have a fast and a slow axis of propagation, as shown in the top view on the molecules in Figure 6.15. The speeds add up pair-wise to an equal speed $V_{fast} + V_{slow}$ in both directions if the pairs are chosen, as indicated by equal numbers, in Figure 6.16. As a consequence, the non-addressed pixels of both displays form an isotropic path for light incident perpendicularly to the glass plates. This results in a perfect black state independent of the wavelength, and hence is free of any coloration. For the addressed pixel on the right-hand side of Figure 6.15, the linearly polarized light reaches the output of the upper cell unchanged, and is rotated by the lower cell towards the crossed analyser. The cell appears white.

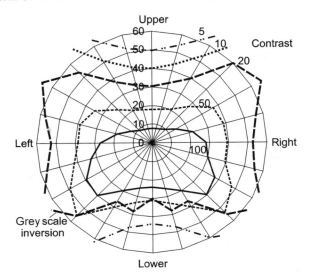

Figure 6.14 Isocontrast and limit for grey inversion after compensation with discotic films in Figure 6.13

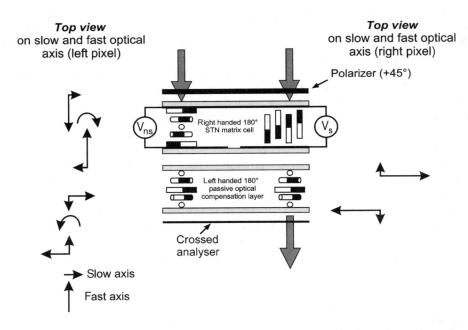

Figure 6.15 Double layer STN cell for compensation of black state and colouration and top view on slow and fast propagation axis

As a rule, the second cell is replaced by a compensation foil which, however, contains only a few layers, mostly two or three, of molecules. This is an acceptable approximation to the exact solution.

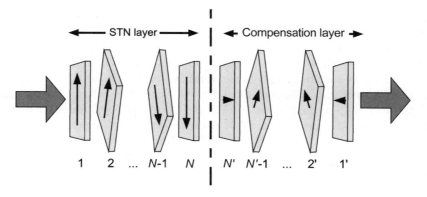

STN layer ——→   ←— Compensation layer —→

1   2   ...   N-1   N   N'   N'-1   ...   2'   1'

Figure 6.16   The pairs of STN layers with opposite twist for compensation in Figure 6.15

6.2.3   Suppression of grey shade inversion and the preservation of grey shade stability

The grey shade inversion and the grey shade stability in Figure 6.6 is related to the minima of luminance shifting to lower viewing angles with increasing voltage V_{LC}. The aim of the compensation is to shift the minima to such large angles that they are out of the viewing range. This places grey shade inversions out of view, and preserves the fairly equal spacing of the grey shades below the minima over the entire viewing range.

To understand the compensation method, we first have to grasp the cause for the minima, which is explained in Figure 6.17 for a TN cell seen in a plane perpendicular to the glass substrates. The larger the voltage V_{LC}, the smaller becomes the angle Θ_e of the midlayer molecules. The optical retardation is lowest for the viewing direction along the director of the midlayer tilt, because the E-field mainly encounters the index $n_0 < n_e$. A low optical retardation causes the polarized optical wave to twist only slightly, resulting in a small amount of light being transmitted by the crossed analyser. To the right and to the left of the direction

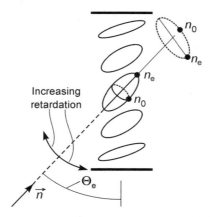

Figure 6.17   Viewing direction for minimum luminance

parallel to the director $\bar{n}$ of the midlayer, the index increases, resulting in increasing retardation and luminance. The lowest retardation can be compensated by adding a positively birefringent layer with its optical axis perpendicular to the optical axis of the midlayer molecules, as shown with dotted lines in Figure 6.17. This positively birefringent compensation layer with an optical axis oblique to the surface of the glass plates is called a *positive 0-layer*. For the compensation, a medium tilt of the midlayer of 45° is chosen; the optical axis of the 0-plate is then at 45° to the glass plates (Yeh and Gu, 1999). For reasons of symmetry, this axis also lies in the y-z-plane of Figure 4.1. To preserve the viewing performance perpendicular to the glass plates ($\Theta_e = 0$), we have to avoid adding optical retardation in this direction by inserting a positive a-plate with its optical axis perpendicular to the 0-plate, as already encountered as an alternative to the Fuji film mentioned earlier. As a result, the minima of the luminance can be shifted for most practical applications out of the viewing range.

6.2.4 Fabrication of compensation foils

Positively bi-refringent a-plates are fabricated by stretching uniaxial polymer films such as polyvinyl alcohol (PVA); a-plates with a negative birefringence consist of uniaxially aligned discotic materials such as triphenylene derivates. Polymer films with a homeotropic alignment of nematic LC materials form positively birefringent c-plates; c-plates with a negative birefringence are fabricated by pressing uniaxially oriented polymer films with random azimuthal angles, by spin-coated polyimide films, or by using homeotropically oriented discotic compounds. Finally, stratified media consisting of isotropic thin layers with alternating different refractive indices are birefringent, and may be used for c-plates with the optical axes perpendicular to the layer interfaces. The thickness of the layers has to be much smaller than the wavelength of the light used. This effect is called form birefringence (Yeh and Gu, 1999).

6.3 Electro-optic Effects with Wide Viewing Angle

So far we have investigated viewing angle limitation and problems with grey shades in TN cells. The key issue in those cells is the minimum luminance for a viewing direction parallel to the director of the tilted mid-layer molecules. This is a basic cause for all the shortcomings of a display associated with an oblique viewing direction. These shortcomings do not show up in displays with multidomain pixels, in non-twisted displays such as displays with In-Plane Switching (IPS) or in a π cell, which is also called a cell with bend-alignment or an Optically Compensated Bend cell (OCB cell).

6.3.1 Multidomain pixels

A further means to suppress inversion and maintain grey shade stability whilst enhancing the viewing angles for which the contrast ratio exceeds 10:1 is the use of multidomain pixels, as shown in Figure 6.18 for the case of four domains in each pixel (Yang, 1991; Iimura and Kobayoshi, 1994). These domains possess two different surface tilts combined with two different twist senses of rotation for the helixes. Let us consider one of the helixes and the

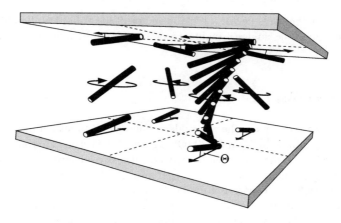

Figure 6.18   A pixel with four domains

direction of the mid-layer tilt at which, as explained in Figure 6.17, a minimum luminance occurs. In this direction, the luminances of the other three helixes do not exhibit their minimum. Therefore, the entire pixel does not generate a minimum luminance in this direction. This is true for all three remaining directions with minimized luminance of the other helixes in a pixel. As a result, the point of low luminance has been shifted to larger angles Θ_e. This effect, we already know, results in avoiding inversion, in maintaining grey shade stability and additionally in enhancing the viewing angle.

The use of more than four domains does not lead to a noticeable improvement, whereas two domains also work, but not so efficiently as four domains.

The alignment of the domains with an adjustable tilt angle on the orientation layer is achieved by a Linear Photo-Polymerization (LPP) (Schadt *et al.*, 1996; Schadt, 1999) of polyvinyl 4-methoxy cinnamate photopolymer with UV-radiation. Liquid crystal alignment occurs as shown in Figure 6.19, at the intersection of the surface of the photopolymer with

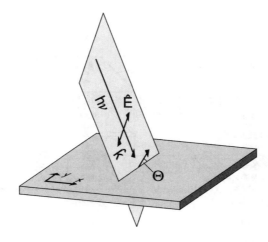

Figure 6.19   The photo-induced alignment and adjustment of the tilt angle Θ of LC molecules on top of the orientation layer

the plane defined by the wave vector $\vec{k}$ and the vector of the E-field of the radiation. The tilt angle Θ depends upon the angle of incidence of $\vec{k}$, and can be adjusted in the entire range $\Theta \in [0, \pi/2]$.

6.3.2 In-Plane switching

In the IPS mode, the directors of the molecules remain at all times parallel to the surface of the glass plates, as outlined in Figure 6.20 (Oh-e *et al.*, 1995). They are rotated by an electric field applied parallel to the glass plates between the two electrodes in each pixel. This rotation generates grey shades similar to the operation of a Fréedericksz cell in Section 3.2.1.

The normally black cell works as follows: in the field-free state in Figure 6.20, the LC molecules are oriented by the orientation layer all in parallel to the direction of the polarizer. Hence, the incoming linearly polarized light does not experience birefringence, reaches the back plate with an unchanged polarization, and is thus blocked in the crossed analyser, as in the Fréedericksz cell. If an electric field is applied, the LC molecules rotate and tend to align parallel to the field. The molecules anchored on the orientation layer and in its vicinity do not rotate, resulting in a twist angle dependent on the distance z in Figure 6.20. Only for a large electric field are all the molecules rotated parallel to the field independent of z besides two thin layers on top of the orientation layers. For this configuration, a straightforward calculation with Jones vectors provides the transmission T through the crossed analyser as

$$T = \frac{1}{2} \sin 2\beta \sin^2 \gamma, \tag{6.49}$$

with the retardation $\gamma = (2\pi/\lambda)\Delta nd$ and the twist angle β of the molecules from the field-free state. The largest transmission $T = 1/2$ is reached for $\gamma = \pi/2$ and $\beta = \pi/4$. In the field-free state with $\beta = 0$, we obtain $T = 0$.

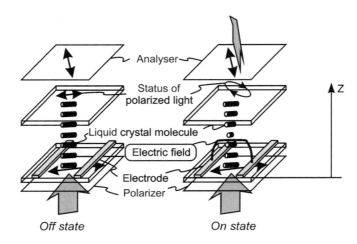

Figure 6.20 Principle of the in-plane switching mode

Parallel polarizers define the normally white mode. The big advantage of the IPS mode is the previously mentioned absence of a minimum in transmission for an oblique viewing angle. This absence, as previously provided by compensation foils or by mulitdomain pixels, is now offered by a different optical system, with widened viewing angles as demonstrated in Figure 6.21. The fact that the directors of the LC molecules are always parallel to the plane of the polarizers suppresses, for all viewing angles, the damaging effect of tiltet molecules in Figure 6.17 and in Equation (6.31) on the viewing angle. Therefore, IPS belongs to the LCDs with the best viewing characteristics. Two shortcomings of IPS also have to be mentioned, even though work to remedy them is under way. The two electrodes in the pixels shrink the aperture ratio to about 40 percent. The torque needed for the rotation of the molecules in plane is larger than for rotating them perpendicular to the plane, as stronger intermolecular forces have to be overcome. Therefore, response time is increased to 60 ms at $V_{LC}=7\,V$, too slow for TV pictures.

6.3.3 Optically compensated bend cells

This cell is the same as the π cell, the fast switching of which was introduced in Section 3.2.7. Besides fast switching, the OCB cell also exhibits a wide viewing angle (Bos and Rahman, 1993; Yomaguchi, Miyashita and Uchida, 1993). Before we embark on this, we have to complete Equation (3.98) for the retardation R by including the refractive index n_f in Equation (6.29). This index applies if light propagates at an angle Θ_e to the director, as shown in Figures 6.22(a) and (b) for the known configuration of the directors in a non-addressed π cell. The retardation for on-axis viewing in Figure 6.22(a) is, with Equations (3.98) and (3.100),

$$R = \int_{z=0}^{d} \Delta n(z)dz = \int_{z=0}^{d} (n_f(\Theta_e(z)) - n_0)dz, \tag{6.50}$$

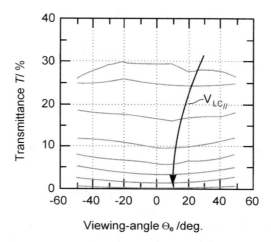

Figure 6.21   Transmittance of various grey levels versus the off-axis Θ_e for an IPS display

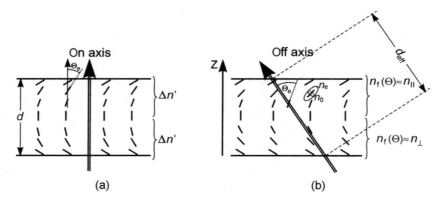

Figure 6.22 The self-compensation of the retardation R of light propagating at an oblique angle through the π-cell, (a) Light propogation in the direction of the cell normal; (b) propagation at an oblique angle

with n_f in Equation (6.29) and $\Theta_e(z)$ the angle changing with z in Figure 6.22. The π cell is a $\lambda/2$ plate, yielding

$$R = \frac{\pi}{2} = \int_{z=0}^{d} (n_f(\Theta_e(z)) - n_0)dz, \tag{6.51}$$

from which the thickness d can be calculated if $n_f(\Theta_e(z))$ is known. With Equation (3.101), an effective anisotropy Δn_{eff} is introduced, yielding

$$R = \Delta n_{\text{eff}}\, d = \int_{z=0}^{d} (n_f(\Theta_e(z)) - n_0)dz \quad \text{or}$$

$$\Delta n_{\text{eff}} = \frac{1}{d} \int_{z=0}^{d} (n_f(\Theta_e(z)) - n_0)dz. \tag{6.52}$$

As a rule, the integrals are evaluated numerically. The retardation for the viewing under the oblique angle in Figure 6.22(b) is

$$R = \int_{z=0}^{d_{\text{eff}}} (n_f(\Theta_e(z)) - n_0)dz. \tag{6.53}$$

The aim is to shift this retardation as close to $\pi/2$ in Equation (6.51) as possible, which renders the viewing under angle Θ_e equal to the viewing perpendicularly.

In the lower portion of the cell, $n_f(\Theta_e)$ in Equation (6.53) is close to $n_f = n_0 = n_\perp$ as $\Theta_e \approx 0$ in Equation (6.29), and in the upper portion $n_f(\Theta_e)$ is close to $n_f = n_e = n_\parallel > n_\perp$ as $\Theta_e \approx \pi/2$. So the excess retardation in the upper portion is compensated by the smaller retardation in the lower portion. This self-compensation of the π cell shifts R in Equation (6.53) close to $R = \pi/2$ in Equation (6.51), which expands the viewing range considerably as an inherent property of the π cell. The isocontrast curve of a π cell is contained in Figure 6.23 (Bos and

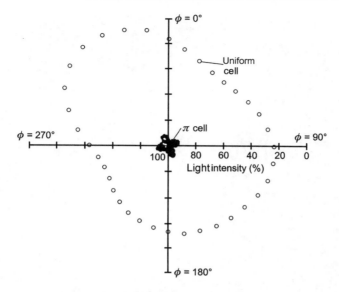

Figure 6.23 Isocontrast curves of a π cell

Koehler, 1984). The viewing angle can be further enhanced by adding a negatively bi-refringent compensation foil with the self-compensating director configuration (Bos and Rahman, 1993). This renders the π cell one of the most interesting LCDs.

6.4 Polarizers with Increased Luminous Output

The issue of increasing the luminance of LC cells that use polarized light is also of some urgence. A linear polarizer allows only 50 percent of the light to pass, the s-wave, because it has to absorb the remaining 50 percent of the light, the p-wave, which is polarized perpendicular to the s-wave. Two 'loss-less' polarizers which recycle the otherwise lost p-wave have been developed. They enhance luminance.

It is worth mentioning a further effect of polarizers which can reduce the viewing angle of a display. The two orthogonal axes of a polarizer for the s-wave and p-wave no longer appear orthogonal if viewed from an oblique angle. This causes leakage of light from the p-wave thus spoiling the black state and the grey shades for larger viewing angles.

6.4.1 A reflective linear polarizer

The 3M company (Wortman, 1907) has developed the reflective polarizer shown in Figure 6.24, which consists of a three layer laminate. Each layer contains an isotropic and a birefringent film. The refractive indices for the ordinary beam in the birefringent film and for the isotropic film are equal. Thus, the stack of layers is isotropic for linearly polarized light in the direction of the ordinary beam. This beam can pass. The complimentary beam polarized in the direction of the extraordinary beam is reflected if the periodicity of the stack and the wavelength are matched. The reflected polarized light is again reflected by a

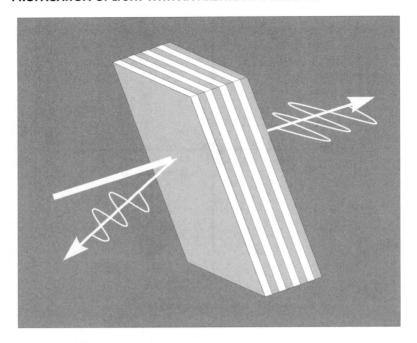

Figure 6.24 The three-layer laminate of the 3M reflective polarizer

depolarizing diffuser, upon which this light is again treated the same way as the previous incoming beam. This transmission and reflection of beams is repeated until, theoretically, all the light has been transmitted in linear polarization. Measurements reveal that practically 85 percent instead of 50 percent of the light is transformed into a linear polarization state, reflecting in a 70 percent gain in luminance.

6.4.2 A reflective polarizer working with circularly polarized light

The Merck company (Coates *et al.*, 1996) introduced the reflective polarizer shown in Figure 6.25, in which unpolarized light is incident on a birefringent foil with a helical molecular structure like in an STN display. The helical structure is generated by a cholesteric film. The helix only allows light circularly polarized in the opposite sense of the helix to pass. The light circularly polarized in the same sense as the helix is reflected. This is shown in the stages 1 and 2 in Figure 6.25. The reflected light incident on the diffuser is depolarized by scattering (stage 3 in Figure 6.25). The component with a circular polarization in the opposite direction as the helix is transmitted, whereas the reflected component undergoes depolarization by scattering. This play is repeated indefinitely. The circular polarization at the output of the cholesteric film is transformed into a linear polarization by a $\lambda/4$-plate, as discussed in Section 5.2. Finally, a linear polarizer suppresses spurious portions of the *p*-wave which has leaked through. This solution is able to deliver 80 percent of the incoming light into the LC cell.

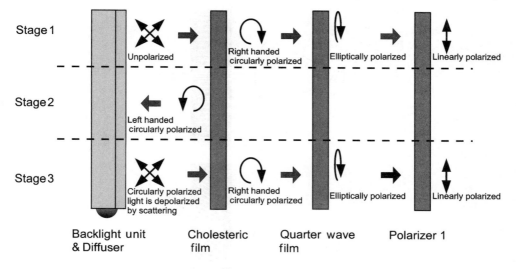

Figure 6.25 Merck's reflective polarizer based on a cholesteric film

6.5 Two Non-birefringent Foils

The Brightness Enhancement Foil (BEF) is a ridged transparent plate, with a cross-section as shown in Figure 6.26. The changes in refraction indices at the solid-air interface causes refractions and reflections, as indicated in Figure 6.26. The reflected light is recycled by a

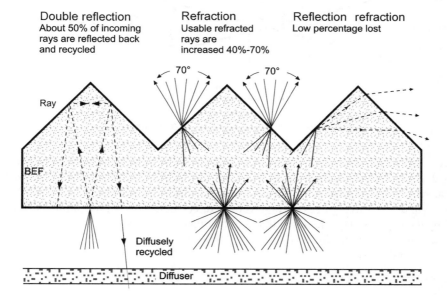

Figure 6.26 The operation of Brightness Enhancement Film (BEF)

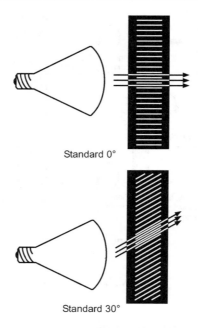

Standard 0°

Standard 30°

Figure 6.27   Optical waveguides

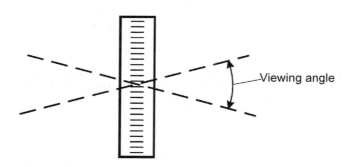

Viewing angle

Figure 6.28   The viewing cone of an optical waveguide

diffuser. The overall effect is the collimation of the incident light into a cone with an angle of $\pm \varphi_0$ degrees. In the example of Figure 6.26, the angle is $\pm 35°$. Within this reduced angular range, the luminance of a display is considerably enhanced.

The cylindrical wave guides in Figure 6.27 direct light into a predetermined direction given by the angle of the cylindrical light guiding tubes. The tubes generate a viewing cone, as shown in Figure 6.28. Such light control films are used to provide a specific viewing direction of a display.

7

Modified Nematic Liquid Crystal Displays

Two modifications of nematic LCDs are in use. Polymer Dispersed LCDs (PDLCDs) consist of droplets of nematic liquid crystals embedded in a polymer. Guest-Host Displays (GHD) contain nematic liquid crystals as host and dopants to influence the properties as guests.

7.1 Polymer Dispersed LCDs (PDLCDs)

7.1.1 The operation of a PDLCD

Droplets of nematic liquid crystals sized $0.3\,\mu m$ to $3\,\mu m$ in diameter are embedded in a polymer binder, and enable operation as a transmissive or reflective display without polarizers. Hence, the displays will be very bright as loss of light in a polarizer is avoided. The droplets usually contain the LC molecules in a bipolar configuration, as shown in Figure 7.1. The polymer binder has the index n_p. In the off-state of a transmissive display in Figure 7.2, the droplets are oriented at random. The incident unpolarized light encounters the index $n_f(\Theta_e)$ of the molecules according to Equation (6.29) with $n_f \neq n_p$, and is hence refracted and reflected, resulting in scattering of the light in a wide angle towards the viewer. This represents the black state. Obviously, it is a poor black state which has a milky white, opaque or translucent appearance. In the saturated on-state in Figure 7.2(b), the directors of the LC molecules orient themselves parallel to the electric field, exposing the refractive index $n_\perp = n_0$ to the light. The polymer is selected to exhibit an index $n_p \approx n_0$. In this index matched state, as a consequence, light can pass providing a very bright white state, the brightest white state offered by any type of LCD. The transparent state can show some haze stemming from the incoming diffuse light, and from the fact that not all molecules, especially those close to the inner surface of the droplets, are aligned perfectly in parallel to the E-field.

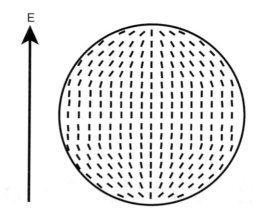

Figure 7.1 The bipolar configuration of the LC molecules in a PDLC droplet

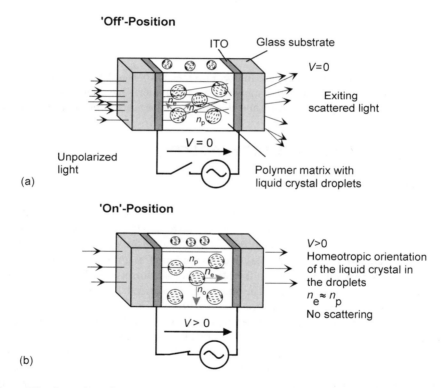

Figure 7.2 Operation of a transmissive PDLC cell (a) in the off-state, and (b) in the on-state

After switching off the voltage, the droplets relax back to the random state, if they are spherical, more slowly than if they are ellipsoidal. The droplets are generated either by a micro-encapsulation process, called NCAP (Fergason *et al.*, 1986), or by phase separation (Doane *et al.*, 1986).

In the NCAP process, a polymer such as polyvinyl alcohol (PVA) is mixed with nematic liquid crystals in an aqueous emulsion. This emulsion is coated onto a conductive transparent substrate and allowed to dry. Then the film can be laminated to a transparent substrate. The droplets often vary in size.

Polymerization induced phase separation starts with a homogeneous solution of prepolymer liquid crystals and a curing agent which catalyses polymerization. The droplet formation is based, for instance, on a photo initiated polymerization, where phase separation is induced by a UV irradiation (Voz *et al.*, 1987). Phase separation may also be thermally or solvent induced. Gelation of the polymer locks in the droplet morphology. Parameters for the fabrication of a high contrast film in a PDLCD light valve determined experimentally are (Ginter *et al.*, 1993): 19wt percent of polymer in an LC material with the high $\Delta n = 0.22$ which causes strong scattering. The PDLCD was produced by polymer induced phase separation at $17\,\text{mW/cm}^2$ UV power in a projection system this formulation produced (according to Figure 7.3) the high contrast of $40:1$ on screen. The transmission versus voltage curve in Figure 7.4 shows that saturation was reached at the relatively low voltage of 12 V.

The switching voltage and the response time of a HPDLC depend upon a variety of factors, but most heavily on the resistive and dielectric properties of the liquid crystal and polymer. These properties influence the electric field inside the droplets, which is different from the field outside imposed by the external voltage. These two fields determine the alignment of the molecules and, as a consequence, the voltage required for a given contrast and the response time. Further factors are the shape and size of the droplets, the viscous torque and elastic torque which returns the droplet to its field-free equilibrium. The contrast increases with the thickness of the PDLC layer, and with the density of the droplets. However, the thicker the PDLC layer, the larger is the voltage required.

n_0 decreases with decreasing temperature, whereas n_e increases; hence, $\Delta n = n_e - n_0$ becomes larger, and so does the scattering. As a consequence, the transmittance in the off-state decreases with falling temperature, which means the display becomes darker.

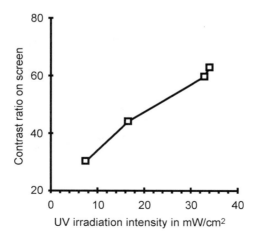

Figure 7.3   Contrast ratio on screen versus power density of UV irradiation of a $10\,\mu\text{m}$ thick PDLC cell

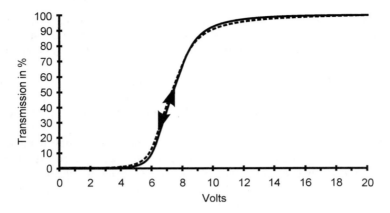

Figure 7.4 Transmission versus voltage of a transmissive PDLC light valve

To enhance contrast, PDLCs can be doped with pleochroic dyes (West *et al.*, 1989; Drzaic *et al.*, 1989). Those dyes absorb a spectral band dependant on the orientation of their long axis. As shown in Figure 7.5, dichroic dyes absorb the band mainly if the E-field of the light oscillates parallel to the long axis, whereas negative dichroic dyes absorb mainly along the axis perpendicular to the long axis. The dichroic ratio D is defined as

$$D = \frac{\text{absorption parallel}}{\text{absorption perpendicular}}. \tag{7.1}$$

The dyes are mixed with the LC molecules in the droplets, and are forced to rotate with the directors. Thus, a pleochroic dye exhibits minimum absorbance in the on-stage, where the molecules allow the light to pass. In the randomly oriented off-state, the dyes exhibit maximum absorbance, substantially enhancing the dark state. To optimize this effect, it is desirable to have as much of the dye in the droplets and as little in the polymer as possible.

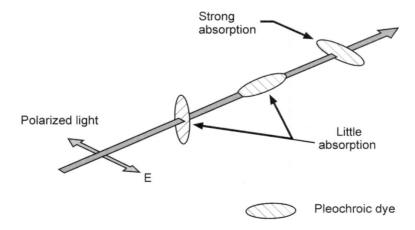

Figure 7.5 The absorption of light by pleochroic dyes

The dark state and the colour, and hence contrast, is further influenced by the order parameter S_T of the vector of the transition moment T of the dye absorption, which exhibits an angle Θ_T to the director of the LC molecules. For fluctuating Θ_T, the order parameter is (Voz *et al.*, 1987)

$$S_T = \frac{1}{2} \langle 3 \cos^2 \Theta_T - 1 \rangle. \tag{7.2}$$

The best black state, and hence the best contrast, is reached for $\Theta_T = 0$.

The PDLCDs with dyes are also called Polymer Dispersed Guest-Host Displays (PDGHD), where the dyes play the role of guests and the liquid crystal the role of hosts.

7.1.2 Applications of PDLCDs

PDLCD cells are very economic devices. They do not require expensive polarizers, and are thus very bright. Since the polymer solidifies the PDLC film, the cells do not need the tight sealing of a TN cell. Further, the orientation layer and its rubbing are not required. These properties also lend themselves to large area displays.

The transmissive PDLCD, with the PDLC layer sandwiched in between the glass plates carrying the ITO electrodes, has been presented in Figures 7.2(a) and 7.2(b). Applications are windows in offices, homes and cars with an adjustable transparency, and bright displays with a wide viewing angle.

The operation of the reflective PDLCD in Figure 7.6 is the inverse of the transmissive display. In the field-free state, the droplets backscatter the light and the pixels appear bright. The bright state becomes brighter with both an increasing thickness of the cell and a larger density of the droplets. In the field-on state the PDLC layer is transparent, and the viewer can see the black absorber at the back of the display. The pixels exhibit a rather good black, which is only slightly spoiled by LC molecules not perfectly aligned parallel to the E-field.

In a reflective display with colour and a doping by black dyes, the absorber is replaced by a pixellized colour filter on top of a reflector. In the on-stage, the colour pixels are visible

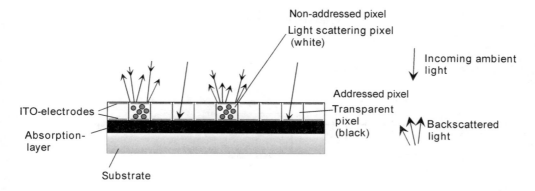

Figure 7.6 A reflective black and white PDLCD cell operating with ambient light

with high brightness and colour purity. In the off-state, the pleochroic dyes provide an acceptable dark state. In the grey shades in between, the colour is brightened quite rapidly by the scattering of white light from the droplets on top of the colour filter.

Haze-free PDLCDs may be designed by replacing the polymer binder by a liquid crystal polymer which is permanently oriented parallel to the E-field of the on-state. For the indices, $n_0 = n_{op}$ and $n_e = n_{ep}$ must hold, where n_{op} and n_{ep} are indices of the LC polymer.

The use of PDLCDs as light valves for bright projectors is discussed in Chapter 24.

7.2 Guest-Host Displays

7.2.1 The operation of Guest-Host displays

The role of dyes as a guest in liquid crystals as the host has been explained in Section 7.1.1. In this section, we investigate further examples of this technique that do not involve dispersion of the LC in polymers.

The basic cell from which all others are derived is the Heilmeier cell in Figure 7.7 (Heilmeier and Zononi, 1968), which uses one polarizer. In the off-state in Figure 7.7(a), a spectral band of the linearly polarized white light is absorbed by the pleochroic dye. The remaining spectrum exits the cell as coloured light. In the on-state in Figure 7.7(b), the dyes allow the light to pass only slightly attenuated. The output appears white.

In the Double-Guest-Host LCD (DGH-LCD) (Uchida *et al.*, 1980) in Figures 7.8(a) and 7.8(b), the polarizer can be omitted as the second cell with an orientation of the LC

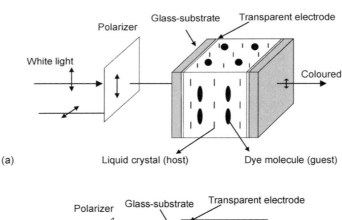

(a)

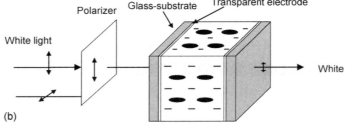

(b)

Figure 7.7 The Heilmeier Guest-Host cell (a) in the off-state, and (b) in the on-state

molecules rotated by 90° with respect to the first cell. Thus the double cell attenuates both components of the incident light. In the off-state, the directors are aligned parallel to the surface of the alignment layer. In the on-stage of the DGH-LCD in Figure 7.8(b), both cells are transparent due to $\Delta\varepsilon > 0$ providing white light at the output. Figures 7.9(a) and 7.9(b) demonstrate an equivalent system in which in the off-state with white output the molecules are homeotropically aligned. The on-state provides (due to $\Delta\varepsilon < 0$) the alignment parallel to the glass plate associated with coloured output.

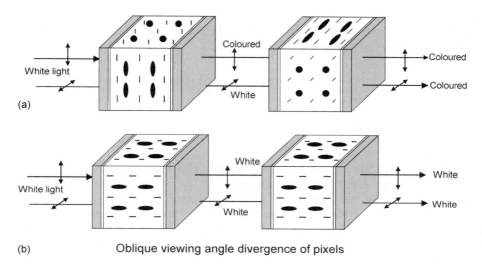

Figure 7.8 The Double Guest-Host cell with $\Delta\varepsilon > 0$ (a) in the off-state and (b) in the on-state (Uchida *et al.*, 1980)

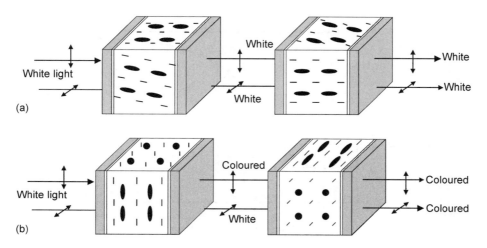

Figure 7.9 The Double Guest-Host cell with $\Delta\varepsilon < 0$ (a) in the off-state and (b) in the on-state (Uchida *et al.*, 1980)

The absorbances in the off- and on-states of the DGH-LCD in Figures 7.9(a) and 7.9(b) are

$$A_{\text{off}} = (a_{\|} + a_{\perp})cd \tag{7.3}$$

and

$$A_{\text{on}} = 2a_{\perp}cd, \tag{7.4}$$

where $a_{\|}$ and $a_{\perp}$ are, respectively, the absorption coefficients parallel and perpendicular respectively to the long axis of the dye, c is the dye concentration and d is the thickness of the cell. The colour contrast is, with Equation (7.1),

$$\frac{A_{\text{off}}}{A_{\text{on}}} = \frac{1}{2}\left(1 + \frac{a_{\|}}{a_{\perp}}\right) = \frac{1}{2}(1 + D). \tag{7.5}$$

The colour contrast is approximately half of the dichroic ratio D. Figure 7.10 depicts the absorbance versus voltage V_{LC} of a pleochroic dye at the wavelength $\lambda=582$ nm of maximum absorption.

The Phase Change Guest-Host Display (PC-GHD), which is also called after the inventors a White–Taylor type display (White and Taylor, 1974), in its off-state in Figure 7.11(a) renders a polarizer obsolete, as does a DGH display, but this time only with one cell and by exposing the incoming light to all orientations of the directors, and hence the dyes. A cholesteric host is used which aligns the dyes along a helix. In this way, both perpendicular components of light undergo the absorption leading to coloured light at the output. In the on-state in Figure 7.11(b), the dyes exhibit minimum absorption, and hence white light. The axis of the helix may be parallel to the surface of the substrates, as in Figure 7.11(a), or perpendicular to it as in Figure 7.12(a). To establish this direction of the axis, the display in Figure 7.11(a) requires homeotropic anchoring of the LC molecules on the orientation layer,

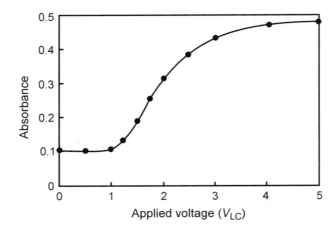

Figure 7.10 Absorbance versus voltage V_{LC} of a pleochroic dye for maximum absorbance at $\lambda=582$ nm

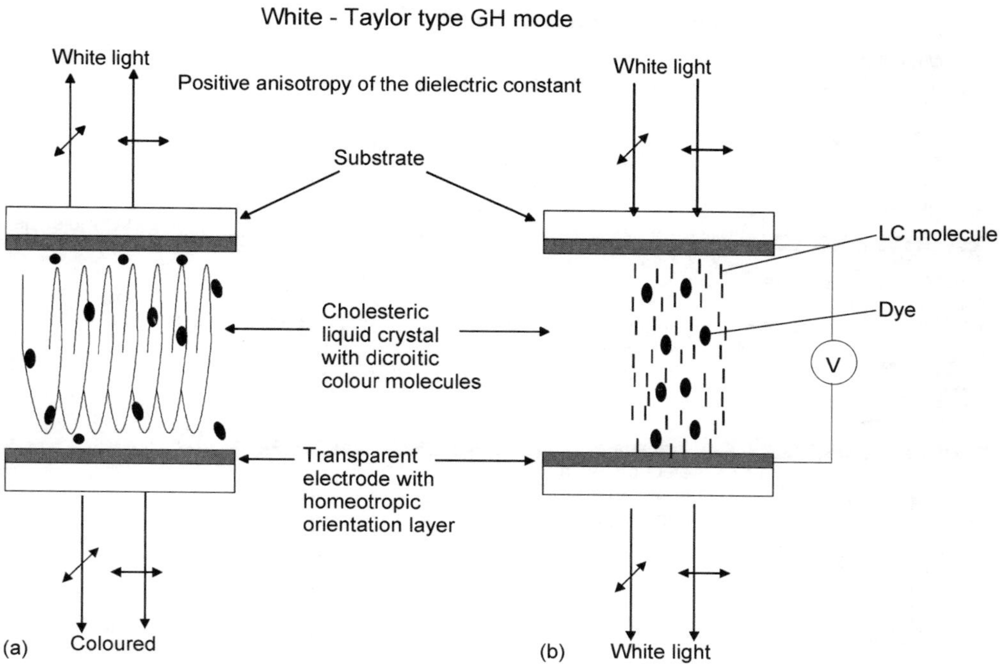

White - Taylor type GH mode

White light

Positive anisotropy of the dielectric constant

White light

Substrate

LC molecule

Cholesteric
liquid crystal
with dicroitic
colour molecules

Dye

V

Transparent
electrode with
homeotropic
orientation layer

(a)   Coloured

(b)   White light

Figure 7.11 Cholesteric nematic phase change guest-host display with the axis of the helix parallel to the glass plates (a) in the off-state, and (b) in the on-state

and in Figure 7.12(a) homogeneous anchoring parallel to the surface of the orientation layer. In the on-state, the cell in Figure 7.11(b) exhibits the same configuration as in Figure 7.12(b). The cholesteric-nematic PC-GHD does not allow grey shades, because its transmittance-voltage curve exhibits hysteresis, especially if the twist angle of the helix exceeds 270°. This is already known from the behaviour of STN cells in Figure 4.7.

In all guest-host displays, the dissolution of dyes in the LC molecules increases viscosity, resulting in an addressing voltage up to 5 V higher and a response time up to 10 percent higher than for a nematic LCD. The viewing angle for a contrast 10 : 1 is very wide, reaching $\pm 90°$. The colours offered by guest-host displays can be complementary, namely cyan, magenta and yellow, as in each case one spectral band is suppressed by absorption. The primary colours red, green and blue are generated by allowing one band to pass and by suppressing the bands with wavelengths above and below. Reflective guest-host displays have found more applications than transmissive ones. They are discussed in the following section.

7.2.2 Reflective Guest-Host displays

For many applications, reflective displays should possess the properties of white paper with black or coloured information and reflectance should exceed 60 percent for the display to appear paper-white, whereas contrast should be greater than 8:1 for legibility. These are the

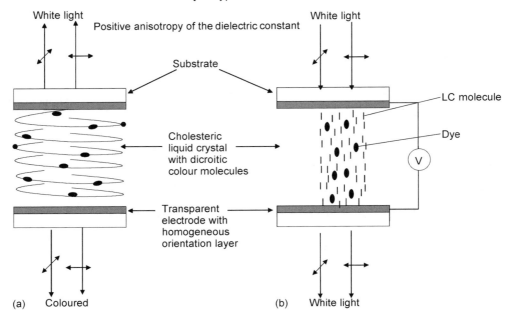

Figure 7.12 Cholesteric nematic phase change guest-host display with the axis of the helix perpendicular to the glass plates (a) in the off-state, and (b) in the on-state

characteristics of a newspaper. As reflective displays rely on ambient light, the loss of light in the LCD, a crucial issue, has to be minimized. As the greatest loss of light is incurred by the use of polarizers, of which at least one is needed in reflective nematic and twisted nematic displays, attention has been drawn to polarizer-free reflective displays. These are PDLCDs, DGHDs and, if only one cell is allowed, PC-GHDs.

We first investigate the reflective black and white PC-GH-cell in Figure 7.13 (Seki *et al.*, 1996; Koisnai *et al.*, 1995). In the off-state, the dyes embedded in the cholesteric-nematic helix absorb incoming light, yielding a dark but not a perfect black state. In the on-state, the transparent GH layer allows the smooth metal electrode to reflect almost 100 percent of the light, which is forward scattered by the front scattering film. This provides a homogeneous luminance over the display area, and a wide viewing angle. The $100\,\mu$ thick scattering film contains plastic particles with a diameter of $60\,\mu$ in a polymer matrix. An alternative to the smooth mirror and scattering film is a scattering mirror with a rough surface. This, however, may cause problems with the orientation layer.

The reflectance in Figure 7.14 exceeds 60 percent in the on-state. Contrast at $10\,V$ is $5:1$. Contrast being still a crucial issue may be enhanced by a larger dicroic ratio D in Equations (7.1) and (7.5), by a larger order parameter S_T of the transition moment of the dye absorption in Equation (7.2), by a smaller pitch of the helix, and an increased thickness of the cell. The latter may increase the voltage above $10\,V$.

Contrast can be considerably enhanced by using a DGH cell. To do this more economically, the two cells can be combined into one single cell in Figure 7.15 (Lowe and Cox, 1981; Lowe and Hasegava, 1992), where the crossed guest-host configuration of the double

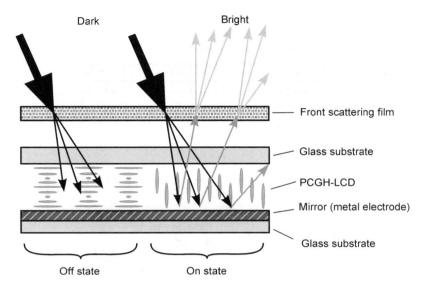

Dark

Bright

Front scattering film

Glass substrate

PCGH-LCD

Mirror (metal electrode)

Glass substrate

Off state   On state

Figure 7.13   A reflective black and white PC-GH display

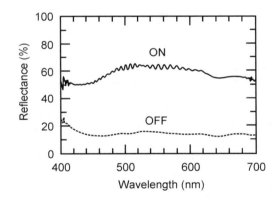

Figure 7.14   Reflectance of a PC-GH LCD versus wavelength

cell is separated by a thin Mylar film only 1.3 µ thick. The easily alignable pixels of the two halves of the cell are addressed by the same TFTs. To keep addressing voltage low, the plastic film ought to be very thin, minimizing the voltage drop across it. This display reached a reflectance of 60 percent and a contrast increased from the single cell GHD to 8 : 1.

The reflective PC-GH colour display in Figure 7.16 (Kanoh *et al.*, 1998) generates the colour by a conventional R-G-B colour filter underneath the top substrate (Uchida, 1984). The operation of the cell follows the description given for the previous PC-GH displays. The colour perceived by the eye consists of the addition of the colour stimulation of three colour dots in the eye. We have to remember that the additive colour generation requires three pixels. The AM-LCD panel belonging to Figure 7.16 had a 28 cm diagonal and

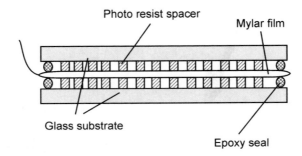

Figure 7.15   A single cell DGH display with a plastic separation layer

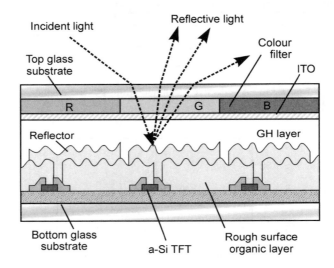

Figure 7.16   A GH display with a conventional colour filter for R, G and B

1200 × 1600 RGB pixels (UXGA) with a pixel pitch of 177 dpi, which was considered to be necessary to produce legibility similar to that of a newspaper.

Colour generation with complementary colours is explained by the spectra in Figures 7.17(a) and 7.17(b). The spectra of the primary colours red, green and blue are shown in Figure 7.17(a), and those for the complementary colours yellow, magenta and cyan in Figure 7.17(b). The stacked three-layer PC-GH display (Uchida, 1984) in Figure 7.18 (Sunohara *et al.*, 1996) generates colour without using a colour filter, by relying on the absorbance of the dyes in the three GH displays. For panels with $\Delta\varepsilon < 0$ in the off-state, the uppermost display allows magenta, the middle one cyan and the lowermost yellow to pass. For the generation of the primary colours R, G and B, the transmittances of two stacked layers have to be added to obtain a primary colour. Thus magenta added to yellow provides red, magenta added to cyan results in blue, and yellow added to cyan provides green. In all combinations, one layer out of the three is not required. It has to be switched into the transparent state. For black all three layers must be in the absorbing and for white in the transparent state. The

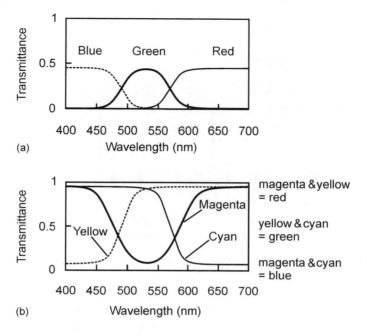

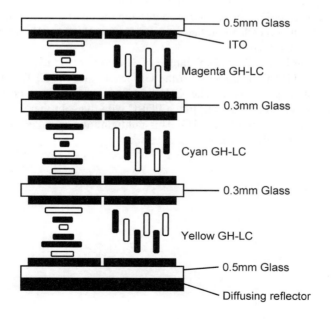

Figure 7.17 Transmission spectra (a) of the primary colours R, G and B, and (b) of the complementary colours yellow, magenta and cyan

Figure 7.18 Structure of a stacked three-layer PC-GH display

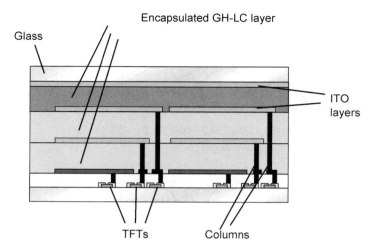

Figure 7.19   The three-layer GH display with stacked films of encapsulated liquid crystals

sequence of the layers is arbitrary; experiments, however, show that contrast is largest for a sequence starting with magenta on top and yellow on the bottom. The colour display in Figure 7.18 exhibits a reflectance of 40 percent and a contrast of 5.3 : 1.

A most severe problem of the three-layer stack is parallax, which can be made tolerable with glass plates as thin as 0.3 mm. Plastic substrates could be even more advantageous. A further problem is loss of light, as an incoming light beam has to pass twelve layers of ITO electrodes in Figure 7.18 with their associated loss of around 8 percent. Therefore, the light budget merits special care. For active matrix addressing the aperture ratio must be enhanced by placing the TFTs underneath the reflector, as shown in Figure 7.16. This increases the aperture to values beyond 90 percent.

A further solution in Figure 7.19 (Sunohara *et al.*, 1998) to avoid parallax and to increase luminance places the ITO films directly on the GH layers and the TFTs for the three GH layers on one common glass plate. The GH layers are realized by encapsulated liquid crystals with dyes dispersed in polymer, like a PDLC with dyes. In that way, the GH layers are made solid enough to support the ITO. This configuration is called Stacked Films of Encapsulated Liquid Crystals (SFELIC).

8

Bistable Liquid Crystal Displays

A configuration of LC molecules is bistable if it can assume two different states at zero electric field. Which state appears depends upon the history of the configuration before it is switched to the field-free state. The two stable states are characterized by minima in the energy stored in these states, which is composed in general of mechanical, dielectric and electric energy terms. The switching from one stable state into the other is performed by feeding energy, as a rule by the addressing circuit, into a stable state, lifting it over the energy barrier between the two states into the second stable state. A stable state maintained without energy consumption represents a storage or a memory effect. Thus, bistable displays are able to realize an image memory at zero field, similar to a page of a book. All displays discussed so far exhibit only one stable state at zero field, which was imposed on the configuration of the directors by the molecules on top of the orientation layers.

Bistable displays utilizing ferroelectric, cholesteric or nematic liquid crystals are described in the following sections.

8.1 Ferroelectric Liquid Crystal Displays (FLCDs)

Ferroelectric liquid crystals consist of smectic A or smectic C materials which are intrinsically chiral or have added chiral components. A chiral C phase is termed smectic C^*. The chiral additions make the molecules align along a helix, for which symmetry considerations require that they exhibit a spontaneous polarization P_s (Meyer, 1974). The LC molecules possess the layered structure in Figure 8.1. The layer boundaries are parallel planes which, as will be explained later, are not always perpendicular to the glass plates of the display. As a rule, the directors of the molecules in each layer are parallel. Only one molecule is shown in Figure 7.20. The two stable positions of the molecules lie on a cone with an angle $2\Theta(T)$,

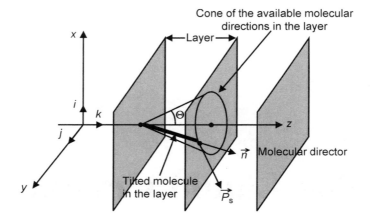

Figure 8.1   Position of a molecule on a cone in each layer

dependent on temperature T. The polarization $\overrightarrow{P_s}$ is perpendicular to $\vec{n}$ and tangential to the circle of intersection of the cone with the boundary plane of the layer. Figure 8.2 depicts the position of the molecules on a helix in a sequence of layers, together with the polarization $\overrightarrow{P_s}$ of each molecule. Within one pitch p of the helix, all directions of $\overrightarrow{P_s}$ occur, adding all $\overrightarrow{P_s}$ vectors to a zero sum. From the outside the material appears to be non-polarized. To be able to use the polarization in the operation of a display, the helix must be suppressed. Clark and Lagerwall (1980) achieved this by placing two narrowly spaced glass plates only $1-2\,\mu m$ apart with the orientation layers in Figure 8.3(a) parallel to and in Figure 8.3(b) slightly inclined to the layer normal. With this concept they opened up a new area of LC technology.

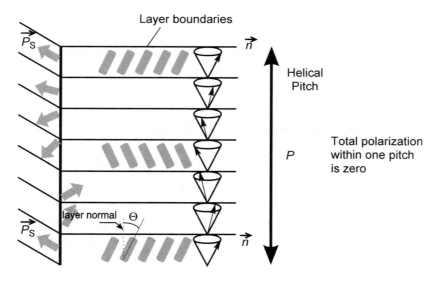

Figure 8.2   The helix in a layered structure of chiral smectic C* liquid crystal

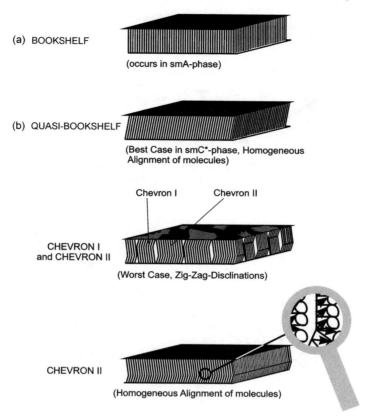

(a) BOOKSHELF

(occurs in smA-phase)

(b) QUASI-BOOKSHELF

(Best Case in smC*-phase, Homogeneous Alignment of molecules)

Chevron I   Chevron II

CHEVRON I and CHEVRON II

(Worst Case, Zig-Zag-Disclinations)

CHEVRON II

(Homogeneous Alignment of molecules)

Figure 8.3 Textures of LC molecules between the substrates of an FLC cell

The anchoring of the molecules tries to impose the constraint for all directors in a layer to be parallel, which is supported by the small thickness $d < p$ of the cell. This is called a Surface Stabilized Ferroelectric Liquid Crystal (SSFLC) cell. The small cell gap implies the need to handle small cell thicknesses economically. The configuration in Figure 8.3(a) is called the *bookshelf texture* and that in Figure 8.3(b) the *quasi-bookshelf texture*.

Before we go on with a more detailed discussion of the textures, and with the introduction of more textures, we can already calculate the propagation of light through the layers, each of which contains LC molecules with parallel directors. Figure. 8.4(a) shows one of the parallel molecules. The polarizer is parallel to $\vec{n}$, and hence the crossed analyser blocks the light. The transmissive cell is black. The molecule is switched into the second stable position in Figure 8.4(b) which rotates the molecules by an effective switching angle β_{eff}. If the axis of the cone is parallel to the plane of the polarizer, we obtain $\beta_{\text{eff}} = 2\Theta$, the angle of the cone in Figure 8.1. As a rule, however, this will not be the case, as investigated later; if the axis of the cone is inclined to the plane of the polarizer, we have to project the angles into the plane of the polarizer, leading to an effective switching angle β_{eff}. The linearly polarized light in Figure 8.4(b) enters the cell at an angle β_{eff} to $\vec{n}$. This is the case of the Fréedericksz cell, for which the Jones vector $J_{x'z}$ for the analyser at $z = d$ is given in Equation (3.40). The

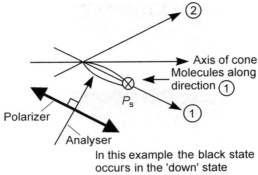

In this example the black state
occurs in the 'down' state

(a)

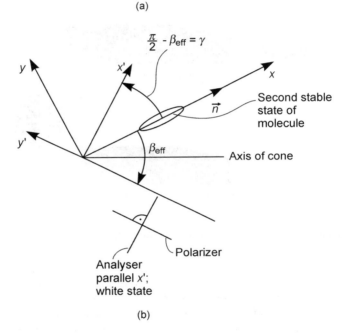

(b)

Figure 8.4   (a) The black and (b) the white state associated with the two stable states of an FLC

coordinates $x-y$ and $x'-y'$ for Equation (3.40) in Figure 3.8 are also drawn in Figure 8.4(b). From these coordinates we learn, for the angles α and γ in Equation (3.40),

$$\alpha = -\beta_{\text{eff}} \quad \text{and} \quad \gamma = \pi/2 - \beta_{\text{eff}}.$$

Introducing these angles in to Equation (3.40) provides, for $z=d$,

$$J_{x'd} = E_{\xi 0}e^{-ik_x d} \cos \beta_{\text{eff}} \sin \beta_{\text{eff}} \left(1 - \cos \frac{2\pi d \Delta n}{\lambda} + i \sin \frac{2\pi d \Delta n}{\lambda} \right)$$

and the intensity

$$I = |J_{x'd}|^2 = E_{\xi 0}^2 (\cos \beta_{\text{eff}} \sin \beta_{\text{eff}})^2 2 \left(1 - \cos \frac{2\pi d \Delta n}{\lambda}\right),$$

or

$$I^2 = E_{\xi 0}^2 \sin^2 2\beta_{\text{eff}} \sin^2 \frac{\pi d \Delta n}{\lambda}. \tag{8.1}$$

The intensity reaches a maximum for $\beta_{\text{eff}} = \pi/4$ and for $d\Delta n = \lambda/2$. As the angle Θ of the cone, related to β_{eff}, is a constant depending on the FLC material, we shall later introduce a free parameter which allows us to reach the optimum for β_{eff}.

The smA-phase exhibits the bookshelf texture in Figure 8.3 in which, as shown in Figure 8.5, the directors are either parallel to the layer normal with tilt angle $\alpha = 0$, or are off the normal with $\alpha > 0$. At temperatures below those for the smA-phase in Figure 2.2 the SmC* phase appears. In this phase more textures are observed (Ricker *et al.*, 1987), to which Clark and Lagerwall's (1980) idea for the suppression of the helix can also be applied. Those textures are the quasi-bookshelf texture in Figure 8.3b, the Chevron I (CI) texture and the Chevron II (CII) texture in Figures 8.3, 8.5 and 8.6 as well as the zig-zag defect texture in

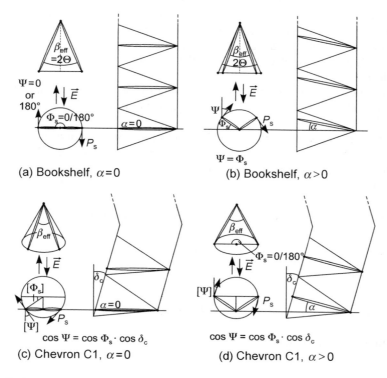

(a) Bookshelf, $\alpha = 0$

(b) Bookshelf, $\alpha > 0$

(c) Chevron C1, $\alpha = 0$
$\cos \Psi = \cos \Phi_s \cdot \cos \delta_c$

(d) Chevron C1, $\alpha > 0$
$\cos \Psi = \cos \Phi_s \cdot \cos \delta_c$

Figure 8.5 Various textures of SSFLCDs with pretilt α and layer tilt δ_c. The angles in brackets [] are shown as projections in the drawing plane

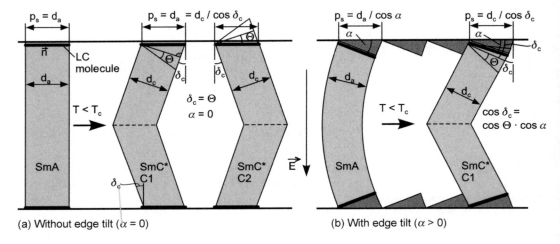

(a) Without edge tilt ($\alpha = 0$)   (b) With edge tilt ($\alpha > 0$)

Figure 8.6 The angles in the Chevron textures

Figure 8.3. In the latter, both the CI and the CII textures appear separated by disclinations. The area of disclinations contains unordered molecules not controllable by the addressing circuit. This is unacceptable for a display. If the tilt angle δ_c of the layers in Figures 8.5 and 8.6 points towards α, we obtain the CI texture, whereas in the CII texture δ_c points away from α. The Chevrons CI and CII may have a tilt $\alpha = 0$ or $\alpha > 0$, as shown in Figures 8.5 and 8.6. In Figure 8.5, ϕ_s is the angle between the directors of the stable states and a diameter of the circle which is the base of the cone; ψ is the angle between the polarization P_s and the electric field $\vec{E}$.

By application of a periodic rectangular or sinusoidal E-field of about 30 Hz and an amplitude of 30 V,..., 40 V, it is possible to switch the Chevrons into a quasi-bookshelf texture.

We now have to visualize the tilt α and the layer tilt δ_c for the various textures in Figures 8.5 and 8.6 in some more detail, and derive a design formula for the FLCD (Buerkle, 1997) based on the bookshelf and the quasi bookshelf textures. For the smA-phase with a layer thickness d_a in Figure 8.6, the layer pitch p_s in the bookshelf texture is $p_s = d_a$, where d_a is the length of a molecule; d_a is also the length of the side of the cone.

In the case of $\alpha \neq 0$, we obtain from Figure 8.6(b) for the SmC*-phase

$$p_s = d_a / \cos \alpha \tag{8.2}$$

or

$$p_s = d_c / \cos \delta_c. \tag{8.3}$$

For all textures, the equation

$$d_c = d_a \cos \Theta \tag{8.4}$$

holds.

Dividing Equation (8.2) by Equation (8.3) and inserting Equation (8.4) provides

$$\cos \delta_c = \cos \Theta \cos \alpha. \tag{8.5}$$

This equation was derived for textures with $\delta_c \neq 0$ and is also only valid for this condition. From Equation (8.5) it follows for $\Theta \varepsilon(0, \pi/2)$ that $\delta_c \geq \alpha$. In Figure 8.7 δ_c (α) is plotted for $\Theta = 20°$. $\delta \geq \alpha$ can be seen. For $\alpha = 0$ Equation (8.5) yields $\delta_c = \Theta$, which is visible in Figure 8.6.

We now investigate the pretilt α, the layer tilt δ_c, the angle 2Θ of the cone and their influence on the effective switching angle β_{eff} and on the torque $\vec{M}$ applied by the electric field $\vec{E}$ in the cell to the molecules with polarization $\vec{P_s}$. The torque is

$$\vec{M} = \vec{P_s} \times \vec{E} \tag{8.6}$$

with

$$|\vec{M}| = |\vec{P_s}||\vec{E}| \sin \psi, \tag{8.7}$$

where Ψ is the angle between $\vec{P_s}$ and $\vec{E}$ as shown in Figure 8.5.

Figures 8.5 and 8.6 show the geometrical situation for the FLC layers in the bookshelf and the Chevron C1 texture with $\alpha = 0$ and $\alpha \neq 0$, together with $\vec{E}$. The effective switching angle β_{eff} is the perpendicular parallel projection of the switching angle β of the molecules in Figure 8.7 into the plane of the polarizers. The closer β_{eff} is to $\pi/4$, the larger is the intensity in Equation 8.1, and hence also the contrast. On the other hand, the closer Ψ is to $\pi/2$, the larger is the initial torque applied to the molecules in Equation (8.7), resulting in a fast switching. Small angles Ψ reflect in a delaying threshold in the switching operation. The bookshelf structure in Figure 8.5(a) with $\alpha = 0$ exhibits a large β_{eff}, but a zero or very small initial torque. The bookshelf in Figure 8.5(b) with $\alpha \neq 0$ provides a larger initial torque but a smaller β_{eff}. Figures 8.5(a) and 8.5(d) are used to derive a design equation for β_{eff}.

The quasi-bookshelf texture has the same angles as the CI- and CII-textures, but they stay unchanged through the thickness of the cell, and do not experience twist where the angles change. The following derivation of design rules for β_{eff} in Equation (8.1) applies only to the

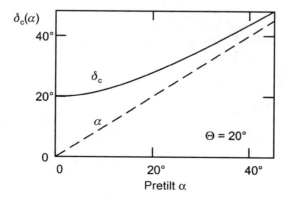

Figure 8.7 The dependency of layer tilt δ_c on the pretilt α in SSFLCDs with the quasi-bookshelf and Chevron textures

quasi-bookshelf and the bookshelf textures, as unchanged angles α and δ_c through the cell are assumed.

The dependence of β_{eff} on Θ, δ_c and α is derived from the Figures 8.5(b), 8.8(a) and 8.8(b). Figure 8.8 depicts the cone with the two stable positions I and II of the molecules, and with the axis d_c of the cone rotated by an angle δ_c from the plane of the polarizers. From Figure 8.8(b) we obtain

$$\cos \Theta = d_c/m, \tag{8.8}$$

and from Figure 8.8(a)

$$\cos (\delta_c - \alpha) = \frac{d_c}{s}. \tag{8.9}$$

For the switching angle β of the molecules, Figure 8.8(a) provides $\cos (\beta/2)=(s/m)$, for which (with s/m from the Equations (8.8) and (8.9)) we obtain

$$\cos \frac{\beta}{2} = \frac{\cos \Theta}{\cos (\delta_c - \alpha)}, \tag{8.10}$$

or

$$\beta/2 = arc \cos \frac{\cos \Theta}{\cos (\delta_c - \alpha)}. \tag{8.11}$$

Further, Figure 8.8(a) yields

$$\tan \frac{\beta}{2} = \frac{b}{s}. \tag{8.12}$$

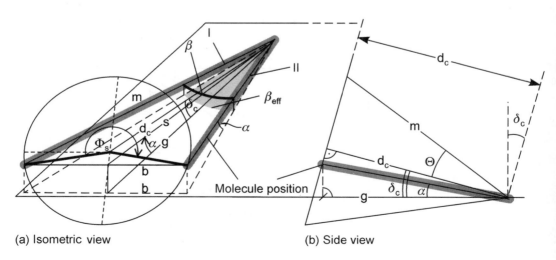

(a) Isometric view   (b) Side view

Figure 8.8 The cone with the two stable positions of the molecules rotated by angle δ_c from the plane of the polarizers: (a) Isometric view, (b) side view

Figure 8.8(a) or 8.8(b) shows

$$g = s \cos \alpha, \tag{8.13}$$

whereas Figure 8.8(a) indicates, with the Equations (8.13), (8.12) and (8.10),

$$\tan \frac{\beta_{\text{eff}}}{2} = \frac{b}{g} = \frac{b}{s \cos \alpha} = \frac{\tan \beta/2}{\cos \alpha} = \tan \left(arc \cos \frac{\cos \Theta}{\cos (\delta_c - \alpha)} \right) \frac{1}{\cos \alpha}. \tag{8.14}$$

The maximum for β is $(\beta/2)=\Theta$, half of the cone angle, which in Equation (8.11) is reached for $\delta_c = \alpha$. According to Equation (8.14), the angle β_{eff} can exceed β, and becomes larger with decreasing α. In Figure 8.9 β_{eff} in Equation (8.14) is plotted versus α for $\Theta=20°$ and $\delta_c(\alpha)$ in Equation (8.5) and Figure 8.7. For $\alpha=0$, and hence $\delta_c=\Theta$ in Equation (8.5), we obtain $\beta_{\text{eff}}=0$ which, according to Equation (8.1), results in an unswitchable dark display. Hence for SSFLCDs with inclined layers, a pretilt $\alpha>0$ is essential for proper functioning. The optimum pretilt yielding $\beta_{\text{eff}}=\pi/4$ for a bright and therefore a high contrast display is reached in Figure 8.9 for $\alpha=32°$. These large pretilts are feasible with rubbed polyimide orientation layers, with evaporated or sputtered SiO_2-orientation layers deposited at an angle of 80° to the normal (Bunz et al. ,1997), or by linear photopolymerization of the orientation layer (Schadt et al., 1996), as discussed in Chapter 6.

A further method which was discovered experimentally is to fabricate Chevron textures with a pretilt of about 5° to 7° and transform it into a quasi-bookshelf texture by applying a 30 Hz 40 V electric field at a 1 μm thick cell. This provides an almost optimum $\beta_{\text{eff}}=41°$.

Polarization causes ionic effects. As they directly influence the addressing of SSFLCDs, they will be discussed together with the addressing schemes in Chapter 13.

8.2 Chiral Nematic Liquid Crystal Displays

Chiral nematic, or as they used to be called, cholesteric liquid crystals (Sage, 1992; Booth, 1998; Colos, 1998), are composed either of intrinsically chiral nematic materials or of

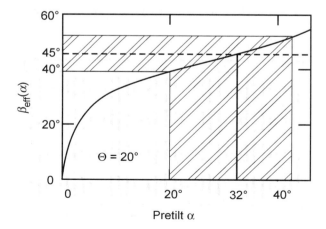

Figure 8.9 The effective switching angle $\beta_{\text{eff}}(\alpha)$ with $\delta_c(\alpha)$ in Figure 8.7

non-chiral nematics with the addition of chiral dopants. As the cholesteric esters are now seldom used, materials scientists prefer the name chiral nematic phase, abbreviated to N*-phase. The chiral components cause the molecules to spontaneously align on the surface of a helix, as shown in Figure 8.10. The helices in a display are stabilized by the anchoring forces of the orientation layer. If the molecules are anchored parallel to the surface of the orientation, as in Figure 8.10, the helices form parallel columns with the axis normal to the glass plates. This texture is called *planar* or *Grandjean*. The display is a Surface Stabilized Cholesteric Texture Display (SSCTD). The planar texture can also be stabilized by a network of polymer chains, resulting in Polymer Stabilized Cholesteric Textures (PSCT). As SSCTs are easier to build, they are now the most widely used.

Anchoring with zero pretilt may be achieved by a rubbed PVA orientation layer, or by friction deposited PTFE (polytetrafluorethylene). Further means are obliquely evaporated or sputtered SiO-layers (Bunz *et al.*, 1997), microgrooves or linear photopolymerization (Schadt *et al.*, 1996). A homeotropic alignment gives rise to the formation of helices with axes parallel to the orientation plane, as shown in Figure 8.11. This configuration is called the *fingerprint texture*. Homeotropic alignment is provided by orientation layers with surfactants such as lecithin, quarternary ammonium, silane derivates, HTAB or Cr-complexes. The focal conic texture in Figure 8.12 in the absence of a polymer network is generated by a rapid cooling from the isotropic to the chiral nematic phase, or by applying a voltage V_{LC} at

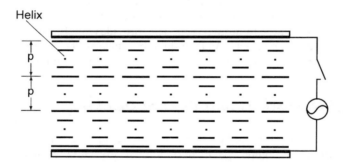

Figure 8.10 The planar texture chiral nematic displays with the axes of the helixes normal to the glass plates

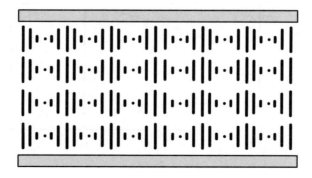

Figure 8.11 Homeotropicly aligned molecules and the fingerprint texture of chiral nematic displays

**Scattering state
(Focal conic)**

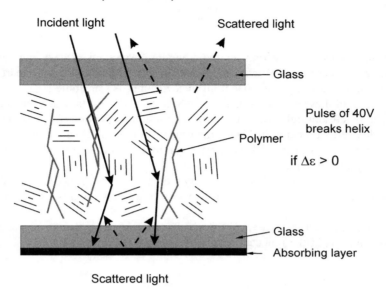

Figure 8.12 The focal conic texture of SSCTDs without and for PSCTDs with the polymer network

a cell with the planar texture. If $\Delta\varepsilon > 0$ the electrical field breaks the helices into the randomly oriented domains in Figure 8.12. The focal conic texture is metastable; in a slow relaxation which may last several months, it tends to realign to the planar state induced by the planar anchoring of the molecules on the orientation layer. The focal conic texture may be stabilized by the polymer network mentioned previously, and can then last for years at zero field. This network is generated from an added monomer. After a UV curing at typically $4\,\mathrm{mW/cm^2}$ for 1 h, the monomer is polymerized into a network, as shown in Figure 8.12. A fourth state occurs at relatively high voltages of 30 V, where for $\Delta\varepsilon > 0$ the molecules align parallel to the electric field. This state is transparent to light, but is only maintained in the presence of an E-field. A high content of chiral components gives rise to a large rotational, twisting power, which is measured in degrees/mm.

The planar texture has the remarkable property of reflecting a component of light incident perpendicularly to the glass plates. This component is circularly polarized with the same sense of polarization as the sense of the helix. The opposite handedness of circularly polarized light is transmitted. The center wavelength λ_0 of the reflected light is

$$\lambda_0 = p\bar{n} = p\frac{1}{2}\left(n_\parallel + n_\perp\right), \tag{8.15}$$

whereas the bandwidth $\Delta\lambda$ is

$$\Delta\lambda = p\Delta n, \tag{8.16}$$

with

$$\frac{\Delta\lambda}{\lambda_0} = \frac{\Delta n}{\bar{n}}. \tag{8.17}$$

This represents a selective reflection centred around λ_0. Any light of wavelength outside the range reflected as described is transmitted by the cell. In other words, apart from the reflected circularly polarized component all incident light is transmitted.

The derivation of these equations (Colos, 1998; Kats, 1971) starts with a representation of incident light by left-and right-handed polarized light in the Equations (5.31) and (5.32), and by solving Maxwell's Equation (6.13) for the propagation of the E-vector through the helix. The dielectric properties of the N^*-phase are given by a tensor for the permittivity which is dependent on the twist angle.

Other approaches to a derivation are based on the constructive interference of wavefronts reflected at different layers similar to the Bragg-reflection (De Gennes and Prost, 1993).

For light incident at an off-axis angle Θ_1 which is viewed under the off-axis angle Θ_2, the centre wavelength λ_0 of the reflected light is (Colos, 1998; Fergason, 1966)

$$\lambda_0 = \bar{n}p \cos \frac{1}{2}\left(arc\sin\frac{\sin\Theta_1}{\bar{n}} + arc\sin\frac{\sin\Theta_2}{\bar{n}}\right). \tag{8.18}$$

Figure 8.13 shows the reflectance R of the planar state versus wavelength. For the centre wavelength in Figure 8.13, $\lambda_0=480\,\text{nm}$ and for $\bar{n}=1.48$ from Table 2.1, Equation (8.15)

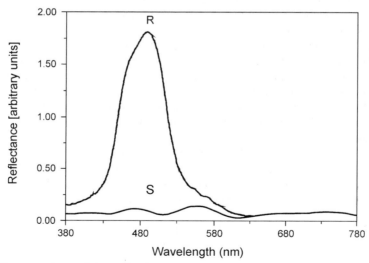

R: spectrum of the reflecting state
S: spectrum of the scattering state

Figure 8.13 The spectra of reflected light R and of scattered light S in an SSCTD

provides a pitch of $p = 324$ nm, reflecting in a rotational power of $1.1 \cdot 10^6$ degrees/mm. This value determines the content of chiral components. The reflected light exhibits a colour given by λ_0. Frequent choices for displays are yellow or green.

In the focal conic texture in Figure 8.12, most of the light can pass the cell and is absorbed at the bottom if there is an absorption layer, as in Figure 8.12. A small portion of the light is scattered forward. The chiral nematic cell in Figure 8.12 can be used as a reflective display. The forward scattered light S is also plotted in Figure 8.13. The levels of the scattered and reflected light define the contrast of the reflective display. The focal conic state, especially if polymer stabilized, is stable at zero field, as is the planar state, as an additional torque from an external electric field is needed to reassemble the broken helices into the helical columns in Figure 8.10. This will be discussed further in Chapter 13.

The peak reflectance at λ_0 can at most reach 50 percent, because the circularly polarized light with a handedness opposite to the helix is lost by transmission. There are two ways to recover this lost portion. The first is an arrangement of two chiral nematic cells with the same pitch but opposite handedness placed on top of each other. The first cell reflects the circular polarized light of the same handedness, and allows the opposite handedness to pass. This portion is reflected by the second cell, and can now pass the first one, thus theoretically doubling the reflectance. The scattering in the focal conic state is also enhanced, but not doubled, due to the increased forward scattering losses of the lower cell. The pitches of the two cells may be temperature dependent. If their shifts with temperature do not match, an embarrassing divergent shift in reflected colours will occur.

The second approach is to place a $\lambda/2$-plate in between two chiral nematic cells with the same pitch and handedness. The circularly polarized light passing the first cell is converted to the opposite handedness by passing the $\lambda/2$-plate, is then reflected by the second cell, changes its handedness again by passing the $\lambda/2$-plate a second time, and can now pass through the first cell. This again theoretically doubles reflectivity. The temperature matching of the shift in the pitches is easier, since both cells have the same handedness and the same LC and chiral materials. The $\lambda/2$-plate, however, renders this solution more expensive, so it is seldom used in cost sensitive display applications. It should be noted that the temperature matched shift of the pitches does, of course, not suppress a colour shift, the shift is only not divergent.

Since both the reflecting planar texture state providing a colour and the focal conic state providing a black state are stable, chiral nematic displays are able to operate as image storage devices, as known from SSFLCDs. They also share the inability to provide grey shades in their basic operation considered so far, as only two stable states are available. For both, SSCTDs and SSFLCDs, grey shades require more elaborate addressing schemes.

As both the helices in the planar texture and the randomly oriented broken helices in the focal conic texture look, to a large degree, alike from all viewing directions, one can expect good viewing angle properties. This is proven by the isocontrast curves in Figure 8.14, which are almost perfectly circular; contrast drops from its maximum of 8.5:1 at a normal viewing direction to 4:1 at an off-axis angle of 50°. As a rule, contrast is higher than in STN displays. Contrast of chiral nematic displays can be enhanced up to 25:1 by using an orientation layer with SiO_2 evaporated at an oblique angle of 80° to the normal (Bunz et al., 1997).

In a stacked configuration of three displays (Okada et al., 1997; Davies et al., 1997), as depicted in the right stack in Figure 8.15, each display has a pitch p tuned to reflect in the planar texture one of the three primary colours red, green and blue. For example, If red is desired the two other non-red displays are switched into the focal conic state, through which

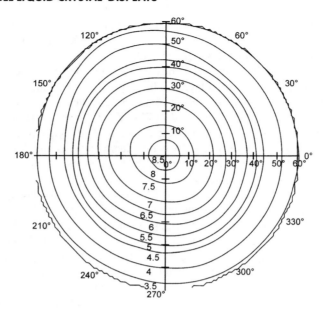

Figure 8.14   Isocontrast curves of a reflective SSCT display

the red light can pass. Thus, in the three-stack reflective display in Figure 8.15, ambient light incident from the top is reflected as one of the three colours. If two displays in the stack are in the planar state, the addition of the two colours appears at the output, whereas the addition of all three colours produces white. All three displays in the focal conic state allow ambient light to pass upon which it is absorbed in the black absorber at the bottom of the stack. This produces black slightly lightened up by the forward scattered light S in Figure 8.13. Hence, six colours in addition to black and white are feasible. The approach is costly as three full fledged displays are required. A more economic solution is represented in the two stacks shown on the left hand side of Figure 8.15. One of the two displays reflects a primary colour (e.g. blue), and the second one the complementary colour, which for blue is yellow. The addition of both provides white; if both displays are focal conic, that is transparent, the black absorber is visible. This arrangement offers two colours in addition to black and white. The most simple solution to generate white is to use only one yellow reflecting display with a blue background. Black is not possible in that way, however. Attempts have been made to build a single display with all three colours (Chien *et al.*, 1995) by providing different pitches in the pixels for different colours. This could be done by a photosensitive chiral component with a rotational power adjustable by the power of UV light. The mixing of the three different N^*-phases would have to be avoided by walls around the pixels, which complicates cell building.

The inherently metastable focal conic texture exhibits a high sensitivity to pressure caused by a lateral flow of the LC material out of the pressure zone. This gives rise to the formation of the stable planar texture, thus destroying the picture content. This sensitivity, together with the need for high addressing voltages in the range of 30 V, are the main shortcomings of chiral nematic displays. The lateral flow could be prevented by spacer walls, a more costly process that limits pixel density.

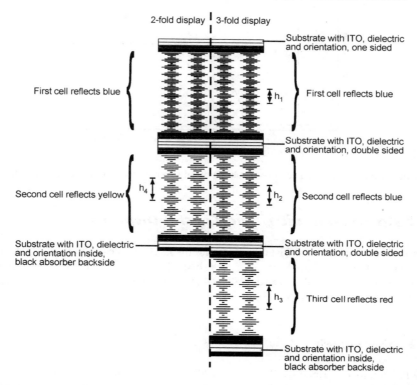

Figure 8.15 Three (resp. two) stacked reflective chiral nematic displays for the generation of colours

In addition to the regular mode of chiral nematic cells described so far, there is an inverse mode cell, which is no longer a bistable device (Yang *et al.*, 1992). It will be described as a transmissive display.

The planar texture for the inverse mode is designed to reflect in the infrared by choosing a pitch in the μm-range. This planar state is transparent for wavelengths in the visible. In addition, the planar state is strongly polymer stabilized. This destabilizes the focal conic state, as the polymer network favours a quick relaxation into the planar state. Thus, only the planar state is stable at zero field. The focal conic texture reached by applying an increased voltage, as shown in Figure 8.16, is scattering, resulting in a low transmittance. The transition between the high and low transmittance is rather flat, which allows us to generate grey shades. This is the main advantage of the inverse mode.

Applications for chiral nematic displays are mainly in reflective bistable displays. As they do not require a power source for storing and displaying information, they are suited for portable systems. Inverse mode displays are more seldom used but, are fit for displaying moving pictures at reduced voltages.

The pitch p is temperature dependent in the form $p(T) \sim 1/T$. This thermochromism can be used to indicate temperature differences on a surface by colour shifts, e.g. to detect areas of larger temperatures by a shift to 'blue tomography'. On the other hand, the independence of $\Delta\lambda/\lambda_0$ in Equation (8.3) of the temperature dependent pitch is useful for many optical instruments.

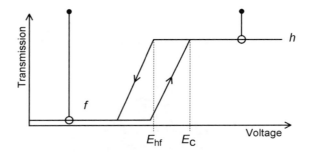

Figure 8.16 Transmittance versus voltage of an inverse mode chiral nematic display

8.3 Bistable Nematic Liquid Crystal Displays

There are three varieties of bistable nematic displays which exhibit the potential for applications in the near future. In all cases, bistability is imposed by a special surface alignment leading to Bistable Twist Cells (BTCs), grating aligned nematic devices and monostable surface anchoring switching.

8.3.1 Bistable twist cells

Bistable twist cells described by Bos *et al.* (1999) are similar to STN cells using a nematic material with a chiral additive and surfaces rubbed in opposite directions. The pretilt angle on both surfaces have the same rotational sense, thus favoring the three stable states in Figure 8.17 with a twist of 0°, 360° and 180°. The LC molecules of the 0° and of the 360° twist exhibit the same topological orientation on the upper and the lower surface, which is no more true for the 180° twist. All three states are stable, given by a local minimum of their free energy which is lowest for the 180° twist. Using crossed polarizers, the 0° twist is designed to show a black state, whereas the 360° twist realizes a white state as long as the

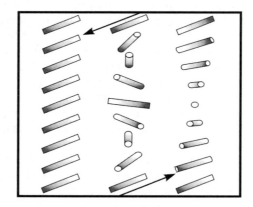

Figure 8.17 The three stable states of a nematic cell with 0°, 360° and 180° twists

180° twist is not present. This state is separated from the other two by disclination lines stemming from nucleation sites. These sites form at glue lines, at the edge of the cell, and at spacers; they tend to expand into the cell. It has been found (Holie and Bos, 1998) that polymer walls suppress the nucleation sites. They are generated from reactive liquid crystals which were UV cured in the presence of 20 V across the cell. This technique allows us to build bistable nematic cells operating at a low power consumption.

Further bistable modes in TN-LCDs are discussed in Tang *et al.* (1994).

8.3.2 Grating aligned nematic devices

Figure 8.18 shows the grating of a zenithal bistable device, in which one alignment is homeotropic (Bryan Brown *et al.*, 1997). Investigations of the free energy revealed that the two director configurations drawn in Figure 8.18 exhibit minimum energy, and are hence stable. The splayed state *b* possesses a flexoelectric polarization *P*. A positive voltage pulse, together with *P*, switches the molecules from state *a* to state *b* in Figure 8.18 whereas a negative pulse favours the homeotropic alignment of state *a*. Contrast can be enhanced by using a 90° twisted nematic configuration with a negative $\Delta\varepsilon$ (Jones *et al.*, 1998). In this configuration, shown in Figure 8.19, the homeotropic alignment on one surface in Figure 8.18 is replaced by a planar alignment of the molecules. The polarizer next to the planar alignment is parallel to the directors, whereas the analyser is crossed, that is perpendicular to the grating grooves. Figure 8.19 shows the switching by a negative pulse $-V$ from the splayed state to the twisted, and on one side, homeotropicly aligned state. A positive pulse switches in the opposite direction. The device operates as a normally white cell. Addressing voltage is as low as 3 V; switching time is several hundred μs; contrast reached is as high as 100:1. The device has the low power consumption needed for portable systems.

8.3.3 Monostable surface anchoring switching

The upper surface in Figure 8.20 (Martinot-Lagarde *et al.*, 1997) provides a strongly tilted anchoring of the molecules throughout all operations, whereas the weak planar anchoring of the lower surface in Figure 8.20(a) is broken by a large field generated by a short pulse *V*

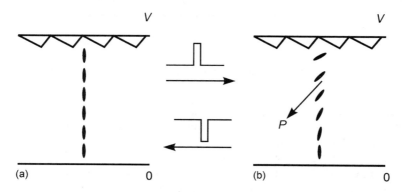

Figure 8.18 Grating and alignment of a zenithal bistable device

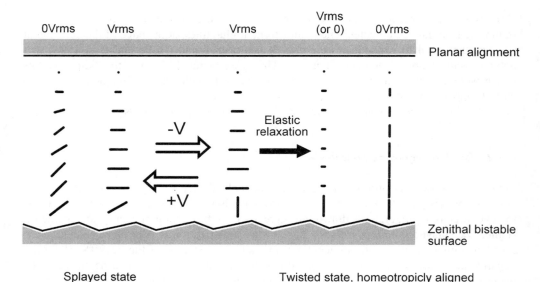

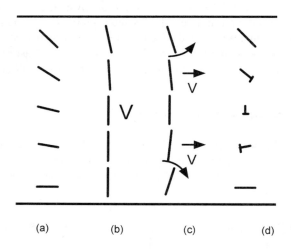

Figure 8.19 Switching of the twisted zenithal bistable device

Figure 8.20 Switching of the monostably anchored device from stable state (a) to stable state (d)

(Figure 8.20(b)). After the pulse, the strongly tilted anchoring relaxes into its initial position inducing the shear flow in Figure 8.20(c), forcing the planar anchoring to tilt in the opposite direction (Figure 8.20(c)), thus generating the twisted texture in Figure 8.20(d). The textures in Figures 8.20(a) and 8.20(d) are stable. The switching from a to d requires a 3 μs pulse of 17 V or a 1 ms pulse of 5 V. The switching from state d back to a (Dozov et $al.$, 1997) is performed by a pulse $22\,V > V > 20\,V$, the texture d is erased and one induces a d to a transition.

9

Continuously Light Modulating Ferroelectric Displays

All versions of ferroelectric displays excel in fast switching due to the high switching torque originating from the interaction between polarization and external E-field. The high switching speed is the main reason for a continued interest in modifications of FLCDs.

9.1 Deformed Helix Ferroelectric Devices

The Deformed Helix Ferroelectric LCD (DHF-LCD) (Beresnev *et al.*, 1989) in Figure 9.1 with a Sm C* material has (without a field applied) a helical pitch p_0 with $p_0 \ll d$, where d is the thickness of the cell. The helix axis is parallel and the FLC layers are perpendicular to the plane of the substrates. The incoming polarized light passes through an aperture $a \gg p_0$ and propagates parallel to the planes of the LC layers. The LC molecules exhibit a tilt angle Θ_0 in Figure 9.1 and a rotation angle φ, a twist, around the helix axis (not shown). An electric field $\pm E$ modulates the angle φ dependant on the location z on the helix, as depicted in the Figure 9.2 for $\cos \varphi$ as a function of the reduced $z' = 2\pi z/p_0$ for positive and negative E-values. This is true as long as $|E| < E_u$, with

$$E_u = \frac{\pi^2 K q_0^2}{16 P_s},\tag{9.1}$$

where $K = K'\Theta_0^2$ is the effective elastic coefficient, $q_0 = 2\pi/p_0$ and $P_s = \eta_\perp \eta_p \Theta_0$ is the piezo-electric polarization with $\eta_\perp$ the susceptibility and η_p the piezoelectric coefficient.

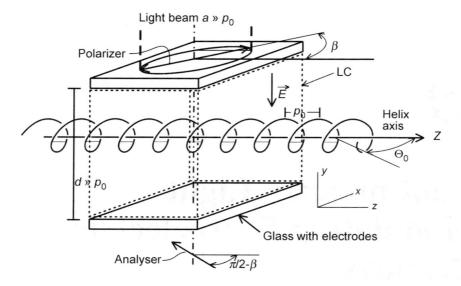

Figure 9.1   The deformed helix ferroelectric LCD

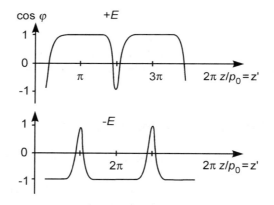

Figure 9.2   Variation of the director twist φ dependent on the reduced location z' on the helix and on the field $+E$ and $-E$

If E approaches E_u, the helix is unwound, as in the case of SSFLCDs, and the device becomes bistable.

In the field free case, one axis of the refractive index ellipsoid is parallel to the helix axis. With increasing and decreasing $\pm E$, the axis of the index rotates to $\pm \Theta_0$ for $E = \pm E_u$. Also, $|\Delta n|$ changes. Depending on the angle β of the polarizer, the effect is a linear or a quadratic change of the transmittance. Thus voltage controlled grey shades can be produced.

The time constant τ_c of the light modulation is

$$\tau_c = \frac{\gamma_\varphi}{K' \Theta_0^2 q_0^2},$$ (9.2)

where γ_φ is the viscosity for the twist φ. The time constant can be as small as $250\,\mu s$ (Beresnev *et al.*, 1989), which puts DHF-LCDs in the category of fast switching displays at voltages as low as 1 V to 2 V. High contrasts of 100 : 1 were reached (Fünfschilling and Schadt, 1994).

9.2 Antiferroelectric LCDs

The configuration of the molecules of an AntiFerroelectric LCD (AFLCD) is called the Sm C_A^* phase shown in Figure 9.3(a), together with the Sm C^* phase of a ferroelectric texture in Figure 9.3(b) (Miyachi and Fukuda, 1998). The tilt angle in the antiferroelectric layers with the normal $\hat{z}$ changes between $+\Theta$ and $-\Theta$ from layer to layer. The spontaneous polarization $\vec{P}$ is given by

$$\vec{P} = P\frac{\hat{z} \times \vec{n}}{|\hat{z}\vec{n}|}, \tag{9.3}$$

and is also depicted in Figure 9.3(a).

Antiferroelectricity leads to the tristate switching in Figure 9.4 (Migachi and Fukuda, 1998). At zero field we encounter the AntiFerroelectric (AF) Sm C_A^* state. Increasing E results in the pretransitional state in Figure 9.4(c), and finally, in the Ferroelectric (F) Sm C^* state. The pertinent tilt angle Θ and the transmittance T for these field-induced phase transitions are plotted in Figures 9.4(a) and 9.4(b). A similar phase-transition occurs for $E < 0$. Figure 9.4 demonstrates that antiferroelectric LCDs are not bistable at zero field. The use of a positive E-field in one frame and of a negative E-field in the next frame provides a wider viewing angle than in SSFLCDs; the quasi-book shelf structure results in a contrast

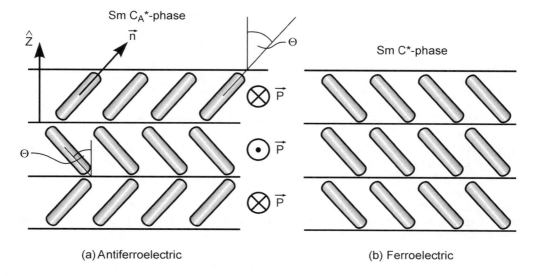

(a) Antiferroelectric   (b) Ferroelectric

Figure 9.3 The texture of the antiferroelectric and of the ferroelectric LC material

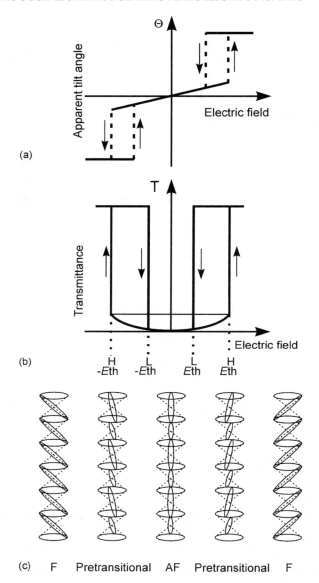

Figure 9.4 The switching characteristics of antiferroelectric LCDs. (a) Tilt angle Θ (E); (b) transmittance T (E); (c) the textures depending on E

ratio of up to $30:1$; response time is as low as $50\,\mu s$, a distinct advantage of LC materials with a spontaneous polarization. Finally, self-recovery from alignment damage caused by mechanical or thermal shocks is observed during operation of the AFLCD. The pretransitional effect in Figure 9.4 limits contrast to about $40:1$. Suppressing this effect by rendering it narrower until it disappears provides the V-shaped switching curve in Figure 9.5. This transmission versus voltage curve represents a thresholdless AFLC, in which the positive or

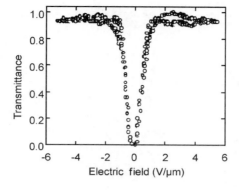

Figure 9.5 The V-shaped switching curve of antiferroelectric LCDs

negative amplitude of the pixel voltage is directly translated into grey shades. A low driving voltage of ± 3 V and a good contrast is feasible. The tristate switching curve in Figure 9.4(a) is driven by mono- or bipolar pulses. As the grey level achieved depends on the initial grey shade, a reset to the zero level is needed. AFLCDs still need further development before they are accepted by the market.

10

Addressing Schemes for Liquid Crystal Displays

The addressing circuit has the task of imposing a given grey shade at a prescribed time onto an individual pixel without causing discernible crosstalk to other pixels. The addressing of the LCDs may be performed electronically, optically or by plasma columns, where each physical effect realizes a switch allowing it to feed in the grey shade information (Kaneko, 1987; Lueder, 1998a, Scheffer and Nehring, 1998; Luo, 1990; Kaneko, 1998).

The grey shades are defined by the linear portion of the curve in Figure 2.13 representing the transmitted luminance versus the voltage V_{LC} across the LC cell. Assuming that the linear portion in Figure 2.13 belongs to a swing of 2 V, then 256 grey shades translate into a change of 7.8 mV per grey shade. This demonstrates how accurately the voltage across the LC cell has to be maintained in order to display an unadulterated picture. The range below the reduced threshold voltage V_{th}/V_0 in Figure 2.13 is required to keep parasitic crosstalk voltages from changing the luminance, and hence the grey shades. As a consequence, the chosen addressing schemes have to limit the sum of all crosstalk voltages to values below V_{th}.

The electronic addressing is subdivided into three sub-groups. The direct addressing in Figure 10.1 connects each portion of the display called a segment, which is supposed to exhibit various grey shades, to an external electrode carrying the voltage applied with respect to the ground electrode common for the entire display area. To avoid an unrealizably large number of connecting links, however, this approach, is limited to a small number of fixed segments, such as the seven segments of a digit in Figure 10.1.

A more versatile addressing is matrix addressing, which uses only the external contact pads in Figure 2.9 at the ends of rows and columns to reach the individual pixels. The Passive Matrix addressed LCD (PMLCD) in Figure 10.2 uses the row address voltages V_r at the row electrodes and the column address voltages V_c at the individual column electrodes which carry the video information to determine the individual grey shade of each pixel.

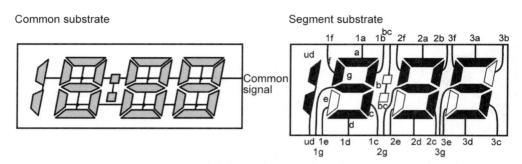

Figure 10.1   The direct addressing of the seven segments of a digit

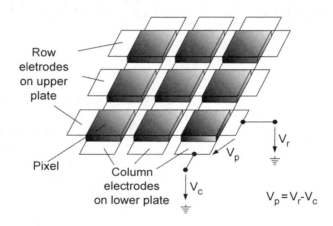

Figure 10.2   The Passive Matrix LCD (PMLCD) with row and column electrodes

As we shall see, an even more powerful control of the grey shades is realized by Active Matrix addressed LCDs (AMLCDs) in Figure 10.3, where each pixel possesses an electronic switch allowing the video information to reach the pixel.

Optical addressing involves the transfer of optical image information of a camera or on a display screen directly into an LCD. As this LCD is not pixellized at all, a high resolution picture may be obtained.

The Plasma Addressed LC (PALC) display uses an ignited and a turned-off plasma column as a switch for the addressing of large area LCDs with a diagonal, at present, up to 42".

There is a variety of electronic switches. Most important are Thin Film Transistors (TFTs) with a-Si, poly-Si or CdSe as the semiconductor. TFTs are Field-Effect-Transistors (FETs). Their manufacture and application for LCDs is well established, especially for a-Si-TFTs. TFTs with CdSe as the semiconductor now play a minor role (Brody *et al.*, 1973; Lee *et al.*, 1991; Lueder, 1994b).

Somewhat more frequent is the use of Metal-Insulator-Metal (MIM) devices as a switch. Semiconductor diodes as switches were investigated for a long time, but have now virtually disappeared in the display field. Nonlinear voltage dependent resistors, called *varistors*,

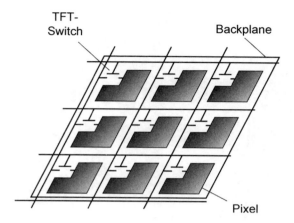

Figure 10.3 The Active Matrix addressed LCD (AMLCD) with a TFT as pixel switch

could not withstand the onslaught of the economically fabricated a-Si-TFTs, and are hence not discussed here.

Waveforms and addressing circuitry are very similar for all nematic LCDs, the PDLCDs and the guest-host displays, whereas for bistable devices such as ferroelectric LCDs and chiral nematic LDCs, different waveforms and addressing circuits are mandatory. As bistable displays do not require a control of continuous grey shades, the expensive AM addressing is not needed. Therefore, the addressing of bistable displays is done with PM schemes.

11

Direct Addressing

All the segments of the 3 1/2 digits in Figure 10.1, which could be used for watches, pocket computers, telephones or price labels, are individually connected to contact pads along the edges of the display. We need as many connections as there are segments. The example in Figure 10.1 has 23 of those connections, which is about the largest number for which the substrate has area for connecting. The backplate possesses the same segments but inter-connected to form a common electrode. Figure 11.1 shows how the voltage V_c at the common backplate and voltage V_s at a segment are applied to form the voltage $V_p = V_c - V_s$ across the LC material between the segments. In Figure 11.2, the voltages V_c and V_s at the off-segments are the same, whereas the voltage V_s at the on-segments are shifted by $T/2$ towards V_c, as also shown in Figure 11.2. The voltage at the off-segments add up to $V_p = 0$ and at the on-segments to $V_p = V$, as a dc-free square wave in Figure 11.2. As liquid crystals react to rms voltages, the on-pixels with the square wave are black and those with zero voltage are white in a normally white display. The reason why LCDs respond to an rms voltage is the fact that the response time τ of an LC display is much larger than the address time τ_A of the electronic circuit, i.e.

$$\tau \gg \tau_A. \tag{11.1}$$

Therefore, the LCD cannot follow the waveform of the voltage V_p, and hence must react to the energy fed into it, which is proportional to the square of the rms voltage averaged over time T

$$V_{rms}^2 = \frac{1}{T} \int_0^T V^2(t)dt. \tag{11.2}$$

The dc free addressing voltage V_p in Figure 11.2 is required for all LCDs in order to prevent decomposition of the liquid crystals due to electrolytical effects in a dc field.

The number of external connections can be decreased by reconfiguring the 23 electrodes in Figure 10.1 into a matrix display with four rows and six columns, shown in Figure 11.3.

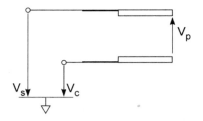

Figure 11.1 The voltage for a directly addressed LCD

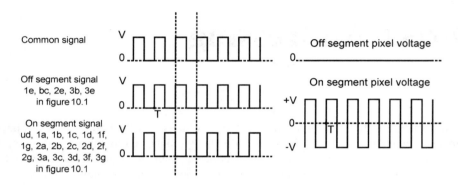

Figure 11.2 The waveforms for V_c, V_s and V_p of a direct addressed LCD

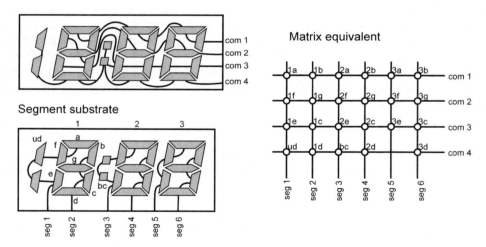

Figure 11.3 The reconfigured addressing of the display in Figure 10.1 into a matrix addressing with only 10 external connections

This array is addressed by the PM addressing discussed in Chapter 12. It reduces the external leads from 23 in Figure 10.1 to 10 in Figure 11.3.

12

Passive Matrix Addressing of TN Displays

12.1 The Basic Addressing Scheme and the Law of Alt and Pleshko

For the PM addressing in Figure 10.2, the voltage V_p across the pixel is

$$V_p = V_r - V_c. \tag{12.1}$$

In the normally black mode, the voltage across an off-pixel, that is a black pixel, is

$$V_p < V_{th}, \tag{12.2}$$

and across an on-pixel, that is a white one, it is

$$V_p > V_{th}, \tag{12.3}$$

where the grey shades in Figure 2.13 are realized by

$$V_p \in [V_1, V_2] > V_{th}. \tag{12.4}$$

The simplest way to achieve this is explained by a display in Figure 12.1, with three rows and two columns, where the pixels designated by '1'○ are supposed to be white, and those designated by '0'● are black (Scheffer and Nehring, 1998). The row selection or scanning voltage $V_r = S$ is shown in Figure 12.2(a); outside the row selection time $\Delta t = T/N$, the row is grounded with potential zero. T is the frame time needed to write in one picture, and N is the number of rows. The column voltage $V_c = F < V_{th}$, also called signal, data or video voltage, is depicted in Figure 12.2(b), whereas Figure 12.2(c) shows the pixel voltage V_p across the

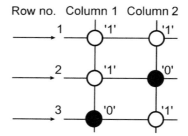

Figure 12.1   A display with 3×2 pixels with the desired video information, $\bigcirc$ white and $\bullet$ black

pixels in row 1/column 1 and row 2/column 2. The voltages in Figure 12.2(c) during the row address time are

$$\text{across the on-pixel:} \quad V_p = S + F > V_{th}; \tag{12.5}$$

$$\text{and across the off-pixel:} \quad V_p = S - F < V_{th}; \tag{12.6}$$

during the non-addressed state they are:

$$V_p = \pm F, \tag{12.7}$$

with

$$F \geq 0. \tag{12.8}$$

The inequalities make sure that the on-pixels are indeed white and the off-pixels black. In the non-addressed state, the parasitic crosstalk voltage across a pixel is $\pm F$, which is of no consequence as long as

$$|F| < V_{th}. \tag{12.9}$$

For grey shades the on-voltage $S + F$ lies between the two boundaries in Figure 2.13 that is

$$S + F \geq V_1 \geq V_{th} \tag{12.10}$$

and

$$S + F \leq V_2 > V_1, \tag{12.11}$$

resulting in the range for the signal voltage

$$F \geq V_1 - S \tag{12.12}$$

and

$$F \leq V_2 - S. \tag{12.13}$$

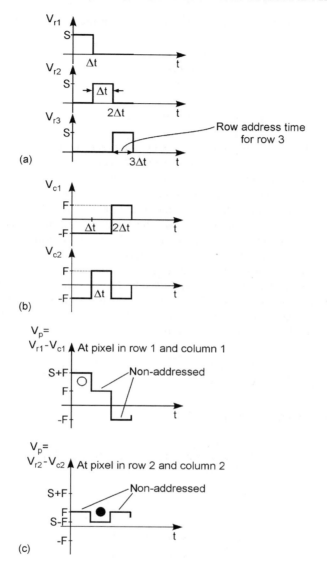

Figure 12.2 (a) Waveforms V_{r1}, V_{r2}, V_{r3} in rows 1, 2 and 3; (b) waveforms V_{c1} and V_{c2} in column 1 and 2; (c) waveforms at pixels in row 1/column 1 and row 2/column 2

The inequalities (12.8), (12.9), (12.12) and (12.13) provide

$$0 \leq F \leq V_{th} \tag{12.14}$$

and

$$V_1 \leq S \leq V_2 - V_{th}. \tag{12.15}$$

The example in Figures 12.1 and 12.2 reveals the rule for the selection of the signal voltage V_c as

$$\text{for an on-pixel:} \quad V_c = -F,$$

$$\text{for an off-pixel:} \quad V_c = +F.$$

The row-voltages V_r are $+S$ if the row is selected and zero otherwise.

In the time slot Δt in which a row is selected, we can apply the signal voltages for all pixels in this row simultaneously. Hence, all the pixels of this row simultaneously show their appropriate grey shades. This procedure is then repeated in the next time slot for the next row. This scheme is called the addressing of a row-at-a-time. As the LC cell reacts to the rms voltage, the signs of the addressing voltages V_r and V_c can be inverted without consequences for the grey shades. This inversion (e.g. after each frame or even after each row) is mandatory for the long term operation of the LC material.

The optimum ratio S/F providing the maximum contrast between an on-pixel and an off-pixel is now calculated according to Alt and Pleshko (1974) (see also Kawakomi, 1976). As we know, during the row-select time T/N an on-pixel carries the voltage $V_p = S + F$ and the off-pixel $V_p = S - F$. During the remaining unaddressed time $(T/N)(N-1)$, the pixel voltage is $V_p = \pm F$. Hence, the rms voltage for an on-pixel is

$$V_{on} = \frac{1}{\sqrt{T}} \sqrt{(S+F)^2 \frac{T}{N} + F^2 \frac{T}{N}(N-1)} = \frac{1}{\sqrt{N}} \sqrt{(S+F)^2 + F^2(N-1)}, \qquad (12.16),$$

whereas the rms voltage for an off-pixel is

$$V_{off} = \frac{1}{\sqrt{T}} \sqrt{(S-F)^2 \frac{T}{N} + F^2 \frac{T}{N}(N-1)} = \frac{1}{\sqrt{N}} \sqrt{(S-F)^2 + F^2(N-1)}. \qquad (12.17)$$

The ratio

$$\frac{V_{on}}{V_{off}} = \sqrt{\frac{(S/F+1)^2 + (N-1)}{(S/F-1)^2 + (N-1)}}, \qquad (12.18)$$

and hence contrast V_{on}/V_{off}, becomes a maximum for

$$\frac{S}{F} = \sqrt{N}, \qquad (12.19)$$

and as a consequence,

$$\frac{V_{on}}{V_{off}} = \sqrt{\frac{\sqrt{N}+1}{\sqrt{N}-1}}. \qquad (12.20)$$

Inserting Equations (12.19) into Equations (12.16) and (12.17) yields

$$V_{on} = F\sqrt{2\frac{\sqrt{N}+1}{\sqrt{N}}} \qquad (12.21)$$

and

$$V_{off} = F\sqrt{2\frac{\sqrt{N}-1}{\sqrt{N}}}. \qquad (12.22)$$

The circuit designer needs the row and column voltages S and F for the case of the optimum selection ratio V_{on}/V_{off}. We select

$$V_{off} = V_{th}, \qquad (12.23)$$

where a pixel is certainly off. Then Equation (12.22) yields

$$F = V_{th}\sqrt{\frac{\sqrt{N}}{2(\sqrt{N}-1)}}, \qquad (12.24)$$

for which Equation (12.19) provides

$$S = V_{th}\sqrt{\frac{N\sqrt{N}}{2(\sqrt{N}-1)}}. \qquad (12.25)$$

The optimum ratio V_{on}/V_{off}, and hence the optimum contrast obtained from Figure 2.13 for a given N, is determined by Equation (12.20), to which the optimum ratio S/F in Equation (12.19), the rms voltages V_{on} and V_{off} in Equations (12.32) and (12.22) and row and column

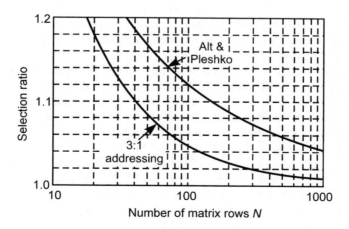

Figure 12.3 The selection ratio V_{on}/V_{off} in Equation (12.21) as a function of the number N of rows

voltages F and S in Equations (12.24) and (12.25) belong. This solution gives better contrast ratios than the $1/3$ bias optimized amplitude selective addressing scheme (Kaneko, 1998), also called $3:1$ addressing, known before Alt and Pleshko (1974). Solving Equation (12.20) for N results in

$$N = \left(\frac{(V_{on}/V_{off})^2 + 1}{(V_{on}/V_{off})^2 - 1}\right)^2 = N_{max}. \tag{12.26}$$

The obtainable selection ratio V_{on}/V_{off} for a given N in Equation (12.20) is plotted in Figure 12.3. Obviously, for increasing N the V_{on}/V_{off} (and hence the pertinent contrast) decreases. For a large enough N in Equation (12.20), V_{on}/V_{off} approaches 1, for which the contrast goes to zero. This is the main limitation of PM addressing.

As a rule, about $N=80$ rows of a TN display are addressable with the above PM addressing scheme. For $N=80$, Equation (12.19) yields $S \approx 9F$; for $F=2V$ we need the rather high row selection voltage of 18 V. Obviously, the larger is N in Equation (12.19), the larger is S, and the shorter the row address time $\Delta t = T/N$. There are further factors limiting the feasible values of N.

The consideration leading to Equation (12.20) does not take into account the steepness of the optical response in Figure 2.13. For a given ratio V_{on}/V_{off}, contrast can be enhanced by increasing the steepness of the optical response in Figure 2.13, because a given voltage swing from V_{on} to V_{off} then covers a larger range in the transmitted luminance. An enhanced steepness is obtained by using STN displays, for which the acceptable N can reach $N=250$. As N is such a crucial number, the matrix arrangement of a non-square display is, if possible, selected such that the number of rows is smaller than the number of columns.

12.2 Implementation of PM Addressing

A dc-free addressing could be realized by also using the addressing voltages V_r and V_c in Figures 12.2(a) and 12.2(b) with a reversed sign. However, this would enlarge the voltage swing of the row drivers to $2S$, extending from $-S$ to $+S$. Since ICs for such a large voltage range are more expensive, the modified addressing in Figure 12.4 has been introduced (Kawakami et al., 1976; Yong et al., 1994). The two left columns in Figure 12.4 depict the regular and reversed polarity for V_r, V_c and V_p. In the new scheme, the regular polarity is offset by F and the reversed polarity by S, resulting in the same regular and reversed polarity V_p in the old and new schemes. However, the voltage swing of the row and column drivers has been reduced from $2S$ to $S+F$. The change between the regular and reversed polarity could be done after each frame, after each row, or with still good dc-free results at other intervals (e.g. for a 240 row display) every seventh row.

The required voltages 0, $2F$, $S+F$ and $S-F$ in Figure 12.4 for a column driver are derived from the voltage source $S+F$ by the circuit in Figure 12.5. A buffered voltage divider is realized by the unit-gain operational amplifiers. The resistors are designed such that the outputs to ground provide the desired voltages. With the optimum ratio $S/F = \sqrt{N}$ in Equation (12.19), the resistor in the middle assumes the value $(\sqrt{N} - 3)R$ or in terms of the bias ratio $B = (F/(S+F))$ found in data sheets it has the value $(1/B - 4)R$. The circuit for a row or a column driver in Figure 12.6 starts with a shift register receiving the high H and

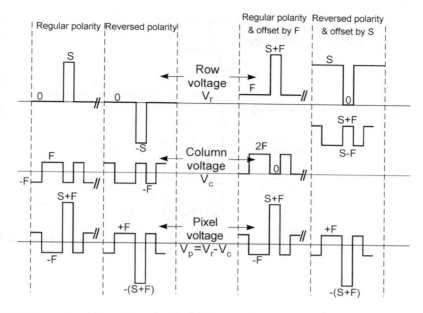

Figure 12.4 The reversal of polarity for row and column voltages in the addressing scheme without and with offset

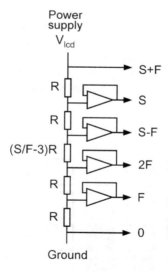

Figure 12.5 The generation of the voltage needed for a column driver

low L data for simultaneously setting the switches. The output provides the voltages for the regular offset polarity, and after activating the switch reverse the voltages for the reverse offset polarity of the column driver. A similar circuit realizes the row drivers. Grey shades may be realized by amplitude modulated column signals, as used in Equation (12.4). Very

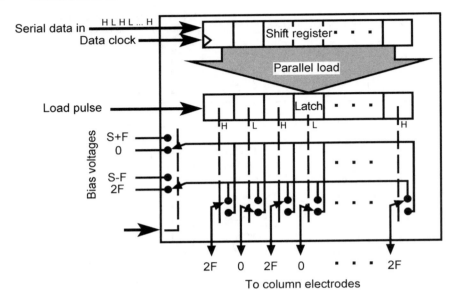

Figure 12.6   The circuit of a column driver for PM addressing

often grey shades are implemented either by Frame Rate Control (FRC) (Suzuki *et al.*, 1983), or by Pulse Width Modulation (PWM) (Kawakami *et al.*, 1980). FRC uses a super-frame in which the pixels in consecutive frames define a grey shade by exhibiting either a black or a white status. This is shown in Figure 12.7, where a superframe consisting of four frames is able to generate five grey shades.

For PWM the voltage waveform of a pixel is given in Figure 12.8 by the signal voltages of the on-pixel $V_c = -F$ during time f, of the off-pixel $V_c = +F$ during time $1 - f$, and of the

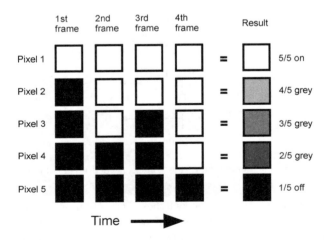

Figure 12.7   A superframe of four frames for the generation of five grey shades

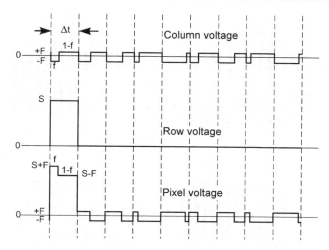

Figure 12.8 Pulse Width Modulation of the signal voltage V_c for the generation of grey shades

unaddressed pixel $V_c = \pm F$, leading to the rms voltage across the pixel for the grey shade

$$V_{\text{grey}} = \sqrt{\frac{f(S+F)^2 + (1-f)(S-F)^2 + (N-1)F^2}{N}}, \qquad (12.27)$$

with $f \,\varepsilon\,[0,1]$, which lies between the rms voltages of the on-pixel and of the off-pixel.

There are two causes of crosstalk. The non-ideal frequency response crosstalk is caused by the shift of the pixel-voltage for 50 percent optical transmission. This shift stems from low-pass filtering of the addressing voltage before it reaches the pixels. Periodically reversing the voltages V_r and V_c reduces this crosstalk. Further crosstalk is introduced by capacitive coupling of steps in V_c onto the row voltage V_r and then on V_p. This may be reduced by counter-impulses compensating the crosstalk caused by V_c, as discussed in Chapter 14.

Matrix architectures for PM addressed LCDs are represented in Figure 12.9. The basic architecture in Figure 12.9(a) shows the ICs for the row and column drivers placed along the edges of the displays and bonded to the contact pads at the end of the rows and columns. Special ICs were designed which are only about 1 mm wide and up to 20 mm long, with up to 60 pins; they require an additional area around the edges only 5 mm wide. For high resolution displays with a small pitch of columns and rows, the modifications in Figures 12.9(b) and 12.9(c) were generated. In the dual scan display in Figures 12.9(c) and 12.9(d), the columns are divided along the middle of the panel; this allows us to simultaneously address the upper and lower part of the display in Figure 12.9(d) with only $N/2$ row-drivers. This halves the time required to write in one frame, and either provides (according to Equation (12.20) and Figure 12.3) a larger selection ratio $V_{\text{on}}/V_{\text{off}}$, or allows us to double the number N while keeping the same selection ratio as in an undivided display. The penalty is the doubling of the number of column drivers. The number of row drivers in the dual scan scheme can also be halved by the set-up in Figure 12.9(e). Each original column is split into two columns, one each for the pixels in rows a and b. This configuration fits to the incoming stream of time sequential pixel data, serving first the even numbered rows and then the odd

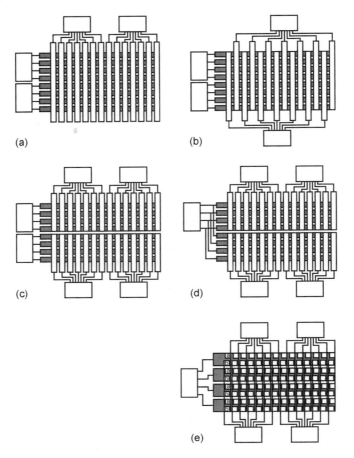

(a)

(b)

(c)

(d)

(e)

Figure 12.9 The various matrix architectures for PM-addressed LCDs (a) and (b) depict modifications of the basic architecture; (c), (d) and (e) show modifications of the dual scan scheme

numbered ones, whereas in the dual scan displays in Figure 12.9(d), a frame buffer for half of the display is required, if the two half pictures are displayed simultaneously. The pitch enhancing in Figure 19.9(b) can also be applied to the row-drives.

12.3 Multiple Line Addressing

12.3.1 The basic equations

To satisfy the rms condition in Equation (11.1), the response time τ of an LCD must be much larger than the address time τ_A of the electronic circuit. This was as a rule met by using slow LCDs which, however, precluded the use of PM addressed displays for fast moving pictures, such as are present in TV. A fast LCD would try to follow the pulsed addressing waveform shown in Figure 12.10. This effect is called *frame response*, because it occurs in the time intervals T_f in Figure 12.10 imposed by the pixel voltages which coincide

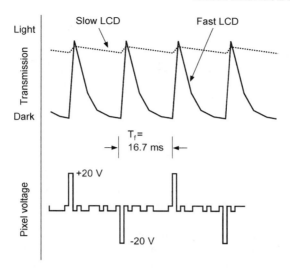

Figure 12.10 The phenomenon of 'frame response' in fast LCDs

with the period T_f of the frames. To render fast LCDs applicable, addressing based on short pulses has to be abandoned and replaced by addressing waveforms spread over the entire frame time T_f. This was achieved by the Active Addressing proposed by Scheffer and Clifton (1992) and Scheffer and Nehring (1998), and by the independently developed Multiple Line Selection by Ihara *et al.* (1992). Their work was based on Nehring and Kmetz's (1979) contribution. The new addressing scheme is now called Multiple Line Addressing (MLA).

The analysis of MLA starts with an LCD with N rows and M columns. The binary information in a pixel in row i and column j is given by I_{ij}, where $I_{ij}=1$ stands for an off-pixel, that is black, and $I_{ij}=-1$ for an on-pixel, which is white. Grey shades are dealt with later. Each row $i=1,2,\ldots,N$ driven by a row voltage $F_i(t)$ with period T, and all columns with a column voltage $G_j(t)$, $j=1,2,\ldots,M$ also with period T. The column signal is

$$G_j(t) = c \sum_{i=1}^{N} I_{ij}F_i(t),$$ (12.28)

where c is a constant independent of I_{ij} (Nehring and Kmetz, 1979). $G_j(t)$ is the sum of the dot product of the information states I_{ij} and all row signals $F_i(t)$ at time t.

The voltage $U_{ij}(t)$ across a pixel located at i,j is

$$U_{ij} = F_i(t) - G_j(t).$$ (12.29)

The LCD reacts to the rms value of U_{ij}, that is to

$$\langle U_{ij} \rangle = \frac{1}{\sqrt{T}} \sqrt{\int_0^T U_{ij}^2(t)dt}.$$ (12.30)

Substituting Equations (12.28) and (12.29) into Equation (12.30) provides

$$\langle U_{ij} \rangle = \frac{1}{\sqrt{T}} \sqrt{\int_0^T F_i^2(t)dt - 2\int_0^T F_i(t)c\sum_{i=1}^N I_{ij}F_i(t)dt + c^2\int_0^T \left(\sum_{i=1}^N I_{ij}F_i(t)\right)^2 dt}. \quad (12.31)$$

Now we require the signals $F_i(t)$ to be orthogonal, leading to

$$\frac{1}{\sqrt{T}}\sqrt{\int_0^T F_i(t)F_k(t)dt} = \begin{cases} F & \text{for } i = k, \\ 0 & \text{for } i \neq k. \end{cases} \quad (12.32)$$

Further, we need in the second term of Equation (12.30) the information states I_{ij} in all rows $i = 1, 2, \ldots, N$. Therefore, the expression requires some time to be calculated. Assuming that at the moment we calculate Equation (12.31) all I_{ij} are zero besides the one in row i, we obtain from Equation (12.31) by inserting Equation (12.32)

$$\langle U_{ij} \rangle = F\sqrt{1 - 2cI_{ij} + c^2N}. \quad (12.33)$$

The assumption made is proven to be correct in Section 12.3 after Equation (12.62).

The rms voltages $\langle U_{ij\,on} \rangle$ for the on-pixel with $I_{ij} = 1$ and $\langle U_{ij\,off} \rangle$ for the off-pixel with $I_{ij} = 1$ are

$$\langle U_{ij\,on} \rangle = F\sqrt{1 + 2c + c^2N} \quad (12.34)$$

and

$$\langle U_{ij\,off} \rangle = F\sqrt{1 - 2c + c^2N}, \quad (12.35)$$

with the selection ratio

$$\frac{\langle U_{ij\,on} \rangle}{\langle U_{ij\,off} \rangle} = \sqrt{\frac{1 + 2c + c^2N}{1 - 2c + c^2N}}. \quad (12.36)$$

The maximum value of the selection ratio in Equation (12.36) occurs for

$$c = c_{opt} = 1/\sqrt{N}, \quad (12.37)$$

leading to

$$\frac{\langle U_{ij\,on} \rangle}{\langle U_{ij\,off} \rangle} = \sqrt{\frac{\sqrt{N} + 1}{\sqrt{N} - 1}} \quad (12.38)$$

and to $G_j(t)$ in Equation (12.28) as

$$G_j(t) = \frac{1}{\sqrt{N}} \sum_{i=1}^{N} I_{ij} F_i(t). \tag{12.39}$$

Equation (12.38) demonstrates that the law of Alt and Pleshko also holds for the general row signals $V_{ri} = F_i(t)$, $i = 1, 2, \ldots, N$ and the pertinent column signals $V_{cj}(t) = G_j(t)$, $j = 1, 2, \ldots, M$ in Equation (12.39), where $F_i(t)$, $i = 1, 2, \ldots, N$ are orthogonal.

The remaining tasks are to identify easily realizable orthogonal functions for the row selection with a minimum of frame response, the generation of the column signal, the hardware implementation of row and column drivers, and finally, the realization of grey shades.

12.3.2 Waveforms for the row selection

Any set of orthogonal functions $F_i(t)$ may be applied to drive the columns at the optimum selection ratio in Equation (12.38). Practical row functions are either bilevel with amplitude $+F$ or $-F$, or trilevel with amplitude $+F$, 0 and $-F$.

A most practical set of bilevel functions easily realizable by digital circuits are the Walsh functions in Figure 12.11 (Clifton *et al.*, 1993; Scheffer and Nehring, 1998). The order of Walsh functions is 2^s, s integer, meaning that there are 2^s time intervals in which the value is 1 or -1. The 32nd Walsh function in Figure 12.11 has the order $2^5 = 32$. To address N

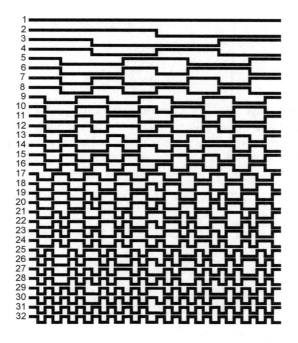

Figure 12.11 The Walsh functions of order 1 to 32

rows simultaneously, we require $2^{s-1} < N \le 2^s$. A 240 row display would require 256 Walsh functions of order $2^8 = 256$. The 1st Walsh function should not be used as it is not dc-free. These bilevel row functions can be represented by a Hadamard matrix, which is square and orthogonal with the elements -1 and $+1$ of the Walsh functions, indicating a pulse $+F$ and $-F$. The N rows of the Hadamard matrix correspond to the signals applied to the N rows of the display, whereas the columns correspond to the N time intervals of the frame period T_f each $\Delta t = T_f / N$ long.

There are other orthogonal functions which have $J > N$ time intervals per frame. Walsh functions exhibit the minimum number of time intervals per frame period reflecting in $J = N$.

The selection intervals are those time intervals in which the selection pulses are non zero. Their number is denoted by K. Thus, the Walsh functions having no zero pulses exhibit $K = J = N$ selection intervals, whereas in trilevel functions $K < J$. To avoid frame response, the intervals with zero pulses should be spread most evenly over the frame time, thus preventing the optical state of the pixel from significantly decaying before the next pulse arrives.

So far we have assumed that all N rows are addressed simultaneously. Computational complexity will, however, be reduced if only $L < N$ rows are addressed at a time. These L row signals must contain $K \ge L$ selection intervals, where $K = L$ is preferred for ease of computation.

For all sets of orthogonal row functions $F_i(t)$, we work with a normalized rms voltage of the off-pixel $\langle U_{ij\,\text{off}} \rangle$ in Equation (12.35) requiring for a general value of c $\langle U_{ij\,\text{off}} \rangle = F\sqrt{1 - 2c + c^2 N} = 1$ or

$$F^2 = 1/(1 - 2c + c^2 N). \tag{12.40}$$

With $c = c_{\text{opt}}$ in Equation (12.37) Equation (12.40) yields

$$F = \sqrt{\frac{\sqrt{N}}{2(\sqrt{N} - 1)}}. \tag{12.41}$$

In a display the rms voltage $\langle U_{ij\,\text{off}} \rangle = V_{\text{th}}$ as the pixel with $V_p = V_{\text{th}}$ is off. Hence, to convert all normalized voltages to actual voltages, they must be multiplied by V_{th}. For $N = 240$ we obtain $F = 0.53$. F is the magnitude of the selection pulses in a frame period if this period contains $J = N$ selection pulses. If there are fewer selection pulses, i.e. if $K < J$, then the magnitude of the pulses has to be increased to

$$S = \sqrt{J/K}\, F \tag{12.42}$$

to maintain the same rms value as for $K = J$. For $J = N$ and $K = L$, Equation (12.42) with F from Equation (12.41) provides

$$S = \sqrt{\frac{N\sqrt{N}}{2(\sqrt{N} - 1)} \frac{1}{\sqrt{L}}}. \tag{12.43}$$

S is the normalized magnitude of the row pulses for MLA with an optimized selection ratio and N rows as a function of the number of $L < N$ rows addressed at a time. As $L < N$, trilevel drivers are required. In Figure 12.12 (Clifton *et al.*, 1993; Scheffer and Nehring, 1998), S is plotted versus L for $N=256$. Obviously, the larger the number L of rows addressed at a time, the smaller the magnitude of S that is required.

12.3.3 Column voltage for MLA

If L rows are selected at a time Equation (12.39) is modified to

$$G_j(t) = \frac{S}{\sqrt{N}} \sum_{i=1}^{L} I_{ij} A_i(t), \qquad (12.44)$$

where $A_i(t) = \pm 1$ are elements of the Hadamard row addressing matrix A in row i at time t, which together with S define the orthogonal function $F_i(t) = \pm S$. The maximum of $G_j(t)$ reached for all $I_{ij}=1$ and $A_i(t)=1$ is, for all j,

$$G_{j\,\text{max}} = G_{\text{max}} = \frac{SL}{\sqrt{N}} = \sqrt{\frac{J}{K}} \frac{L}{\sqrt{N}} F, \qquad (12.45)$$

where S from Equation (12.42) has been used.

For the previously investigated case $J=N$ and $K=L$, we obtain with F from Equation (12.41)

$$G_{\text{max}} = \sqrt{L}F = \sqrt{L}\sqrt{\frac{\sqrt{N}}{2(\sqrt{N}-1)}}. \qquad (12.46)$$

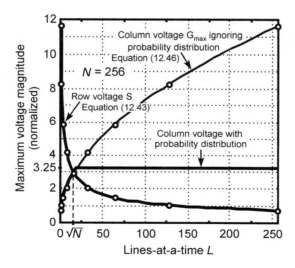

Figure 12.12 The row voltage S in Equation (12.43) and the column voltage G_{max} in Equation (12.46) for $N=256$ as a function of $L<N$ rows addressed at a time

This is plotted again for $N=256$ in Figure 12.12 as a function of L. S in Equation (12.43) is equal to G_{max} in Equation (12.46) for

$$L = \sqrt{N}. \tag{12.47}$$

Obviously, the larger is L the larger is the column voltage G_{max} required.

We now investigate according to Scheffer *et al.* (1993) the probability for the occurance of larger values for $G_j(t)$ in Equation (12.44). To this end, we randomly select about half of the N rows in the row selection matrix A and reverse their polarities. This preserves the orthogonal property of the row vectors, and ensures the system will obey the probability of a value $G=G_j$ exceeding a given value G_0,

$$P(|G| \geq G_0) = 2Erfc(G_0/R), \tag{12.48}$$

where the complimentary error function is defined by

$$Erfc\left(\frac{G_0}{R}\right) = \frac{1}{\sqrt{2}} \int_{G_0/R}^{\infty} \exp\frac{z^2}{2}dz. \tag{12.49}$$

For $G_0=3.25$ and $N=256$ the probability that G exceeds G_0 is $8.57 \cdot 10^{-6}$. Thus, only one out of $116\,000$ column levels exceeds G_0. Therefore, limiting G in Figure 12.12 by $G_0=3.25$ is justified. This limit is independent of L. The limit G_0 is reached at about $L=16$. The probability for the $L+1$ terms for the column voltage levels $G_j(t)$ in Equation (12.44) is depicted in Figure 12.13 for $L=4$, 32 and 256. The distribution is binomial approaching a gaussian with increasing L. Levels beyond ± 2.5 V are very unlikely.

12.3.4   Implementation of multi-line addressing

We have to calculate the voltage $G_j(t)$ of column j of the display $j=1,2,\ldots,M$ in Equation (12.44), and do it for the N time intervals $\Delta t_k=(T_f/N)$, $k=1,2,\ldots,N$, leading to

$$G_j(\Delta t_k) = \frac{S}{\sqrt{N}} \sum_{i=1}^{L} I_{ij}A_i(\Delta t_k), \quad k = 1,2,\ldots,N, \quad j = 1,2\ldots M. \tag{12.50}$$

$A_i(\Delta t_k)$, $i=1,2,\ldots,L$ are the elements in the column k of the normalized row address matrix A, an (L,N) matrix with

$$A = \begin{array}{c} \text{row} \\ (i)\downarrow \end{array} \begin{array}{c} 1 \\ 2 \\ \\ i \\ \\ L \end{array} \left(\begin{array}{ccccc} 1 & & & & \\ 1 & & & & \\ & & & & \\ & \ldots 1 \ldots & & \xrightarrow{} A_i(\Delta t_k) \\ & -1 & & & \\ & 1 & & & \end{array} \right). \tag{12.51}$$

$$\begin{array}{ccccc} \text{time interval} & \Delta t_1 & \Delta t_2 \ldots & \Delta t_k \ldots & \Delta t_N \\ (k)\rightarrow & & & & \end{array}$$

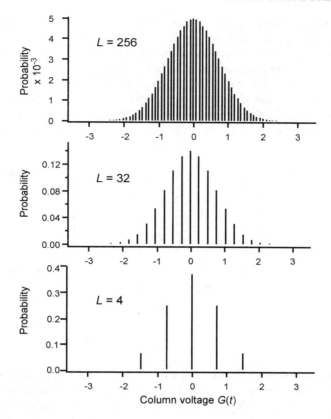

Figure 12.13   Probability for column voltage levels as a function of column voltage for $L=4$, 32, 256 rows addressed at a time

The (L, N)-matrix I is the information matrix, with the elements I_{ij} given as

$$
\begin{array}{cc}
\text{row of display} & 1 \\
& 2 \\
I = & i \\
& L \\
\text{column of display} & 1 \quad j \quad N
\end{array}
\left(
\begin{array}{c}
I_{1j} \\
\\
I_{ij} \\
\\
I_{Lj}
\end{array}
\right). \tag{12.52}
$$

The matrix product $A'I$, where A' is the transpose of A, exhibits as element k, j the sum in Equation (12.50). This leads to a column voltage matrix

$$ C = A'I, \tag{12.53} $$

in which the element k, j represents the voltage at column j of the display during time interval k, as required in Equation (12.50). The matrix multiplication transforms a given image in the space domain (matrix I) into the time domain (matrix C).

The example for a column signal in Figure 12.14 was calculated using Equation (12.53) for $N=256$ time intervals Δt_k and for $L=32$.

The computation of the matrix product C in Equation (12.53) can be done in two ways.

The first way is by combining, as shown by solid arrows in Equation (12.54), row k in A' belonging to time interval k with the column j in I. By doing this for all columns $j=1,2,\ldots,M$ in I, we obtain the elements k, j with $j=1,2,\ldots,M$, that is the row k in C belonging to time interval k for all column voltages. This procedure is repeated for the next time interval $k+1$ (row $k+1$ in A'), etc., and may be called a column sequential update.

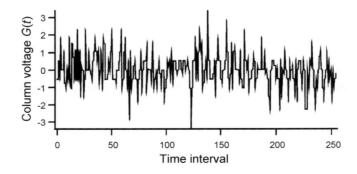

$$= C.$$

$$\begin{array}{ccc} A' & I & C \end{array}$$

$$(12.54)$$

This method has the advantage that the elements in row k of C can directly be connected to the contact pads of the column $1,\ldots,M$ of the display feeding them with the signal for a given time interval $k=1,2,\ldots,N$. The disadvantage is that in the matrix I the information of a column pertaining to all time intervals of a frame are needed, which requires a frame buffer. The second method of matrix multiplication is the combination of a column in A' with a row r in I, as indicated by a dotted arrow in Equation (12.54). By doing this, we obtain column i in C with the column voltage $G_i(\Delta t_k)$ for all time intervals $k=1,2,\ldots,N$. This method requires only one row of I to be presented to the display at the same time, and hence does not require a frame buffer. The column voltage is, however, provided for all time intervals, and therefore has to be stored and composed to a column vector for a given time interval.

Figure 12.14 Waveform of a column voltage with 256 time intervals and 32 lines selected at a time

The block diagram for the hardware for row and column drivers in Figure 12.15 contains the frame buffer memory with the content of matrix I and a row function ROM in which the orthogonal row functions are stored representing the content of the matrix A'. The row of A' belonging to time interval Δt_k, denoted by $A(\Delta t_k)$, is read out and fed into the row drivers. They can be realized as the ICs of STN column drivers.

The multiplication $I_{ij}A_i(\Delta t_k)$ with $i = 1, 2, \ldots, L$ involves a column vector l_j of the matrix I and the row $A(\Delta t_k)$ with time interval Δt_k of the matrix A'. If the row address functions are $\pm F$ and 0, the multiplication reduces to establishing if the product is $+F$, $-F$ or zero, which can be performed by the exclusive or gate (XOR) in Figure 12.15. After the summation, a DAC also performs the multiplication with $S/\sqrt{N}$ in Equation (12.44). The result is fed into the column drivers, which may be realized by the ICs for the data drivers in AMLCDs.

The advantage of MLA is the capability to handle a larger number of rows than imposed by the law of Alt and Pleshko, and to use faster LCDs suitable for fast moving pictures, such as those required for TV and even HDTV. The selection of $L < N$ rows to be addressed at a time reduces the computational complexity.

12.3.5 Modified PM addressing of STN cells

Decreased levels of addressing voltages

The original single line at-a-time addressing of STN and also TN cells requires a swing of the row voltage of $2S$ reaching from $-S$ to $+S$. As a rule, S is around 5 V, and always larger

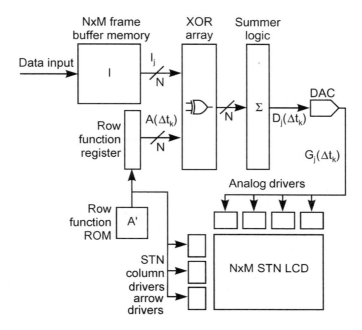

Figure 12.15 Block diagram for row and column drivers of a MLA-LCD

than the data voltage F, lying in the range of 2 to 3 V. Therefore, the addressing ICs must withstand a voltage of $2S$ rendering them more expensive. The modified addressing waveforms in Figure 12.16 are similar to the waveforms in Figure 12.4. They also limit the voltage swing to the smaller values $S+F$. During the first frame time the row addressing in Figure 12.16(a) has a zero voltage during the line address time and the voltage S in the non-select time, whereas the data voltage in Figure 12.16(b) is $S+F$ in the select time and $S-F$ in the non-select time. This leads to the pixel voltage being $-(S+F)$ in the select time and F in the non-select time. In the second frame time, the voltages are as shown in Figures 12.16(c) and 12.16(d), with the pixel voltages the reverse from the first frame time, as required for stability of the LCDs. This scheme reduces the maximum voltage swing to $S+F$, rendering the ICs cheaper.

For MLA the row voltage S and the maximum data voltage G_{max} in Figure 12.12 are dependent on the number L of rows addressed at one time. Also, these voltages can be reduced following a scheme proposed by Kuijk *et al.* (1999). He observed that choosing a smaller number of $N' < N$ lines than the optimum number of N lines in Equation (12.38) leading to the optimum selection ratio $\langle U_{ij\,on}\rangle/\langle U_{ij\,off}\rangle$ yields a larger on-voltage $\langle U'_{ij\,on}\rangle$ than the automatic use of Equation (12.38) would give. This is illustrated in Figure 12.17, where the difference between $\langle U_{ij\,off}\rangle = V_{th}$ and $\langle U_{ij\,on}\rangle = V_{sat}$ and the difference between V_{th} and $\langle U'_{ij\,on}\rangle = V_{on}$ is $1/N$ and $1/N'$. However, the optically effective selection ratio determining the transmission in the linearly rising flank is the same in both cases, namely due to Equation (12.38) equal to $\sqrt{(\sqrt{N}+1)/(\sqrt{N}-1)}$.

With Equation (12.36), this fact results in

$$\frac{\langle U'_{ij\,on}\rangle^2}{\langle U_{ij\,off}\rangle^2} = \frac{1+2c+c^2 N'}{1-2c-c^2 N'} = \frac{\sqrt{N}+1}{\sqrt{N}-1}. \tag{12.55}$$

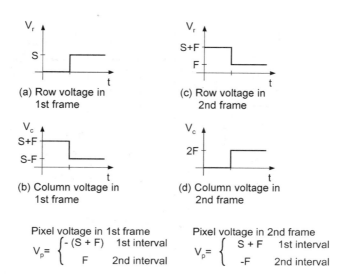

(a) Row voltage in
1st frame

(c) Row voltage in
2nd frame

(b) Column voltage in
1st frame

(d) Column voltage in
2nd frame

Pixel voltage in 1st frame
$$V_p = \begin{cases} -(S+F) & \text{1st interval} \\ F & \text{2nd interval} \end{cases}$$

Pixel voltage in 2nd frame
$$V_p = \begin{cases} S+F & \text{1st interval} \\ -F & \text{2nd interval} \end{cases}$$

Figure 12.16   The decreased levels of row and column voltages for PM addressing

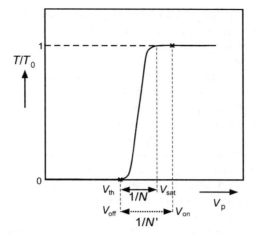

Figure 12.17 The reduced transmission at the pixel voltage V_{th}, V_{sat} and $V_{on} > V_{sat}$ for $N' < N$

This equation is met for

$$c = \frac{\sqrt{N}}{N'} \pm \frac{1}{N'}\sqrt{N - N'}. \tag{12.56}$$

The new c-value coincides with the previous value c_{opt} in Equation (12.37) for $N' = N$.

For c in Equation (12.56) the row voltage F in Equation (12.40) for a single row addressed at a time and the row voltage S in Equation (12.42) for L rows addressed at a time become for actual voltages

$$FV_{th} = \frac{V_{th}}{\sqrt{1 - 2(\sqrt{N}/N') + (N/N')(2(N/N') - 1) - 2(\sqrt{N - N'}/N')((\sqrt{N}/N') - 1)}}, \tag{12.57}$$

and

$$SV_{th} = \sqrt{\frac{N'}{1 - 2(\sqrt{N}/N') + (N/N')(2(N/N') - 1) - 2(\sqrt{N - N'}/N')((\sqrt{N}/N') - 1)}} \frac{V_{th}}{\sqrt{LF}}. \tag{12.58}$$

The maximum column voltage is, according to Equation (12.46) $G_{max} = \sqrt{LF}$ and using F in Equation (12.57), leading to

$$G_{max}V_{th} = \sqrt{L}\frac{V_{th}}{\sqrt{1 - 2\sqrt{N}/N' + (N/N')((2N/N') - 1) - (2\sqrt{N - N'}/N')((\sqrt{N}/N') - 1)}}. \tag{12.59}$$

SV_{th} in Equation (12.58) and $G_{max}V_{th}$ in Equation (12.59) both have relatively small values where their curves in Figure 12.12 intersect at

$$L = \sqrt{N'}. \tag{12.60}$$

It can be shown that the row and column voltages in Equations (12.58) and (12.59) are lower than those for MLA in Equations (12.43) and (12.46) for $L=8$ and lower for single line addressing without and with the modified voltage swing. The lower voltages also ensure a lower power consumption.

As an example, we consider a display with $N'=64$ rows, $V_{th}=1.1$ V and an Alt and Pleshko limit of $N=100$ rows. We calculate for MLA the row voltage and column voltage for $N \neq N'$ as given, and for $N=N'=64$, as well as both voltages for single line addressing.

From Equation (12.60) we obtain $L=8$ rows for MLA. The Equations (12.58) and (12.59) yield $SV_{th}=G_{max}V_{th}=1.52$ V with a voltage level $SV_{th}+G_{max}V_{th}=3.04$ V. For a MLA with $N'=N=64$ rows and the optimum $L=8$, Equations (12.58) and (12.59), or Equations (12.43) and (12.46), provide $SV_{th}=G_{max}V_{th}=2.13$ V and a pertinent voltage level of $SV_{th}+G_{max}V_{th}=4.62$ V. Obviously, the approach using $N \neq N'=64$ provides the lower level. For single line addressing according to Equations (12.24) and (12.25), we obtain $FV_{th}=0.83$ V and the column voltage $SV_{th}=6.65$ V with a voltage level $FV_{th}+SV_{th}=7.48$ V. Hence, single line addressing requires the largest voltage level of all three approaches.

Contrast and grey shades for MLA

The frame response can be diminished if the frame frequency is increased, as transmittance has less time to drop. Thus, a higher frame frequency enhances contrast. This is depicted in Figure 12.18 (Hirai *et al.*, 1995). The penalty is the need of faster signal processing for MLA. Experiments have demonstrated that contrast for $L=3$ to 4 lines addressed at a time is already quite close to the contrast reached in saturation, as is also visible in Figure 12.18.

Grey shades with Frame Rate Control (FRC) are limited to a few shades, as the larger frame frequencies required may cause flicker if a fast responding LCD is used. The fast response, however, is needed for fast moving pictures, as on computer screens and for TV (Scheffer and Nehring, 1998). Pulse Width Modulation (PWM) needs several $G_j(\Delta t_{ik})$ values per information vector l_j, each based on different grey shades. Then the $G_j(\Delta t_{ik})$ signals are time weighted within each time interval to provide a flicker-free grey shade. With PWM only

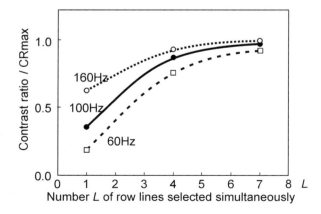

Figure 12.18 Dependence of contrast on frame frequency and on the number L of lines addressed at a time

around 16 grey shades can be realized, because the shorter pulses contain higher frequencies which do not reach the pixels due to the RC low-pass filters formed by the addressing lines (Scheffer and Nehring, 1998).

Pulse amplitude modulation used with TFT addressing is also unacceptable, because the pulses turn out to be too unevenly distributed over the frame time, thus giving rise to frame response. The addressing described in Section 12.1 with levels F of the data voltage dependent on the grey shade represents an amplitude modulation, and is therefore only useful for a few coarse grey levels.

A large number of grey shades beyond 256 is feasible with full-interval Pulse Height Modulation (PHM) (Connes and Scheffer, 1992). The approach is based on MLA in Section 12.3, and contains single line addressing as a special case. The waveforms for the selection of a line i in a display with N lines are the orthogonal functions $F_i(t)$ with $i=1,2,\ldots,N$; the signals $G_j(t)$ with $j=1,2\ldots,M$ for the M columns are given in Equation (12.39) for the case of an optimum selection ratio $\langle U_{ij\,\text{on}}\rangle/\langle U_{ij\,\text{off}}\rangle$ defined by $c=c_{\text{opt}}=1/\sqrt{N}$ in Equation (12.37). Inserting $G_j(t)$ in Equation (12.39) into the rms pixel voltage $\langle U_{ij}\rangle$ in Equation (12.31) yields

$$\langle U_{ij}\rangle = \frac{1}{\sqrt{T}}\sqrt{\int_0^T F_i^2(t)dt - \frac{2}{\sqrt{N}}\int_0^T F_i(t)\sum_{i=1}^N I_{ij}F_i(t)dt + \frac{I}{N}\int_0^T \left(\sum_{i=1}^N I_{ij}F_i(t)\right)^2 dt,}$$

which is simplified by the orthogonality of F_i in Equation (12.32) to

$$\langle U_{ij}\rangle = F\sqrt{1 - \frac{2}{\sqrt{N}}I_{ij} + \frac{1}{N}\sum_{i=1}^N I_{ij}^2}. \tag{12.61}$$

The information elements I_{ij} introduced in Section 12.3.1 so far could assume the values $I_{ij}=\pm 1$ corresponding to black and white pixels. Equation (12.29) for the pixel voltage U_{ij} indicates, together with Equation (12.39), that $I_{ij}=-1$ belongs to the larger and $I_{ij}=+1$ to the smaller pixel voltage. Hence, $I_{ij}=-1$ stands for the black state in a normally white, and for the white state in a normally black, display. Now we introduce intermediate values $-1<I_{ij}<1$ representing grey shades. Equation (12.61) has the shortcoming that the grey shade in pixel i, j is not only dependent on its information element I_{ij}, but also on all information elements in column j. As a remedy, it was proposed (Connes and Scheffer, 1992) to add an additional row $N+1$ with the information elements $I_{N+1,j}, j=1,2,\ldots,M$. This modifies Equation (12.61) into

$$\langle U_{ij}\rangle = F\sqrt{1 - \frac{2}{\sqrt{N}}I_{ij} + \frac{1}{N}\sum_{i=1}^N I_{ij}^2 + \frac{1}{N}I_{N+1,j}^2}. \tag{12.62}$$

Selecting

$$I_{N+1,j} = \sqrt{N - \sum_{i=1}^N I_{ij}^2} \tag{12.63}$$

simplifies Equation (12.62) to

$$\langle U_{ij} \rangle = F\sqrt{2}\sqrt{1 - I_{ij}/\sqrt{N}}, \tag{12.64}$$

in which the pixel voltage U_{ij} only depends upon the information I_{ij} of this pixel. The added row $N+1$ has not to be realized being virtual row. However, to the column signals $G_j(t)$ in Equation (12.39), the term $\sqrt{N}^{-1}I_{N+1,j}F_{N+1}(t)$ with the orthogonal line voltage $F_{N+1}(t)$ must be added, providing

$$G_j(t) = \frac{1}{\sqrt{N}} \sum_{i=1}^{N} I_{ij}F_i(t) + \frac{1}{\sqrt{N}}I_{N+1,j}F_{N+1}(t). \tag{12.65}$$

The formation of this column signal requires the knowledge of all elements I_{ij} which have to be stored in a frame buffer. $F_{N+1}(t)$ is just one more sample of the set of orthogonal functions $F_i(t)$. We consider two special cases. If there are no intermediate grey shades in column j, the elements are $I_{ij} = \pm 1$ rendering $I_{N+1,j} = 0$ in Equation (12.63). This reduces Equation (12.62) to Equation (12.33) for $c = 1/\sqrt{N}$. This also proves Equation (12.33), which was derived by making an assumption of the values of I_{ij}. If all pixels assume the value $I_{ij} = 0$, that means a grey level half way in between the black and the white state, $I_{N+1,j}$ reaches its maximum $I_{N+1,j} = \sqrt{N}$. For $N = 240$, $I_{N+1,j} = \sqrt{240} = 15.5$ lying outside the feasible range from -1 to $+1$. However, this does not matter as row $N+1$ is not realized.

From Equation (12.62) we obtain $\langle U_{ij\,on} \rangle$ for $I_{ij} = -1$ and $\langle U_{ij\,off} \rangle$ for $I_{ij} = +1$ of a normally black display, yielding

$$\frac{\langle U_{ij\,on} \rangle}{\langle U_{ij\,off} \rangle} = \sqrt{\frac{\sqrt{N}+1}{\sqrt{N}-1}}. \tag{12.66}$$

Obviously, the law of Alt and Pleshko still holds also in the presence of a virtual row.

Examples of single row addressing and of MLA are instructive (Connes and Scheffer, 1992). Figure 12.19 shows the row signals for single row addressing, including the virtual row. The column signal between $\pm F$ for the first six real rows realizes the additional intermediate grey shades $-1/2$, 0, 0, and $+1/2$ depicted in Figure 12.19 leading for the six rows with Equation (12.63) to $I_{7,1} = \sqrt{6 - 2.5} = 1.87$.

Equation (12.65) provides the voltage $G_1(t)$ for the only column which is shown in Figure 12.19 versus the time intervals $i = 1, 2, \ldots, 7$. In these intervals, the value of $G_1(t)$ is $(1/\sqrt{6})FI_{i1}$, where I_{i1} are the values -1, $-1/2$, 0, 0, 1/2, 1 and 1.87. Obviously, the virtual row influences the column signal only in the additional time interval 7. It should be noted that this time interval is part of the column signal, though row 7 does not exist.

The same 1-column grey shades as in Figure 12.19 are used as an example for MLA addressing of six rows simultaneously in Figure 12.20. The row signals are Walsh functions, including the virtual row 7. The zero order Walsh function is omitted as it has a dc component. The amplitudes are $\pm F$. The additional information element in Equation (12.63) is the same as in the preceding example, because the information pattern is the same. The column signal computed from Equation (12.65) is shown in Figure 12.20 as solid waveform, where the individual row voltages $F_i(t)$ in Figure 12.20 were used. The dotted waveform

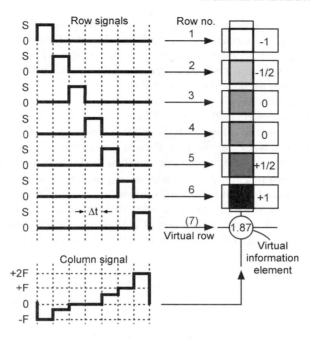

Figure 12.19 Generation of full-interval PHM grey shades for a single row at a time addressed LCD

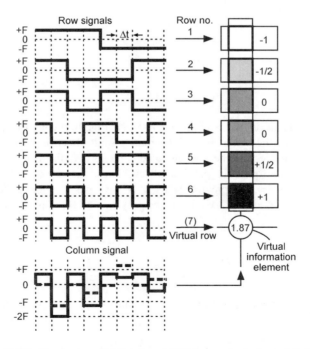

Figure 12.20 Generation of full-interval PHM grey shade in an MLA-LCD

demonstrates the column signal without the signal F_i of the virtual row. In contrast to Figure 12.19, the influence of the virtual row is distributed over the time intervals of the column signal, whereas it has been added as an additional time interval in Figure 12.19. The method described makes use of the full undivided time intervals. An alternative called *split-interval PHM* divides an interval into two parts with different heights of the pulses, and is described in Connes and Scheffer (1992).

The hardware for full-interval PHM published by Clifton and Price (1992) has to provide the row voltages $F_i(t)$ and the column voltages $G_j(t)$ according to Equation (12.65). The row function can be bilevel Walsh functions or Pseudo-Random Binary Sequences (PRBS), as used in communications as a data signal with zero mean (Stein and Kashnow, 1971). Each row in the PRBS system is a cyclically shifted version of the previous row. These functions are orthogonal as required. Trilevel row functions with values $+F$, $-F$ and 0 are generated from small blocks of Walsh functions distributed over the frame period and separated by zero elements. The zeros reduce the number of computations needed for $G_j(t)$ in Equation (12.65). For the explanation of the hardware Equation (12.65) is detailed as

$$G_j(\Delta t_k) = \frac{1}{\sqrt{N}}(I_{1j}F_1(\Delta t_k) + I_{2j}F_2(\Delta t_k) + \ldots, + I_{Nj}F(\Delta t_k) + I_{N+1\,j}F_{N+1}(\Delta t_k)),$$

$$j = 1, 2, \ldots, M, \quad k = 1, 2, \ldots, N. \tag{12.67}$$

An $(N+1)$th virtual row must be added, but not an $(N+1)$th virtual time interval. The evaluation of Equation (12.67) can be carried out with the row address matrix A introduced in Equation (12.51) and the information matrix I in Equation (12.52) by calculating, according to Equation (12.53),

$$C = A'I, \tag{12.68}$$

in which the elements

$$C_{kj} = G_j(\Delta t_k) \tag{12.69}$$

are found. The explanation has already been given together with the Equations (12.50) through (12.53). There the evaluation of the matrix multiplication with column sequential and time sequential update, and the associated advantages and disadvantages, have also been discussed. The extension over this previous discussion is that we now have grey shades in between the two states $I = \pm 1$ previously used. If we have 2^q grey shades between 0 and $2^q - 1$, then the binary coded q-bit word for the grey shades D_{ij} in between those two values is

$$D_{ij} = \sum_{\nu=0}^{q-1} r_{ij\nu}2^\nu \in [0, 2^q - 1], \tag{12.70}$$

where the bits $r_{ij\nu} = 0,1$. We have to translate D_{ij} into $I_{ij} \in [-1, 1]$, resulting in

$$I_{ij} = \frac{2}{2^q - 1}\left(D_{ij} - \frac{1}{2}(2^q - 1)\right) \in [-1, 1]. \tag{12.71}$$

We also translate the row voltages $\pm F$ into binary values R_i, in which 0 stands for $+F$ and 1 for $-F$. This yields

$$F = (1 - 2R_i)|F|. \tag{12.72}$$

Substituting Equations (12.70), (12.71), (12.72) and (12.63) into Equation (12.65) yields, for the time interval $t = \Delta t_k$,

$$G_j(\Delta t_k) = \frac{|F|}{\sqrt{N}} \sum_{i=1}^{N} \left(\frac{2}{2^q - 1} \sum_{\nu=0}^{q-1} r_{ij\nu} 2^\nu - \frac{2^q - 1}{2} \right)(1 - 2R_i(\Delta t_k))$$

$$+ \frac{|F|}{N} \sqrt{N - \sum_{i=1}^{N} \left(\frac{2}{2^q - 1} \left(\sum_{\nu=0}^{q-1} r_{ij\nu} 2^\nu - \frac{2^q - 1}{2} \right) \right)^2} (1 - 2R_i(\Delta t_k)). \tag{12.73}$$

The evaluation of Equation (12.73) can be performed by a correlation (Clifton and Price, 1992), where the correlation score for each time interval Δt_k is proportional to the dot product; one can also use fast transform algorithms such as a fast Fourier transform, where the number of multiplications required is reduced from $N(N-1)$ to $N \log_2 N$; further, the product $r_{ij\nu}$ times R_i may be evaluated as an exclusive-or logical operation. The number of operations is considerably reduced by the zeros in trilevel row functions.

The hardware for the generation of the column signal from the row functions and the information elements I_{ij} is achieved by the circuits in the block diagram in Figure 12.21 (Clifton and Price, 1992), which exhibits similarities to the diagram in Figure 12.15. The main difference is the column voltage generator according to Equation (12.73), shown in Figure 12.22 for $2^q = 64$ grey shades. The correlators generate the individual bits of the column voltage which are then weighted and summed. The voltage of the virtual pixel in Equation (12.73) is computed in the two multiplier-accumulators (MACs) in Figure 12.22.

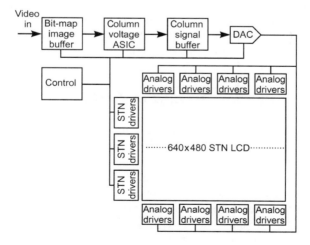

Figure 12.21 Hardware for the generation of row and column voltages for pulse height modulated PHM displays

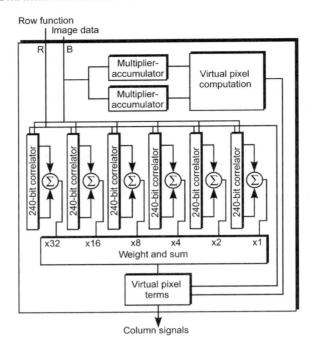

Figure 12.22 The generation of column voltages including the virtual row by correlation for PHM displays

The square root of the result is provided by a look-up table. The virtual pixel term block combines this result with the dot product term, resulting in the unscaled column voltages $G_j(\Delta t_k)$.

If row functions with zeros are used, the realization of the column sequential matrix product may not require too much memory, and can be performed with smaller correlators. In this case, the column buffer in Figure 12.19 is not needed, as the column signals are generated in the time sequence they are required.

12.4   Two Frequency Driving of PMLCDs

The dielectric anisotropy $\Delta\varepsilon=\varepsilon_\parallel-\varepsilon_\perp$ depends upon the frequency, as demonstrated in Figure 12.23, as $\varepsilon_\parallel$ is decreasing with frequency. At the crossover frequency f_c in Figure 12.23 $\Delta\varepsilon$ changes from positive to negative values. As a consequence, addressing with frequencies $f<f_c$ aligns the LC molecules parallel, and addressing with $f>f_c$ perpendicular, to the electric field. This effect is the cause for a change in the transmission of the LC cell, and is used in a two frequency addressing scheme (Dijan, 1990; Koneko, 1987). As introduced in Section 12.1, we apply the voltage S_L to the columns and F_L to the rows. Both voltages have a low frequency $<f_c$, as indicated by the subscript L. A second high frequency $>f_c$ with the voltage U_H is permanently superimposed on the row voltage, leading to the square of the rms voltages of an on-pixel and an off-pixel based on Equations (12.16)

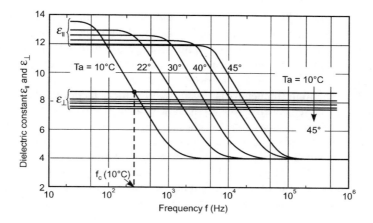

Figure 12.23 Change of the dielectric constants $\varepsilon_\parallel$ and $\varepsilon_\perp$ with frequency

and (12.17):

$$V_{\text{on}}^2 = wU_H^2 + \frac{1}{N}(S_L + F_L)^2 + \frac{N-1}{N}F_L^2 \tag{12.74}$$

$$V_{\text{off}}^2 = wU_H^2 + \frac{1}{N}(S_L - F_L)^2 - \frac{N-1}{N}F_L^2, \tag{12.75}$$

where w is a weight factor accounting for the difference in the electro-optic response to high and low frequency signals.

The dependence of the electro-optic response on the two frequencies is shown in Figure 12.24, in which the 10 percent and 90 percent light transmission is depicted as a function of

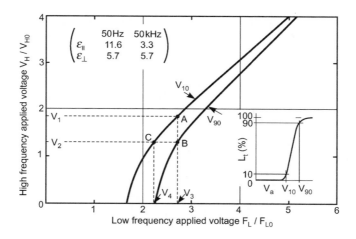

Figure 12.24 The high frequency voltage versus the low frequency voltage for a 10 percent and a 90 percent optic transmission in a two frequency driving scheme

the rms voltage of the high frequency signals and the rms voltage of the low frequency signals. The 10 percent transmission V_{10} corresponds to the threshold voltage (off-pixel) and the 90 percent transmission V_{90} to the saturation voltage (on-pixel). Point A with the high- and low frequency voltages V_1 and V_3 respectively represents an off-pixel. By decreasing the high frequency voltage from V_1 to V_2, the pixel is switched to the on-stage, point B, while V_3 remains constant. The high frequency has been used for the selection of the on-pixel. If we start from point C, increasing the low frequency voltage from V_4 to V_3, while the high frequency voltage V_2 stays constant, switches to the on-stage in point B. Here the low frequency is used for the pixel selection. The permanently applied high frequency voltage also serves as a bias which raises the voltage close to V_{th}, rotating the LC molecules closer to the state of saturation. Therefore, switching into saturation can be achieved rather quickly, rendering the display capable of a larger number of rows with still acceptable grey shades. This is an advantage of two frequency driving. The penalty is the use of two frequencies one of which lies in the 5 KHz range. This is a reason why this addressing scheme is rarely used.

13

Passive Matrix Addressing of Bistable Displays

Bistable displays neither require the more expensive active matrix addressing, nor are they bound by the rule of Alt and Pleshko limiting the number of addressable lines. The bistable devices discussed are ferroelectric, cholesteric and nematic LCDs.

13.1 Addressing of Ferroelectric LCDs

As described in Chapter 8.1, FLCDs possess two stable states I_1 (black) and I_2 (white) of the intensity of light at zero field, shown in Figure 13.1. It was found that FLCDs (in contrast to TN LCDs) respond to the addressing by the area $A = V\tau$ of a pulse, where V stands for the amplitude and τ for the duration. Increasing the area A lifts the device from the minimum energy state I_1 at $A = 0$ into the second minimum energy state I_2, again at $A = 0$, as indicated by arrows in Figure 13.1. Similarly, switching from I_2 to I_1 is performed by a negative pulse, and hence $A < 0$ in the direction of the arrows. Bistability is necessarily associated with a hysteresis, as in Figure 13.1. If the dielectric displacement D of the device were plotted versus E, we would obtain the hysteresis curve of ferroelectric materials, from which the name ferroelectric LCD is derived.

The threshold curves τ_{th} versus V_{th} separating the areas where no switching occurs from the switching areas are shown in Figure 13.2 (Buerkle, 1970) for a low polarization P_s and a large $|\Delta\varepsilon|$, as well as for a high P_s and a small $|\Delta\varepsilon|$. The electrically induced torques driving an FLC molecule are $M_{ferro} \sim \vec{P}_s x \vec{E}$ and $M_{diel} \sim \Delta\varepsilon E^2$ establishing an equilibrium with the mechanical torques M_{visco} and M_{elast} (Dijon, 1990). For a large $P_s > 20\,\text{nC/cm}^2$ and a small $|\Delta\varepsilon|$ the torque M_{ferro} is dominant, and switching occurs for $V_{th}\tau_{th} = const$. This is the hyperbola in Figure 13.2. If M_{diel} also plays a role, then an additional torque is present, which for $\Delta\varepsilon < 0$ as used in LCDs tries to keep the molecules with their directors perpendicular to E. If this is the starting position, then switching can only take place for an

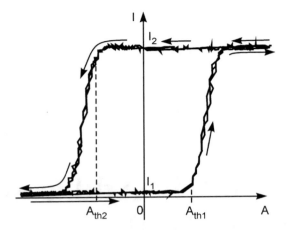

Figure 13.1 The hysteresis of the electro-optic response of an FLCD with intensity I versus pulse area A

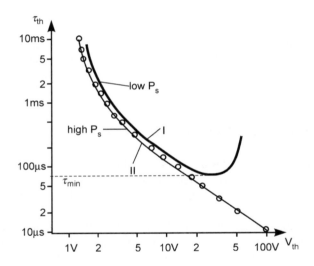

Figure 13.2 Switching thresholds of pulse duration τ_{th} versus pulse amplitude V_{th} of FLCDs

increased M_{ferro}, reflected in an increase of the switching threshold for larger V_{th}, as seen in Figure 13.2. This is shown in the parabola-like curve in Figure 13.2. For this curve, switching can only be achieved for a pulse duration $\tau_{th} > \tau_{min}$, as indicated in Figure 13.2; τ_{min} is the time in which a pulse of a given polarity is applied. Each threshold A_{th1} and A_{th2} in Figure 13.1 results in a pertinent threshold curve of the kind given in Figure 13.2. The addressing schemes will be different for the two threshold curves in Figure 13.2.

13.1.1 The $V-\tau_{min}$ addressing scheme

This method is used for FLC materials with a smaller polarization $P_s < 10 \, nC/cm^2$, where both torques M_{ferro} and M_{diel} play a role and τ_{min} in Figure 13.2 has to be considered (Surguy *et al.*, 1991). Hence, a long pulse with $\tau > \tau_{min}$ initiates switching even if the amplitude is low, whereas a short pulse with $\tau < \tau_{min}$ does not, even if the amplitude is high. The addressing for the $V - \tau_{min}$ scheme is plotted in Figure 13.3. The pixel voltage is the column voltage minus the row selection voltage. This is the inverse sign of the pixel voltage in Chapters 11 and 12 and stems from an inversed sense of the voltage arrows. The pixel voltage for reset is negative and long enough to switch from I_2 in Figure 13.1 to a black pixel I_1, whereas a black pixel remains black. This reset pulse is applied at the beginning of each row addressing, and establishes a common initial state for all pixel information, namely switching to bright or no switching, that is, remaining black.

The addressing waveforms generating a long positive pulse across the pixel switch into the bright position I_2 and the waveforms generating a shorter positive pulse across the pixel do not switch at all after having caused a small disturbance in the optical response. The advantage of this method is the strong selectivity of the addressing voltage as the threshold curve in Figure 13.2 rises rapidly as soon as the pulse duration exceeds τ_{min}. The switching time for an FLC molecule is

$$\tau_s = \eta / P_s E = \eta d / P_s V, \tag{13.1}$$

where η is the rotational viscosity. This time is relatively long for the materials with a low P_s. Increasing P_s is not feasible for the $V - \tau_{min}$ addressing scheme, because the minimum in the threshold curve as seen in Figure 13.2 can only be maintained by simultaneously increasing M_{diel}, or in other words, by increasing E^2 or V^2 / d^2. A large addressing voltage, however, is impractical. A further shortcoming is the strong temperature dependence of τ_{min}.

13.1.2 The $V - 1/\tau$ addressing scheme

This scheme is applied to FLC materials with a high $P_s > 20 \, nC/cm^2$, where the hyperbolic threshold curve in Figure 13.2 applies. This curve also demonstrates that switching is simply performed when the pulse area $A = V_{th} \tau_{th}$ exceeds the threshold curve. For switching from I_1

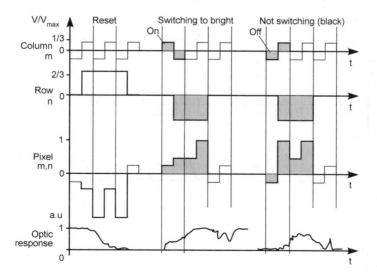

Figure 13.3 The addressing of FLCDs with the $V - \tau_{min}$ scheme

to I_2, a positive A is required, and is obtained by integrating over the positive values of the waveform $V(t)$ of the pixel voltage, leading to

$$A = \int_{t_1}^{t_2} \frac{d_0 V(t)}{d_{pel}} dt > V_{th} \tau_{th},$$ (13.2)

with $V(t) \geq 0$ for $t_1 \leq t \leq t_2$ and d_0 the cell thickness as designed, and d_{pel} the actual thickness at an individual pixel. If d_{pel} fluctuates around d_0 due to fabrication inaccuracies, A in Equation (13.2) must be large enough to compensate for that.

Harada's addressing scheme (Harada *et al.*, 1985) in Figure 13.4 is most often used. The scheme operates either with two time slots per switching period, as in Figure 13.4(a), or with four, as in Figure 13.4(b). An exception is the reset pulse, switching all pixels into the dark state I_1 using all four time slots. The actual switching pulse is negative for the reset and positive for switching from I_1 to the bright state I_2 and in each case is preceded by a pulse of the same amplitude, but opposite sign to maintain dc-free operation. The row address impulse is always the same double pulse in Figure 13.4, and is zero during the frame time outside the row address time. The data voltage at the columns starts with a positive pulse, followed by a negative one for switching in the bright state; the signs of the two pulses are reversed for staying in the dark state, generated by the reset pulses. The conditions for switching into the bright state and for staying in the dark state are, respectively,

$$V_r - (-V_c) = V_p > A/T_p,$$ (13.3)

and

$$V_r - V_c = V_p < A/T_p,$$ (13.4)

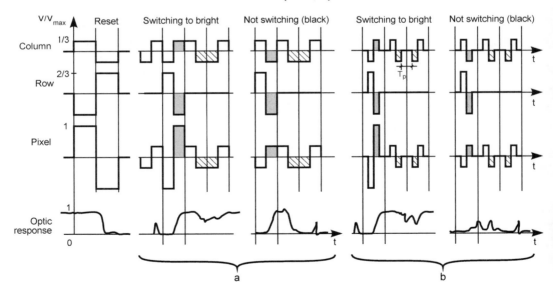

Figure 13.4 The addressing of FLCDs with the Harada scheme with (a) 2 time slots and (b) 4 time slots per pixel information

which is met for the reduced row voltage $V_r=2/3$ and the reduced column voltage $V_c=1/3$. T_p is the time for one time slot. Then $V_p=1\cdot T_p>A$ for switching and $(1/3)T_p<A$ for non-switching is feasible. If switching data is followed by non-switching data, the two hatched negative data pulses in Figure 13.4(a) lie next to each other, forming a long impulse. Even in this case Equation (13.3) provides for $V_r=0$ the value $V_p=(1/3)2T_p<A$, and non-switching is still maintained. However, it may have the disadvantage that the margin for acceptable voltages is small, and the inhomogeneities in the cell thickness can no longer be compensated for by larger voltages. The remedy is the 4-slot addressing in Figure 13.4(b), where the two slots responsible for switching or non-switching are separated by a time slot with zero voltage. This avoids the hatched long pulses. The disadvantages in this case are longer addressing times or, for shorter slots T_p, higher voltages.

An appealing way out is the 2 1/2-slot-scheme in Figure 13.5 (Schweikert *et al.*, 1993). The data pulses are still separated from their neighbours by one time slot on each side; the row pulses are, however, extended over all four time slots, thus contributing more pulse area to the pixel voltage, as in the 4-slot scheme, as can be seen in Figure 13.5. Therefore, the pixel voltage remains on longer while the data voltages still remain separated, resulting in a higher margin for the voltages while still exhibiting the same selectivity as the 4-slot scheme.

13.1.3 Reducing crosstalk in FLCDs

The data pulses travelling down the columns generate a parasitic voltage across each pixel of the column. This voltage causes the LC molecules to oscillate around their stable state, causing transient variations of the grey shade, which decreases contrast. The effect is called *crosstalk jitter*. The data pulses exhibit a higher frequency than the row address signals. At this higher frequency, the negative dielectric anisotropy is even more negative, providing a stronger dielectric torque which tries to keep the molecules perpendicular to the E-field. Thus, the oscillations are attenuated. This so-called *ac-stabilization* enhances contrast, as has already been used for the two frequency addressing scheme in Chapter 12.

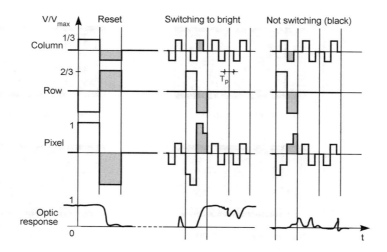

Figure 13.5 The addressing of FLCDs with a $2\frac{1}{2}$ time slot scheme

13.1.4 Ionic effects during addressing

The polarization P_s aligns the electric dipoles of the LC molecules along P_s, as shown in Figure 13.6(a), (Dijon, 1990; Escher *et al.*, 1991). The charges of the dipoles compensate each other in the interior of the LC cell.

There are, however, uncompensated surface charges on the insulated orientation layers of the cells generating a depolarization field E_{dep} oriented against P_s. The LC material contains ions, partly from impurities. The cations and anions are slowly separated in Figure 13.6(a), thus generating the field E_{ion} in the equilibrium compensating E_{dep}. The time constant of the ions is

$$t_{ion} = \frac{d}{\mu E} = \frac{d^2}{\mu V_{pel}}, \tag{13.5}$$

where $\mu = 10^{-10}\,\mathrm{cm^2/Vs}$ is the mobility of the ions. For a cell thickness of $d = 1.5\,\mu$ and $V_{pel} = 10\,V$, we obtain $t_{ion} = 2\,\mathrm{ms}$, meaning that E_{ion} is built up quite slowly until the steady state is reached.

Now we apply an external field E in Figure 13.6(b), rotating the dipoles by $180°$. The time constant τ_s in equation (13.1) of this switching provides for $P_s = 10\,\mathrm{nC/cm^2}$, the rotational

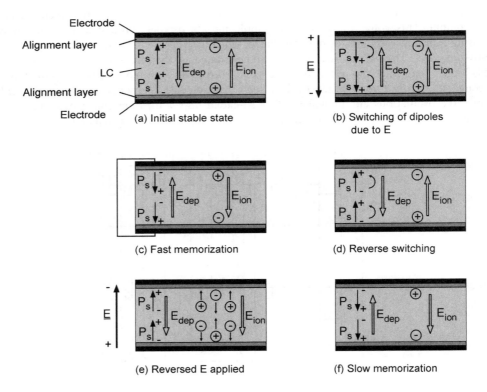

Figure 13.6 The depolarization field E_{dep} and the ionic field E_{ion} during switching of an FLCD

viscosity $\eta = 10^{-7}\,\mathrm{Ws/cm^3}$ and d and E as above $\tau_s = 100\,\mu s$, which would be reduced to $\tau_s = 28.6\,\mu s$ for $P_s = 35\,\mathrm{nC/cm^2}$. Obviously, the ferroelectric switching is much faster than the ionic movement. E_{ion} in Figure 13.6(b) remains virtually unchanged after the switching of the dipoles. At this time, the ionic field is

$$|E_{ion}| = \frac{|P_{ion}|}{\varepsilon_0 \varepsilon_{eff}}, \tag{13.6}$$

with $|P_{ion}| = |\,P_s|$ and $\varepsilon_{eff} = \varepsilon_{AL} d_{LC}/d_{AL} + \varepsilon_{LC}$; ε_{AL} and d_{AL} are the dielectric constant and the thickness of the alignment layer. The ionic field is large if P_s as well as d_{AL} are large, and small otherwise. After turning off E by shorting the input, two responses can occur.

If the ionic field is not large enough to reverse the dipoles of the LC mixture, as in Figure 13.6(c), the positive and negative ions travel driven by $E_{dep} > E_{ion}$ until E_{ion} with its opposite sign neutralizes E_{dep}, and the new memory state of the LC dipoles remains stable. They were switched quickly into this new state with the fast memorization time

$$\tau_M = K\tau_s = K\frac{\eta}{P_s E}. \tag{13.7}$$

$K > 1$ is a constant which is dependent on the anchoring at the orientation layer, the cell technology and the voltages, as well as on the waveforms of the addressing scheme. For fast memorization to take place, we require (with equations (13.5) and (13.1))

$$\frac{t_{ion}}{\tau_s} = \frac{P_s d}{\eta\mu} = \frac{d}{L} > 1, \tag{13.8}$$

with the characteristic thickness

$$L = \frac{\eta\mu}{P_s}. \tag{13.9}$$

Equation (13.8) demonstrates that fast memorization depends on a sufficiently large P_s. If $E_{ion} > E_{dep}$ is large enough to rotate the LC dipoles, which occurs for very large P_s values, then the FLCD switches back to the previous state in Figure 13.6(a). This reverse switching in Figure 13.6(d) destroys bistability, the device is monostable and no memorization takes place.

The second case of no memorization can be turned into a slow memorization by applying, for a few ms (i.e. longer than t_{ion}), a reverse external field E, (as shown in Figure 13.6(e)) to an FLCD residing in the initial state of Figure 13.6(a) before E is applied. During this pulse in Figure 13.6(e), which has no effect on the dipoles, the ions start migrating into the state with reverse polarity to E; the new reversed ionic field E_{ion} in Figure 13.6(d) is strong enough to switch the LC dipoles into the same state as after fast memorization, but only after a longer time determined by t_{ion}. Because switching time is long, slow memorization is seldom used.

For fast memorization to occur, a large $P_s > 20\,\mathrm{nC/cm^2}$ and a low density of ions are required. The ions can originate from the LC material and from the orientation layer. Hence, pure FLC materials and a technology providing orientation layers with few free ions are desirable.

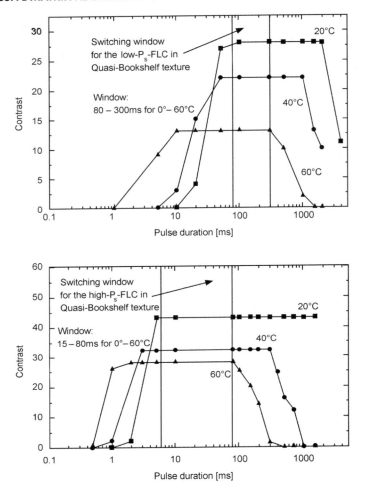

Figure 13.7 Contrast versus duration of addressing impulse as switching window for (a) low P_s FLCDs and (b) higher P_s FLCDs

A further effect stemming from the ionic field is image sticking (De Vleeschouwer *et al.*, 1999). The phenomenon occurs after an image has been presented for a long time, so that the ionic charge separation had time to generate a large E_{ion}. As the time constant E_{ion} in Equation (13.5) is large, E_{ion} disappears only slowly, resulting in a remnant image on the display. Pure LC materials and a low P_s also fight image sticking.

The threshold $A = V_{th} \cdot \tau_{th}$ in Figure 13.2 can be expressed with $\tau = \tau_M$ the fast memorization time in Equation (13.7) and $V = Ed$ as

$$A = V_{th}\tau_{th} = V_{th}\tau_M = K\frac{d\eta}{P_s}.$$

The larger P_s the smaller are the pulse amplitudes V_{th} needed.

The practical consequences of the switching threshold in the last Equation and of the ionic effects on the duration of the switching pulses result in the switching windows in Figures 13.7(a) and 13.7(b) for low and for high polarization FLCDs with a quasi-bookshelf structure (Frey, 2000). The low polarization was $P_s=9.5\,nC/cm^2$ and the high polarization $P_s=36\,nC/cm^2$. In the switching windows, the measured contrast is plotted versus pulse duration τ in ms for a constant pulse amplitude of $V=5\,V$ and for temperatures from 20°C to 60°C.

The two vertical lines indicate a lower and an upper limit for the pulse duration which hold for all temperatures in the two diagrams.

The lower limit of the pulse duration, τ, is for all temperatures 80 ms for the low P_s material and 5 ms for the high P_s material, which is in accordance with τ_M in Equation (13.7). The upper limit for τ stems from the higher E_{ion} resulting from the larger ionic charges accumulated in the longer duration of the pulses. The larger E_{ion} can cause reverse switching which must be avoided by staying below the upper limit of τ. For all temperatures the upper limit is $\tau=300\,ms$ for low P_s and $\tau=80\,ms$ for high P_s. This reflects the fact that a higher P_s generates a larger ionic charge resulting in the constraint of shorter pulses.

The design of the addressing parameters for FLCDs is more complex than for TN displays. However, if properly performed, FLCDs provide a fast and low energy display with memory. In the reflective version there is no energy consumption at all, like for a printed page, when the information content is not changed.

13.2 Addressing of Chiral Nematic Liquid Crystal Displays

As is known from Chapter 8.2, chiral nematic displays can assume the two stable states, the reflective planar state and the focal conic texture, which is slightly back-scattering. (It follows that the FC is also transparent.) The homeotropic texture is not stable at zero field, and is fully transparent. It is generated for $\Delta\varepsilon>0$ by a high field in which the molecules align parallel to the field. There are two addressing schemes: a straightforward, but also a slow, one (Doane *et al.*, 1992; Bunz, 1999); and a more involved but fast scheme (Zhu and Yang, 1997; Huang *et al.*, 1995, 1996).

The slow but simple addressing scheme is explained by Figures 13.8(a) and 13.8(b). In Figure 13.8(a) the reflectivity is plotted versus the pixel voltage. For a 4.5 µ thick cell, typical voltages are $U_1=6\,V$, $U_2=14\,V$, $U_3=30\,V$ and $U_4=38\,V$. Starting in the reflective planar state or in the scattering focal conic state, we can reach the homeotropic state for the high voltage $U>U_4=38\,V$, after having passed through a focal conic state. Figure 13.8(b) is crucial for the addressing scheme; it shows the reflectivity after the voltages required to reach a particular value have been turned off. The only difference to Figure 13.8(a) is that the unstable homeotropic state was transformed into the planar state. Reducing the voltage from $U>U_4$ reveals a hysteresis as the reflectivity follows the dashed line. Figure 13.8(b) demonstrates that the focal conic state can be reached by a pixel voltage $U\varepsilon[U_2,U_3]$ (with a duration longer than 10 ms), and the planar state by $U>U_4$ both from any initial state.

Figure 13.8(b) also indicates that along the two ramps, grey shades seem to be possible. They are, however, not realized, since the large column voltages needed to obtain them would change the grey shades in other pixels due to a large crosstalk. Figure 13.9 shows the bipolar dc-free pixel voltage used for switching from the focal conic state to the planar state

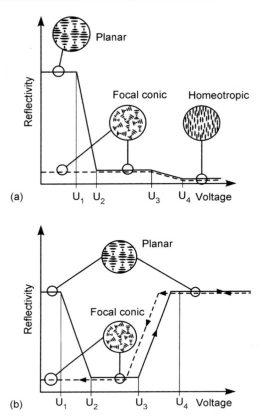

Figure 13.8 The reflectivity of a cholesteric display versus the applied pixel voltage (a) if the voltage is maintained at the pixel, and (b) if the voltage is applied and then switched off

and back. The planar state is only reached after 100 ms, whereas the focal conic state is obtained within 0.5 ms.

The first addressing scheme operates with the row selection voltage U_r, the data voltage U_c at the column, and the ensuing pixel voltage

$$U_p = U_c - U_r,$$

with

$$U_2 < U_r < U_3, \tag{13.10}$$

$$U_1 > U_c > U_4 - U_3 \tag{13.11}$$

and

$$U_r + U_c > U_4. \tag{13.12}$$

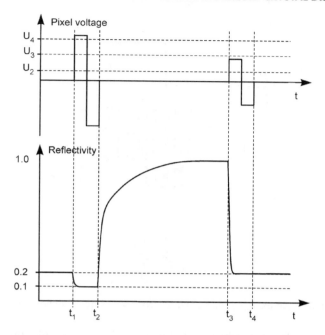

Figure 13.9 Reflectivity and pixel voltage of a cholesteric display versus switching time

U_r within the limits (13.10) is chosen such that together with $U_c=0$, the focal conic state in Figure 13.8 is obtained. The planar texture is realized by the superposition of U_r and U_c with the boundary (13.12). $U_c < U_1$ guarantees that the reflective state is not reduced by a parasitic column voltage U_c; $U_c > U_4 - U_3$ guarantees that the column voltage is large enough for switching into the planar state. An example for switching into the focal conic (fc) and the homeotropic (h) state is shown in Figure 13.10. After turning off the voltage at the homeotropic pixels they revert to the planar state.

A second scheme operates with the following voltages drawn in Figure 13.8:

$$U_3 < U_r < U_4 \tag{13.13}$$

$$\frac{U_4 - U_3}{2} < U_c < U_1 \tag{13.14}$$

$$U_4 < U_r + U_c \tag{13.15}$$

$$U_r - U_c < U_3. \tag{13.16}$$

In this scheme the row voltage lies within the upper ramp in Figure 13.8(b), and is hence larger than in Equation (13.10). The column voltage has to be chosen such that, by its addition to U_r, the planar texture (Equation (13.15)) and by its subtraction, the focal conic

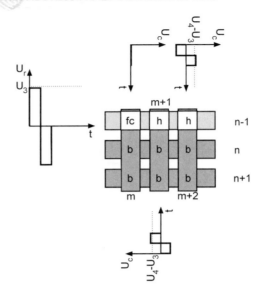

Figure 13.10 An example for passive matrix addressing of a cholesteric display

texture (Equation (13.16)) is reached. U_c in Equation (13.14) is smaller by a factor of 2 than in Equation (13.11) of the first scheme.

Both schemes are simple and easy to implement. Shortcomings are the high voltage levels and the slow operation.

For an explanation of fast addressing (Zhu and Yang, 1997; Huang *et al.*, 1995, 1996) there is a need to delve deeper into the phases encountered during switching. Starting from the focal conic texture in Figure 13.8(a), along the dashed line the homeotropic texture is reached by applying a field

$$E > E_c = \frac{2\pi^2}{h}\sqrt{\frac{\pi K_{22}}{\Delta \varepsilon}}, \qquad (13.17)$$

where h is the pitch of the helix and K_{22} the elastic constant for twist. The threshold E_c corresponds to U_4 in Figure 13.8(b). The homeotropic state becomes unstable by reducing the field below

$$E_{hf} = \frac{2}{\pi}E_c. \qquad (13.18)$$

The thresholds E_c and E_{hf} describe the hysteresis, visible in Figure 13.8(b). E_{hf} may be shifted by the properties of the alignment layer. E_c corresponds to a pixel voltage of 24 V for a cell 4.5 µ thick.

If a field E is reduced, the homeotropic state can relax into two different states. The first case is defined by reducing the field E to

$$E < E_{hp} = \frac{2}{\pi}\sqrt{\frac{K_{22}}{K_{33}}}E_c, \qquad (13.19)$$

where K_{33} is the elastic constant for bend. As $K_{22}/K_{33} < 1$, this field is low compared to E_c. After switching off E the homeotropic state relaxes into the transient planar state, which is unstable and exhibits a pitch by the factor K_{33}/K_{22} larger than the pitch h of the stable planar state. The transition into this state happens with the time constant

$$\tau_{hp} = \frac{\eta h^2}{K_{22}}, \tag{13.20}$$

where η stands for the rotational viscosity.

The second case is defined by reducing the field, E, to

$$E > E_{hf}. \tag{13.21}$$

After switching, the homeotropic state relaxes towards the focal conic state with the time constant

$$\tau_{hf} = \frac{\eta z_0^2}{K_{22}}, \tag{13.22}$$

where $z_0 \geq 5\,\mu m$ is the distance between two growth centres for the focal conic state. From the transient planar state reached with E in Equation (13.19) and the time constant τ_{hp} in Equation (13.20), both the stable planar state and the focal conic state may be reached, which is the essential point of the fast switching discussed later.

The stable planar state is reached for $E < E_{hp}$ by a diffusion process lasting about 100 ms. The ratio

$$\tau_{hf}/\tau_{hp} = \left(\frac{z_0}{h}\right)^2 \tag{13.23}$$

must be $\tau_{hf}/\tau_{hp} \gg 1$ for the planar state to be obtained, because then the process leading to the planar state with time constant τ_{hp} is much faster and dominates. This is the case for $z_0 = 5\,\mu m$ and $h = 100\,nm$. For $E > E_{hp}$ or for $\tau_{hf}/\tau_{hp} < 1$, the focal conic texture is dominant.

Fast switching is based on a proven idea which shifts the status of the director field just short of the point where the crucial change in the field occurs. This way only a little time is required to perform the essential final switching, because the time consuming preparations have been done before. In Zhu and Yang (1997), the five-phase switching scheme in Figure 13.11 with different voltages in each phase of the switching process was introduced. The selection phase exhibits two voltages dependent on the texture desired, either the planar state with $E < E_{hp}$ or the focal conic state with $E > E_{hp}$.

Figure 13.12 shows how switching with the five phases in Figure 13.11, the preparation phase p, the post-preparation phase pp, the crucial selection phase s, the post-selection phase ps and the evolution phase e, progresses from the focal conic texture through the homeotropic and the transient planar textures to the planar texture or back to the focal conic texture. The blocks in Figure 13.11 indicate the phases, the voltages and their duration. This is again listed in Figure 13.12. Only the switching time of 50 µs for the selection phase counts for the switching process, because all other phases can be shared for all pixels, and are performed in a pipeline fashion outside the selection time. Hence, the frame time for a

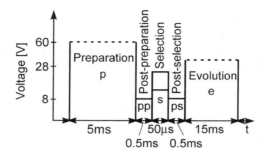

Figure 13.11 The five-phase drive scheme for cholesteric displays

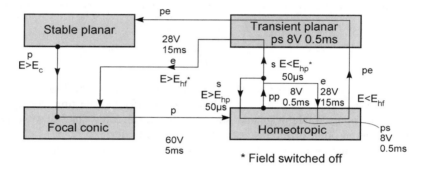

Figure 13.12 The switching of chiral nematic displays with five phase in the drive scheme. p: preparation phase; pp: pose preparation phase; s: selection phase; ps: post selection phase; e: evolution phase; pe: post evolution (only relaxation, no action needed)

display with n lines is

$$T_f = T_p + T_{pp} + P_{ps} + T_e + nT_s, \tag{13.24}$$

where the indices indicate the phases. With this addressing scheme, the display of TV is feasible due to the short selection time of $50\,\mu s$.

The conventional drive scheme takes 20 ms to address a line; with a dynamic drive (Huang *et al.*, 1995) this time was reduced to 1 ms/line, and with a four-phase drive (Huang *et al.*, 1996) to 0.5 ms/line.

The texture present in a cholesteric display can be determined by measuring the capacitance. For $\Delta\varepsilon > 0$ the homeotropic texture has the largest, the planar and the transient planar textures the smallest, and the focal conic texture an intermediate value.

14

Addressing of Liquid Crystal Displays with a-Si Thin Film Transistors (a-Si-TFTs)

An introduction to TFT addressed LCDs has been given in Chapter 2. In this and the following three chapters, we build on the introduction by first looking at the TFT as a component with a-Si and poly-Si semiconductors, its operation and properties, before we turn to the active matrix addressing of LCDs with TFTs. Then, we describe the manufacture of TFTs together with the other layers of an LC cell. Finally, two more AMLCDs with MIMs and diodes will be discussed afterwards.

14.1 Properties of a-Si Thin Film Transistors

TFTs with the voltages and currents in Figure 14.1(a) are Metal-Insulator-Semiconductor Field Effect Transistors (MIS-FETs) which are used most often as bottom-gate, and more seldom as top-gate, transistors with the schematic cross-sections in Figures 14.1(b) and 14.1(c). On a substrate, as a rule usually of glass or more recently plastic foil, the metallic bottom gate in Figure 14.1(b) is deposited and structured, followed by the gate-dielectric, the amorphous-Si semiconductor, the metallic drain and source electrodes, and a protection layer covering the semiconductor channel. The materials used are given in Figure 14.1(b). The top gate TFT in Figure 14.1(c) has an inverted sequence of layers. The main difference to a monocrystalline MIS-FET is the amorphous nature of the semiconductor. This is un-avoidable, since thin film deposition processes are unable to provide crystalline layers. The most one can achieve is to induce crystal growth either thermally or by laser annealing, resulting in crystallized regions separated by grain boundaries. These poly-crystalline Si layers can be used for poly-Si-TFTs.

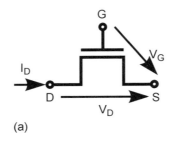

(a)

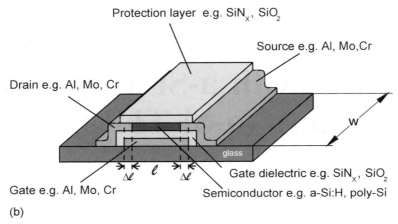

(b)

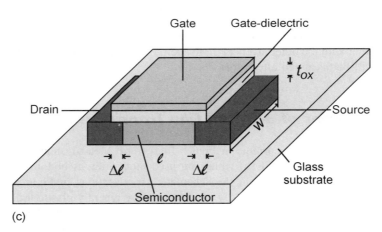

(c)

Figure 14.1 (a) The symbol for a TFT; (b) cross-section of a bottom gate TFT; (c) cross-section of a top gate TFT

The operation of an a-Si-TFT starts with a separation of charge in the capacitor between the gate electrode and the electrode formed by the drain and source, and the distribution of the potential along the upper surface of the semiconductor channel. The charge separation is induced by the gate-source voltage V_G. The voltage V_D between drain and source causes an electric field parallel to the surface of the n-channel of the a-Si, which transports the negative charge to the drain. The drain current I_D in Figure 14.1(a) follows approximately the same well known law as for FETs (Sze, 1981; Borkon and Weimer, 1963; Khakzar, 1991). It is

$$I_D = 0 \quad \text{for} \quad V_G \leq V_{\text{th}} \tag{14.1}$$

$$I_D = \mu C_G \frac{w}{2l}((V_G - V_{\text{th}})^2 - (V_D - (V_G - V_{\text{th}}))^2) \quad \text{for} \quad V_D \leq V_G - V_{\text{th}}, \; V_G > V_{\text{th}} \tag{14.2}$$

and

$$I_D = \mu C_G \frac{w}{2l}(V_G - V_{\text{th}})^2 \quad \text{for} \quad V_D > V_G - V_{\text{th}}, \; V_G > V_{\text{th}}. \tag{14.3}$$

For discussion, an alternative version of the same equations will be helpful, where the sequence of the last two equations is inverted and the inequalities are rewritten:

$$I_D = 0 \quad \text{for} \quad V_G \leq V_{\text{th}} \tag{14.4}$$

$$I_D = \mu C_G \frac{w}{2l}(V_G - V_{\text{th}})^2 \quad \text{for} \quad V_G - V_{\text{th}} < V_D, \; V_G > V_{\text{th}} \tag{14.5}$$

and

$$I_D = \mu C_G \frac{w}{l}\left((V_G - V_{\text{th}})V_D - \frac{V_D^2}{2}\right) \quad \text{for} \quad V_G - V_{\text{th}} \geq V_D, \; V_G > V_{\text{th}}, \tag{14.6}$$

where μ stands for the electron mobility, and w and, respectively, for the width and length of the channel respectively.

$$C_G = \frac{\varepsilon_0 \varepsilon_r}{d} \tag{14.7}$$

is the gate capacitance, where d is the thickness of the gate dielectric, and

$$V_{\text{th}} = -e \, n_0 \, d_{\text{HL}}/C_G \tag{14.8}$$

represents the threshold voltage, where e is the charge of an electron, n_0 the charge density in the semiconductor channel and at the interface to the dielectric before voltages are applied, and d_{HL} the thickness of this channel.

Equations (14.1) and (14.4) describe the ideally blocked TFT. In reality, the channel exhibits a finite sheet resistance R_{CH}, resulting in an off-current

$$I_{\text{off}} = V_D w/R_{\text{CH}} \, l \tag{14.9}$$

The transition region is governed by Equations (14.2) and (14.6), and the saturation region by Equations (14.3) and (14.5). The basis for the derivation of the equations is Shokley's gradual channel approximation (Sze, 1981), which assumes that the electric field in the channel directed from drain to source is much larger than the field perpendicular to it. For a-Si-TFTs the equations, however, imply further approximations as effects such as localized states in the band gaps and traps at the dielectric and in the a-Si channel are not considered. As a consequence of these traps in which the electrons are caught, only about 1 percent to 5 percent of the induced charge contributes a-Si interface to the current I_D. This reflects in a low electron mobility of a-Si in the range of $0.2 \, cm^2/Vs$ to $1.5 \, cm^2/Vs$, whereas monocrystalline Si has more than $1800 \, cm^2/Vs$. Due to these localized states, the power of 2 in Equations (14.2), (14.3), (14.5) and (14.6) is changed into an exponent in the range 2.1 to 2.3.

The reduced output characteristics derived from Equations (14.2) and (14.3), $I_{D0} = I_D/\mu C_G(w/l) = f(V_D)$ with V_G as a parameter, are plotted in Figure 14.2(a). The transition region represents parabolas opened downwards and with the vertex at

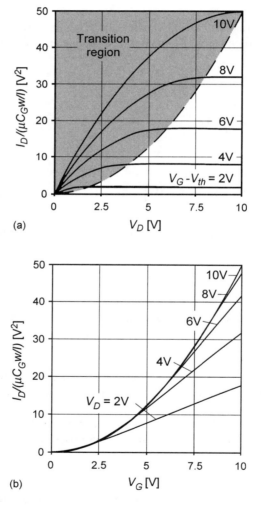

(a)

(b)

Figure 14.2 (a) Output characteristics of a TFT; (b) input characteristics of a TFT; (c) input characteristics of a TFT with $\sqrt{I_D}$ as the ordinate

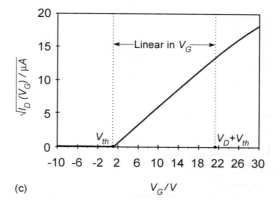

(c)

Figure 14.2   (continued)

$V_D = (V_G - V_{th})$ and $I_{D0} = (1/2)(V_G - V_{th})^2$. In the saturation region, I_{D0} is a constant. The border between the transition region and the saturation occurs for $V_D = V_G - V_{th}$. Substituting this into Equation (14.2) or (14.3) yields

$$I_{D0} = \frac{1}{2} V_D^2.$$ (14.10)

This border is shown as a dashed line in Figure 14.2(a).

The reduced input characteristics derived from Equations (14.5) and (14.6), $I_{D0} = I_D/\mu C_G(w/l) = f(V_G)$ with V_D as a parameter, are plotted in Figure 14.2(b). For $V_G < V_D + V_{th}$, Equation (14.5) provides the parabola

$$I_{D0} = \frac{1}{2}(V_G - V_{th})^2$$ (14.11)

independent of V_D for the saturation region. In the transition region with Equation (14.6) and $V_G \geq V_D + V_{th}$, the curve for I_{D0} is linear in V_G with a steepness proportional to V_D, as shown in Figure 14.2(b).

In Figure 14.2(c) $\sqrt{I_D} = f(V_G)$ derived from Equations (14.5) and (14.6) is plotted. For $V_G < V_D + V_{th}$ in Equation (14.5), we obtain the straight line

$$\sqrt{I_D} = M(V_G - V_{th})$$ (4.12)

with

$$M^2 = \mu C_G \frac{w}{2l}.$$ (14.13)

A measurement of points on the straight line for $\sqrt{I_D}$ provides V_{th} as the intersection of this line with the abscissa and the inclination M, from which we obtain the mobility

$$\mu = M^2/C_G \frac{w}{2l} = \frac{M^2 2ld}{\varepsilon_0 \varepsilon_r w},$$ (14.14)

where C_G in Equation (14.7) was used.

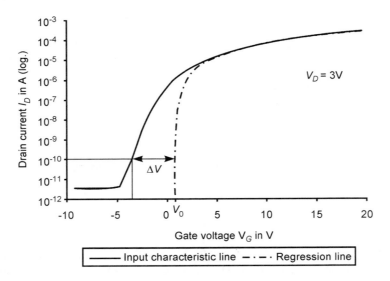

Figure 14.3 Measured (full line) and ideal (dashed line) input characteristics of a TFT with logarithmic ordinate

To determine the property of the a-Si-TFT at small currents we plot the measured data $I_D = f(V_G)$ as the logarithmic curve $\log I_D = f_0(V_G)$ in Figure 14.3. Further, we plot the curve derived from the saturation region in Equation (14.5) as the log I_D of the parabola in Equation (14.5), shown as a dashed line in Figure 14.3. We define the voltage ΔV in Figure 14.3 as the difference between the saturation curve and the measured curve at an off-current of 10^{-10} A. The smaller is ΔV, the better are the blocking properties of the TFT. As a substitute for ΔV, the off-current in Equation (14.9) also describes the blocking performance.

The three quantities μ, V_{th} and ΔV or I_{off} define an a-Si-TFT well enough for our applications, where μ is derived from the saturation region and ΔV or I_{off} from the low current region. Typical values are $\mu = 0.8 \, \text{cm}^2/\text{Vs}$, $V_{th} = 1\text{V}$, $\Delta V = 2.5 \, \text{V}$ and $I_{off} = 10^{-12} \text{A}$ at -3.5 V.

14.2 Static Operation of TFTs in an LCD

The operation of TFTs as switches at the pixels allowing the video voltage at the columns to charge the storage capacitors has been outlined in Chapter 2. The circuit is depicted in Figure 2.16. The charging of the storage capacitors $C_{LC} + C_s$ in a pixel must occur with a small enough time constant T_{on} in Equation (2.18) from which the limit for the on-resistance R_{on} of the TFT follows as

$$R_{on} \leq \frac{0,1 T_f}{N(C_{LC} + C_s)}. \tag{14.15}$$

On the other hand, this resistance can be expressed by a basic equation for current transport in a medium with mobility μ, and the areal charge density σ, that is the density in a unit area with the thickness d_{HL} of the channel with width w and length l, as (Sze, 1981)

$$R_{on} = \frac{l}{\mu \sigma w}.$$ (41.16)

The charge on $C_{LC} + C_s$ defines the grey shade, and hence imposes the constraint in Equation (2.19) on the time constant T_{off} of the discharge. This results in the constraint for R_{off} of the TFT

$$R_{off} \geq \frac{200T_f}{C_{LC} + C_s}.$$ (14.17)

The constraints for R_{on} and R_{off} will be required for the discussion of the performance of a TFT in the environment of a pixel in Figure 14.4. The TFT is embedded in its own voltage dependent parasitic capacitance C_{GS}, C_{GD} and C_{DS}; C_{LC} is the capacitance and R_{LC} the resistance of the LC layer, with the lower electrode being at ground; C_s is an additional thin film storage capacitance enhancing the value of C_{LC}. However, C_s is not connected to ground as is C_{LC}, but to the next gate line, saving a ground line. From the two columns the parasitics C_{c1e} and C_{c2e} shown in Fig. 14.7 couple onto V_p, which is equal to V_{LC}. The voltage V_{FP} at the front-plate will be explained later. The rows r_1 and r_2 are gate lines, and carry the gate pulses V_{r1} and V_{r2} in Figures 14.4 and 14.5, which render the appropriate TFTs conductive during row address time T_r with voltage V_g, and block it with voltage V_0 in Figure 14.5 during the remainder of T_f. The video signal V_{c1} on column 1 is positive during one frame time T_f and, in order to provide an almost dc-free pixel voltage, V_{c1} is negative during the following frame time. Capacitive voltage dividers transmit the steps of the gate voltage onto V_{LC}, where they cause visible changes in the grey shade. Assuming that 256 grey shades are generated by a voltage swing of 8 V, then 31 mV define one grey shade. This demonstrates how precisely V_{LC} has to be maintained.

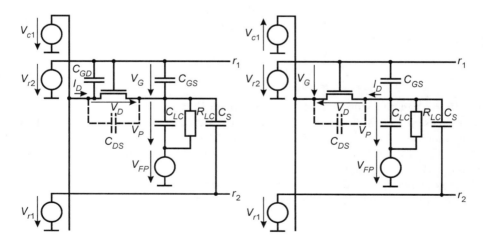

Figure 14.4 The TFT and its environment in a pixel (a) for charging of C_{LC} to a positive voltage and (b) to a negative voltage

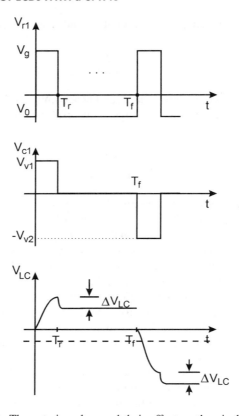

Figure 14.5 The gate impulses and their effect on the pixel voltage V_p

V_{LC} in Figure 14.5 reaches the desired value V_{v1} within T_r. Then the negative step $-(V_g - V_0)$ of the row pulse, that is the falling edge of V_{r1} at $t = T_r$, reaches V_{LC} through the capacitive voltage devider C_{GS} and $C_{LC} + C_s$ as a stepwise change ΔV_{LC}, first published by Suzuki (1987) and later by Lauer (1996), with

$$\Delta V_{LC} = -(V_g - V_0) \frac{C_{GS}}{C_{GS} + C_{LC} + C_s}. \tag{14.18}$$

Also, for the negative video-voltage $-V_{v2}$ in Figure 14.5, the same negative step $V_g - V_0$ of the row pulse V_{r1} again lowers V_{LC} by ΔV_{LC} in Equation (14.18). Some numbers illustrate the importance of Equation (14.18). The typical values $C_{GS} = 20$ fF, $C_{LC} = 80$ fF, $C_s = 60$ fF and $V_g - V_0 = 8$ V result in $\Delta V_{LC} = 1$ V, corresponding to 30 grey shades. Hence, a compensation of ΔV_{LC} is required.

Another derivation of ΔV_{LC} reveals other aspects of its nature (Howard, 1995). The charge Q in the channel during the presence of the gate pulse is

$$Q = \sigma w(l + 2\Delta l) < 0, \tag{14.19}$$

where σ is the electron charge used in Equation (14.16) and Δl is the length of the overlap between drain or source with the gate in Figure 14.1(a). After the drop to zero of the gate impulse, a part k of this negative charge is distributed mainly onto the larger capacitances C_{LC} and C_s, causing the voltage drop

$$\Delta V_{LC} = \frac{kQ}{C_{LC} + C_s} = \frac{k\sigma w(l + 2\Delta l)}{C_{LC} + C_s} < 0. \tag{14.20}$$

Substituting $C_{LC} + C_s$ in Equation (14.20) by $C_{LC} + C_s$ in Equation (14.15), and eliminating R_{on} by Equation (14.16), yields

$$\Delta V_{LC} \leq \frac{Nl(l + 2\Delta l)k}{2 \cdot 0.1 T_f \mu}. \tag{14.21}$$

This equation demonstrates that $|V_{LC}|$ can be decreased by a large mobility and a small channel length, as well as a small overlap Δl. The technology for TFTs has to take note of these requirements. A further means to decrease $|\Delta V_{LC}|$ is to choose a large C_s in Equation (14.18), which is limited, however, by the constraint in Equation (14.15) and the area needed in the pixel. Finally, a small C_{GS} in Equation (14.18) is also helpful. It is achieved by a large thickness of the gate dielectric which, however, reduces C_G and I_D in Equation (14.2).

The shift of V_{LC} by ΔV_{LC} not only causes changes in the grey shades, but also generates a damaging dc voltage. A remedy for both is to lower the potential of the back plate by $|\Delta V_{LC}|$ to the dashed line in Figure 14.5, which requires a voltage source V_{FP} in Figures 14.4(a) and 14.4(b) with

$$V_{FP} = \Delta V_{LC} < 0. \tag{14.22}$$

As C_{GS} and C_{LC} are voltage dependent, in this case they depend upon the video voltage, an average value for ΔV_{LC} has to be chosen.

A further, more costly, remedy is to introduce the compensation impulse V_{com} in Figure 14.6 (Khakzar, 1991). The lines are now addressed starting with the lowermost one, that is

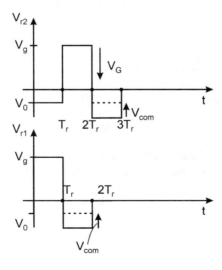

Figure 14.6 The TFT addressing with a compensation impulse

with the row voltage V_{r1}. Then row two is addressed. The falling edge of V_{r2} has increased by V_{com} to $V_g - V_0 + V_{com}$, $V_0 < 0$ resulting in a ΔV_{LC} as

$$\Delta V_{LC} = -(V_g - V_0 + V_{com}) \frac{C_{GS}}{C_{GS} + C_{LC} + C_S}. \tag{14.23}$$

A step V_{com} in the compensation impulse of the previously addressed line r_1 feeds through the capacitive voltage divider C_s and $C_{GS} + C_{LC}$ the voltage

$$\Delta V_{com} = V_{com} \frac{C_s}{C_s + C_{LC} + C_{GS}} \tag{14.24}$$

onto V_{LC} of row r_2, as is shown in Figure 14.6. For compensation of ΔV_{LC} we require

$$\Delta V_{LC} + \Delta V_{com} = -(V_g - V_0 + V_{com}) \frac{C_{GS}}{C_{GS} + C_{LC} + C_s} + V_{com} \frac{C_s}{C_{GS} + C_{LC} + C_s} = 0,$$

providing

$$V_{com} = (V_g - V_0) \frac{C_{GS}}{C_s - C_{GS}}. \tag{14.25}$$

With $V_g - V_0 = 8$ V, $C_{GS} = 20f$ F and $C_s = 60f$ F, we obtain $V_{com} = 4$ V.

Now we are ready to derive the requirements for the values of the row voltages V_g in the on state (Schwarz, 1990; Lauer, 1996) and V_0 in the off state (Schwarz, 1990) in Figure 14.5. We perform this with one of the two corrections of the voltage shift ΔV_{LC}, namely by shifting the potential V_{FP} of the back plate by ΔV_{LC} in Equation (14.18), or by the compensation impulse V_{com} in Equation (14.25). The effect of V_{com} on V_{LC} is the same as the lowering of the front plate potential by ΔV_{LC}. Therefore, introducing the voltage source $V_{FP} = \Delta V_{LC}$ in Equation (14.18) describes both correction methods, and is used in the calculation of V_g and V_0. For the further calculations, we recall that V_g is the row voltage in Figures 14.4 and 14.5 and V_G is the gate-source voltage of the TFT in Figures 14.1(a) and 14.4.

In the on state, we require

$$V_G \geq V_{th}. \tag{14.26}$$

From Figure 14.4(a), we obtain

$$V_g = V_G + V_{LC} + V_{FP} \tag{14.27}$$

for a positive voltage V_{LC}. Figure 14.4(b) provides for negative voltages V_{LC}, where the transistor operates with source and drain interchanged, which is possible due to the symmetry of the TFT,

$$V_g = V_G + V_{c1}. \tag{14.28}$$

The video-voltage V_{c1} is

$$V_{min} \leq V_{c1} \leq V_{max} \tag{14.29}$$

for positive V_{c1}, and

$$-V_{max} \leq V_{c1} \leq -V_{min} \tag{14.30}$$

for negative V_{c1} in Equation (14.28). The voltage across the pixel is, according to Figures 14.4(a) and 14.4(b), with $V_D = 0$ in the conductive state

$$V_{LC} = V_{c1} - V_{FP}. \tag{14.31}$$

V_{min} and V_{max} are chosen such that V_{LC} in Equation (14.31) covers the entire linear interval of the luminance-voltage curve of the LC cell.

From Equation (14.27), we obtain, with Equation (14.26), for the worst case

$$V_g \geq V_{th} + V_{LCmax} + V_{FP}. \tag{14.32}$$

With $V_{LCmax} = V_{c1max} - F_{FP} = V_{max} - F_{FP}$, where Equations (14.31) and (14.29) were used, we finally obtain from Equation (14.32)

$$V_g \geq V_{th} + V_{max} \tag{14.33}$$

for positive voltages V_{LC}. For negative voltages, Equtions (14.28), (14.30) and (14.31) yield similarly

$$V_g \geq V_{th} - V_{min} \tag{14.34}$$

The stronger condition (14.34) has to be used for the design of the electronics.

For the derivation of V_0 in Figure 14.5 (Takahashi et al., 1990), where the pixel is not selected, we require

$$V_G < V_{th}, \tag{14.35}$$

and similarly to Equation (14.32) for the worst case,

$$V_0 \leq V_{th} + V_{LCmin\,a} + V_{FP}. \tag{14.36}$$

For holding the stored charge, Equation (14.36) has to be considered after the steps of all voltages have exercised their effect ΔV_{LCmax}, in all the conditions where V_{LCmin} occurs. This voltage is

$$V_{LC\,min\,a} = V_{LC\,min\,b} + \Delta V_{LC\,max} \tag{14.37}$$

with

$$V_{LC\,min\,b} = V_{c1} - V_{FB}, \tag{14.38}$$

which is the voltage across C_{LC} before the drop caused by the row voltage. In Equation (14.37),

$$\Delta V_{LC\,max} = -(V_g - V_0) \frac{C_{GS}}{C_{GS} + C_{LC\,min} + C_s} \tag{14.39}$$

is the largest drop derived from Equation (14.18).

Inserting Equations (14.37), (14.38), (14.39) and V_g from constraint (14.33) into Equation (14.36) with $V_{c1} = V_{max}$ yields, for the worst case,

$$V_0 \leq V_{th} - V_{max} \frac{2C_{GS} + C_{LC\,min} + C_s}{C_{LC\,min} + C_s}. \tag{14.40}$$

The row voltage V_g in Equation (14.37) in the on-stage and V_0 in Equation (14.40) in the off-stage, as well as the shift of the potential of the front-plate by V_{FP} in Equation (14.22) or the compensation impulse V_{com} in Equation (14.25) determine the row address signals. In the performance of both solutions, the shifted back-plate potential or the compensation impulse are equivalent. The shifted front-plate potential is more often used, because one can use gate line drivers independent of the TFT-specific and pixel design-specific compensation impulse in Equation (14.25).

Not only the falling flanks of the row address signals in Figure 14.5, but also the rising flanks couple a voltage onto V_{LC}. Although it is transitory as the last flank, the falling edge determines the steady state, the effects may be visible as jitter. This adverse influence can be eliminated by introducing a second line to ground, to which C_s is connected, as shown in Figure 14.7. This additional line does not, of course, remove the effect on V_{LC} stemming from the falling flank of the gate impulse. However, it removes the possibility of feeding in the compensation impulse. Hence, the only correction left is the shift of the potential of the back plate.

In addition, the currents for charging and reverse charging the storage capacitors flow in the gate line, changing the potential of the line where charging takes place. This is especially noticeable in large displays with a higher resistance of the lines. It can be significantly reduced by alternating the sign of neighbouring column signals.

Further capacitive couplings in Figure 14.7 feed the video signals V_{c1} and V_{c2} of the two columns bordering a pixel through the parasitic capacitances $C_{c1e} + C_{SD}$ and C_{c2e} onto $C_{LC} + C_S$, causing the pixel voltage to change by

$$\Delta V_{LC} = V_{c1} \frac{C_{c1e} + C_{SD}}{C_{c1e} + C_{SD} + C_{LC} + C_S} + V_{c2} \frac{C_{c2e}}{C_{c2e} + C_{LC} + C_S}. \tag{14.41}$$

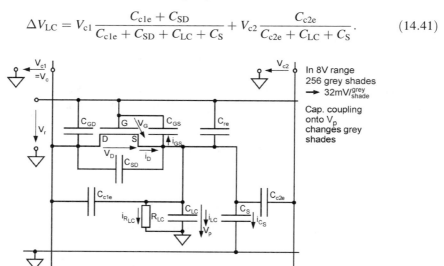

Figure 14.7 The TFT in a pixel with voltages, currents and parasitic capacitances

An exact compensation to $\Delta V_{LC}=0$ is impossible as V_{c1}, V_{c2}, $C_{SD}(V)$ and $C_{LC}(V)$ are dependent on the video voltages. A substantial decrease of ΔV_{LC} is achieved by alternating the signs of V_{c1} and V_{c2} either after each frame or line by line, or even better, from column to column, or as the ultimate, framewise, linewise and columnwise combined.

The capacitive coupling is noticeably reduced by replacing the steep falling edge of the row impulse by the gradual decline in Figure 14.8(a). As this increases the addressing time, an overlap of the two consecutive gate pulses in Figure 14.8 restores the previous speed. Rounding-off the upper portion of the flank in Figure 14.8(b) and then staying with the steep drop has the same beneficial effect. The gate line represents an RC-line, including the crossover capacitances of rows and columns. The step response of a homogeneous RC-line at the end of the line is

$$V_s(t) = V_g \left(1 + \frac{2}{\pi} \sum_{i=1}^{\infty} \frac{(-1)^i}{i - (1/2)} \exp \frac{-\pi^2 (i - (1/2))^2}{R_{tot} C_{tot}} t \right), \tag{14.42}$$

where R_{tot} and C_{tot} are the total resistance and capacitance of the line. The step response exhibits a delay and a decrease of the rise time T_0 from 10 percent to 90 percent of the final values to $T_0 \approx 0.9\, R_{tot} C_{tot}$ (Lauer, 1996). The delay and reduced rise time limit the resolution r for a display with a diagonal D by the relationship (Howard, 1995)

$$r \approx \frac{1}{\varsigma_s D^3 (1 + \alpha)}, \tag{14.43}$$

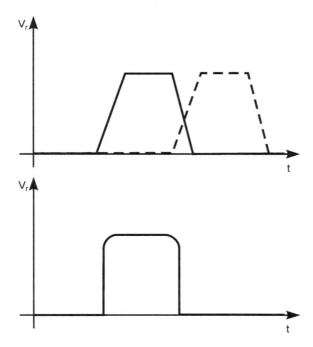

Figure 14.8 Gate pulses for diminished cross talk (a) with trapezoidal and (b) with rounded wave form

where ζ_s is the sheet resistance of the row and $\alpha = f(C_s/C_{LC})$, typically $\alpha \approx 1$. The resolution may be enhanced by reducing the gate line (row) resistance by progressively replacing Mo/Ta by Al and Cu for large diagonal displays.

14.3 The Dynamics of Switching by TFTs

The row address voltages derived in the previous section are steady state values reached after an infinite time. In this section, a differential equation will provide the desired voltages at any time, such as the time T_r with more realistic components such as an LC material with the resistance R_{LC} and a nonlinear TFT.

The differential equation for $V_{LC}(t)$ (Stroomer, 1984; Miyata *et al.*, 1998) is formulated for the ac values in Figure 14.7, and starts with the node equation

$$i_{C_{GS}} + i_{C_{LC}} + i_{C_S} + i_{R_{LC}} = i_D, \tag{14.44}$$

or as a differential equation with $i_C = C\dot{V}$

$$(C_{GS} + C_{LC} + C_S)\dot{V}_{LC} + \frac{V_{LC}}{R_{LC}} = i_D = \mu C_G \frac{w}{l}\left(V_r - V_{LC} - V_{th} - \frac{V_c - V_{LC}}{2}\right)(V_c - V_{LC}), \tag{14.45}$$

where Equation (14.6) for the TFT in the transition region and $V_G = V_r - V_{LC}$ and $V_D = V_c - V_{LC}$ have been used. Reordering of Equation (14.45) provides Riccati's nonlinear differential equation

$$\dot{V}_{LC} = k_1 V_{LC}^2 + k_2 V_{LC} + k_3 \tag{14.46}$$

with

$$k_1 = k/2 \quad \text{and} \quad k = \frac{\mu C_G w/l}{C_{GS} + C_{LC} + C_S},$$

$$k_2 = k(V_{th} - V_r) - \frac{1/R_{LC}}{C_{GS} + C_{LC} + C_S},$$

$$k_3 = kV_c(V_r - V_{th} - V_c/2).$$

As a rule, the simplification $|\mu C_G w/l(V_{th} - V_r)| \gg 1/R_{LC}$ can be applied. The nonlinear transformation of Equation (14.44)

$$V_{LC}(t) = V_c + \frac{1}{\bar{V}_{LC}(t)} \tag{14.47}$$

provides the first order linear differential equation with constant coefficients

$$\dot{\bar{V}}_{LC} = -(2k_1 V_c + k_2)\bar{V}_{LC} - k_1, \tag{14.48}$$

To avoid inconsistancies with the dimension V in Equation (14.47), one would have to divide Equation (14.46) by 1V. For simplicity, the ensuring normalized terms are denoted unchanged V_{LC} including the new variable $\overline{V}_{LC}$.

The general solution of Equation (14.48) is

$$\overline{V}_{LC}(t) = \frac{-k_1}{2k_1 V_c + k_2} + ae^{-(2k_1 V_c + k_2)t}, \tag{14.49}$$

where a is the integration constant. The initial condition $V_{LC}(0) \neq 0$ yields, with (14.47),

$$\overline{V}_{LC}(0) = 1/(V_{LC}(0) - V_c),$$

and with Equation (14.48),

$$a = \frac{1}{V_{LC}(0) - V_c} + \frac{k_1}{2k_1 V_c + k_2}. \tag{14.50}$$

Substituting the constant a in Equation (14.49) finally leads to the solution

$$V_{LC}(t) = V_c + \left[\frac{1}{2q} + \left(\frac{1}{V_{LC}(0) - V_c} - \frac{1}{2q} \right) e^{kqt} \right]^{-1}, \tag{14.51}$$

with

$$q = V_r - V_{th} - V_c. \tag{14.52}$$

The time $t = T_{r-}$, the time immediately before the end of the charging period, gives the pixel voltage to which C_{LC} has been charged. The stepwise change ΔV_{LC} in Equation (14.18) due to the falling flank of V_r occurs immediately afterwards reducing $V_{LC}(T_{r-})$ from Equation (14.51).

Charging $C_{LC} + C_S$ to a negative voltage in Figure 14.4(b) is governed by the node equation

$$-i_{GD} - i_{C_{LC}} - i_{C_S} - i_{R_{LC}} = i_D = \mu C_G \frac{w}{l} \left(V_r - V_{LC} - V_{th} - \frac{V_{LC} - V_C}{2} \right) (V_{LC} - V_C). \tag{14.53}$$

With the same considerations as for the charging to a positive voltage, this leads to Riccati's differential equation

$$\dot{V}_{LC} = k_1' V_{LC}^2 + k_2' V_{LC} + k_3' \tag{14.54}$$

with

$$k_1' = \frac{k'}{2} \quad \text{and} \quad k' = \frac{\mu C_G w/l}{C_{GD} + C_{LC} + C_S},$$

and

$$k'_3 = k'V_c(V_rV_{th} - V_c/2).$$

After the transformation Equation (14.47), we finally obtain the solution

$$V_{LC}(t) = V_c + \left[\frac{1}{2q} + \left(\frac{1}{V_{LC}(0) - V_c} - \frac{1}{2q}\right)e^{k'qt}\right]^{-1} \tag{14.55}$$

for charging C_{LC} to a negative voltage; k' is given below Equation (14.54), and q in Equation (14.52). The results in Equations (14.51) and (14.55) are now used to determine the voltages V_g for the on state and V_0 for the off state of the TFT.

The main requirement for the voltage V_{LC} is that it reaches within T_r the desired voltage V_c, determining the grey shade, and that it remains constant during the remainder of T_f. Reaching V_c within T_r is equivalent to equal voltages $V_{LC}(t)$ at $t=T_{r-}$, T_{r+} and negative equal ones at $t=T_f+T_{r-}$, T_f+T_{r+}. This is expressed by the requirement

$$V_{LC}(T_{r-}) = V_{LC}(T_{r+}) = -V_{LC}(T_f + T_{r-}) = -V_{LC}(T_f + T_{r+}). \tag{14.56}$$

Inserting Equations (14.56) into Equations (14.51) and (14.55) leads to the conditions

$$\left[\frac{1}{2q} - \left(\frac{1}{2V_c} + \frac{1}{2q}\right)e^{kqT_r}\right]^{-1} \to 0 \tag{14.57}$$

and

$$\left[\frac{-1}{2q} - \left(\frac{1}{2V_c} - \frac{1}{2q}\right)e^{k'qT_r}\right]^{-1} \to 0 \tag{14.58}$$

for charging to V_c and to $-V_c$. It is practical to replace 0 by a limit εV_{LC} with $|\varepsilon| \leq 1$, and as close to zero as required by the display performance (Suzuki, 1992). With this constraint, Equations (14.51) and (14.55) yield

$$\left[\frac{-1}{2q} + \left(\frac{1}{2V_c} + \frac{1}{2q}\right)e^{kqT_r}\right]^{-1} \leq \varepsilon V_{LC}, \quad \text{for } V_c > 0 \quad \text{and} \tag{14.59}$$

$$\left[\frac{-1}{2q} - \left(\frac{1}{+2V_c} + \frac{1}{2q}\right)e^{k'qT_r}\right]^{-1} \leq \varepsilon V_{LC}, \quad \text{for } V_c < 0, \tag{14.60}$$

where it was considered that in Equation (14.59) the steady state was reached from below, and in Equation (14.60) from above. Rewriting Equations (14.59) and (14.60) provides

$$kqT_r \geq \ln\left(\frac{1 + 2q/\varepsilon V_{LC}}{1 + q/V_c}\right) \quad \text{for } V_c > 0 \quad \text{and} \quad V_{LC} < 0 \quad \text{and} \tag{14.61}$$

$$k'qT_r \geq \ln\left(\frac{1 + 2q/\varepsilon V_{LC}}{1 + q/V_c}\right) \quad \text{for } V_c < 0 \quad \text{and} \quad V_{LC} < 0. \tag{14.62}$$

According to Equations (14.51) and (14.29) as well as (14.55) and (14.30), the time constants for charging to $+V_c$ and to $-V_c$) are respectively

$$\tau_+ = \frac{1}{\mu C_G w/l} \frac{C_{GS} + C_{LC} + C_S}{V_r - V_{th} - V_c} \quad \text{for } V_{min} \leq V_c \leq V_{max} \tag{14.63}$$

and

$$\tau_- = \frac{1}{\mu C_G w/l} \frac{C_{GD} + C_{LC} + C_S}{V_r - V_{th} - V_c} \quad \text{for } -V_{max} \leq V_c \leq -V_{min}. \tag{14.64}$$

For $C_{GS} = C_{GD}$ and considering $V_c < 0$ in Equation (14.64), we recognize

$$\tau_- < \tau_+.$$

The largest time constant is encountered for $V_c = V_{max}$ in Equation (14.63) reaching the value

$$\tau_+ = \tau_{max} = \frac{1}{k_{max}(V_r - V_{th} - V_{max})}, \tag{14.65}$$

where

$$k_{max} = \frac{\mu C_G w/l}{G_{GS} + C_{LC\,max} + C_S}. \tag{14.66}$$

The constraint in Equation (14.61) is stronger than that in Equation (14.62), because it is associated with the larger time constant τ_+. For the worst case, constraint (14.61) provides

$$k_{max} T_r (V_g - V_{th} - V_c) \geq \ln \frac{1 + 2(V_g - V_{th} - V_c)/\varepsilon V_{LC}}{1 + (V_g - V_{th} - V_c)/V_c}, \tag{14.67}$$

where $V_r = V_g$ for $V_r > 0$ was used.

The voltage-dependent non-linear capacitance $C_{LC}(V_{LC})$ is depicted in Figure 14.9. The minimum value is reached if the electric field is perpendicular to the director with a dielectric constant $n_\perp < n_\parallel$ and the maximum if the field is parallel to the director.

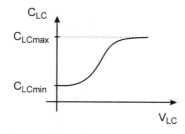

Figure 14.9 The voltage dependent capacitance of liquid crystals

The constraint (14.67) provides, in form of $u = V_g - V_{th} - V_c$, a nonlinear equation for the smallest V_g with given values of V_c, V_{th}, V_{LC}, ε, T_r and k_{max} in Equation (14.66); this equation may be solved by an approximation method such as Newton's. The chosen V_g should be too much above the minimum for V_g, in order to avoid disturbing nonlinearities of the TFT.

For the determination of the blocking voltage V_0, we investigate two cases (Lauer, 1996). First, in the pixel under investigation, $C_{LC} = C_{LC\,min}$ is charged to V_{min}. As a consequence of the linewise sign inversion, the column signals for other pixels travelling down a column are for black pixels in the column $V_c = +V_{min}$ or $V_c = -V_{min}$ in time $T_f - T_r$ during which V_{min} across $C_{LC\,min}$ is maintained. Therefore, the TFT in the pixel with $C_{LC\,min}$ experiences during the time $T_f/2$, that is assuming half the time for a positive and half the time for a negative $V_c = \pm V_{min}$

$$V_{D1} = 0 \tag{14.68}$$

and

$$V_{D2} = 2V_{min}. \tag{14.69}$$

Due to these equations, only during time $T_f/2$ is a current I_D flowing through the TFT.

The ensuing decrease of V_{LC} by dV_{LC1} during the time $T_f - T_r \approx T_f$ degrades the image, because of the parasitic discharge of capacitors by a current I, resulting in

$$dV_{LC1} = \frac{I}{C_{GS} + C_{LC\,min} + C_S} \, dt \approx \frac{I}{C_{GS} + C_{LC\,min} + C_S} T_f. \tag{14.70}$$

With $I = -(I_D/2) - i_{R_{LC}}$ as the currents in Figure 14.7 discharging $C_{GS} + C_{LC\,min} + C_S$, we obtain from Equation (14.70) in the worst case

$$dV_{LC1} = \frac{T_f}{C_{GS} + C_{LC\,min} + C_S} \left(-\frac{I_D(V_G, V_{D2})}{2} - \frac{V_{min}}{R_{LC}} \right). \tag{14.71}$$

The current $I_D/2$ must be considered because it flows only during time $T_f/2$, and not T_f as in Equation (14.71). Our goal is to minimize the degradation of the image by selecting a low enough V_0 blocking the TFT.

In the second case, $C_{LC} = C_{LC\,max}$ is charged to V_{max}. As a consequence of the linewise sign inversion, the column signal travelling along the column during $T_f - T_r$, in which V_{max} is maintained across $C_{LC\,max}$, is $V_c = \pm V_{max}$. Hence, the voltage V_D at the TFT during the time $T_f/2$ is either

$$V_{D3} = 0 \tag{14.72}$$

or

$$V_{D4} = 2V_{max}. \tag{14.73}$$

The ensuing decrease of V_{LC} is

$$dV_{LC2} = \frac{T_f}{C_{GS} + C_{LCmax} + C_S} \left(-\frac{I_D(V_G, V_{D4})}{2} - \frac{V_{max}}{R_{LC}} \right). \tag{14.74}$$

The value with the largest magnitude

$$dV_{LC} = \max(dV_{LC1}, dV_{LC2}) \tag{14.75}$$

is required to be at least by a factor $\varepsilon > 0$ smaller than for instance, V_{min}, leading to

$$dV_{LC} = \max(|dV_{LC1}|, |dV_{LC2}|) = dV_{LCmax} = \varepsilon V_{min} \tag{14.76}$$

This limits the parasitic discharge of C_{LC} in the most severe case when it is charged to V_{min}. Solving Equations (14.71) and (14.74) for I_D, with the boundary in (14.76), yields

$$|I_D(V_G, V_{D2})| \leq 2\left(\frac{C_{GS} + C_{LCmin} + C_S}{T_f}dV_{LCmax} - \frac{V_{min}}{R_{LC}}\right) = I_{D2} \tag{14.77}$$

and

$$|I_D(V_G, V_{D4})| \leq 2\left(\frac{C_{GS} + C_{LCmax} + C_S}{T_f}dV_{LCmax} - \frac{V_{max}}{R_{LC}}\right) = I_{D4}, \tag{14.78}$$

where

$$V_{D2} = 2V_{min} \text{ according to Equation (14.69)},$$

$$V_{D4} = 2V_{max} \text{ according to Equation (14.73)},$$

and

$$dV_{LCmax} \text{ is taken from Equation (14.76)}.$$

In both cases, there is only a solution if the conditions

$$\frac{V_{min}}{R_{LC}} \leq \frac{C_{GS} + C_{LCmin} + C_S}{T_f}dV_{LCmax} \tag{14.79}$$

and

$$\frac{V_{max}}{R_{LC}} \leq \frac{C_{GS} + C_{LCmax} + C_s}{T_f}dV_{LCmax} \tag{14.80}$$

are met. They ensure a positive I_D. If they are not met, V_{min} or V_{max} and ε in Equation (14.76) have to be adjusted accordingly which, however, worsens the performance of the LCD.

The possibility of meeting the conditions in Equations (14.77) and (14.78) are checked by using the measured logarithmic input characteristics in Figure 14.10. The ideal characteristics due to Equations (14.5) and (14.6), shown as a dotted line, do not include the low off-currents caused by the off-resistance of the TFT being finite. We require the measured input characteristics for the drain-source voltages $V_{D2}=2V_{min}$ and $V_{D4}=2V_{max}$ in Equations (14.69) and (14.73) as parameters. The two curves are shown in Figure 14.10.

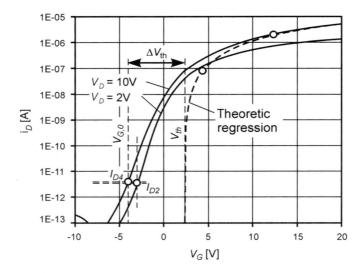

Figure 14.10   The measured logarithmic input characteristics of an a-Si-TFT with the off-currents

The point I_{D2} from Equation (14.76) on the curve with parameter $V_{D2}=2V_{min}$ and the point I_{D4} from Equation (14.78) on the curve with parameter $V_{D4}=2V_{max}$ determine the pertinent gate voltages V_G. The smaller of those two voltages is called V_{G0}, and it represents the more demanding case to be realized in the circuit. This indicates that the real threshold voltage $V_{th0}=V_{G0}$ replaces the ideal threshold V_{th} in Equation (14.40), which now becomes

$$V_0 < V_{G0} - V_{max}. \qquad (14.81)$$

V_{G0} is, by $\Delta V_{th}=V_{th}-V_{G0}$, lower than V_{th}, as is indicated in Figure 14.10.

A practical example will demonstrate that the row voltage $V_r=V_0$ obtained in Equation (14.81) by investigating the dynamics of the addressing circuit can be considerably lower than V_0 in Equation (14.40) based on the ideal behaviour of the TFT.

The following data are given: $V_{min}=1$ V, $V_{max}=5$ V, $C_{LCmin}=40\,fF$, $C_{LCmax}=80\,fF$, $C_S=60\,fF$, $C_{GS}=20\,fF$ and, in Figure 14.10, $V_{th}=2.38$ V. Equations (14.77) and (14.78) provide $I_D \le 3.4 \cdot 10^{-12}$ A for $V_{D2}=2$ V and $I_D \le 3.8 \cdot 10^{-12}$ A for $V_{D4}=10$ V. The smaller of the two pertinent gate-source voltages in Figure 14.10 is $V_{G0}=-4$ V yielding, with Equation (14.81) $V_0 < -9$ V. V_0 in Equation (14.40) would have yielded $V_0=-4.62$ V. The substantial change to Equation (14.81) is due to the corrected input characteristics for low off-currents in Figure 14.10.

14.4   Bias-Temperature Stress Test of TFTs

The measurements which determine μ, V_{th} and ΔV characterize the actual performance of a TFT. For stability over time, the shift ΔV_{th} of the threshold voltage is the most important property for addressing of LCDs. The investigations in Sections 14.2 and 14.3 demonstrate that the appropriate grey shade and the blocking property of a TFT are based on V_{th}.

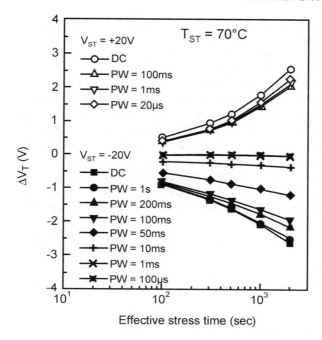

Figure 14.11   The shift of V_{th} during bias temperature stress tests

Equation (14.8) describes how V_{th} depends upon the charge trapped in the semiconductor at the interface to the dielectric and in the dielectric itself.

The Bias Temperature Stress (BTS) test (Chiang and Kaniki, 1996) deals with detrapping and trapping of this charge at an elevated temperature, e.g. at 70°C. A sequence of positive or negative gate-source pulses $\pm V_G$ is applied, with the pulse width as a parameter, while $V_D = 0$ during a given stress time. The measured data in Figure 14.11 shows a larger shift of V_{th} for negative than for positive gate pulses. In each case, the shift increases with the pulse width. A phenomenological explanation is that negative pulses try to empty the traps the more effectively the longer the pulse. The traps are again gradually, but not completely, filled by the intrinsic charge. This results in a lowering of V_{th}. Positive gate pulses fill more of the so far empty traps, resulting in an increase of V_{th}. The gate pulses used for addressing LCDs consist of a sequence of positive and negative pulses with different widths. The superposition of the shifts provides an increase in V_{th}. This increase, measured at an elevated temperature and over a stress time of more than a day, reveals an indication of the life of a TFT under the conditions used in a display.

14.5 Drivers for AMLCDs

The row and column drivers provide the signals (which were discussed in Chapter 13 for PM displays, and in Sections 14.2 and 14.3 for AM displays) to the contact pads. A more detailed discussion will be presented in this section. The next section will be devoted to the processing of the video signal from the picture source to the column drivers.

As a rule, the drivers are ICs placed along the edges of the glass or plastic panel and bonded to the contact pads. Special driver ICs have been developed which require only a narrow stripe along the edges. They are, for example, only 1.8 mm wide, 16 mm long and have 240 pins with a pitch of 64 μm.

The row- or gate-line drivers are shift registers where each stage, as the example in Figure 14.12 shows, provides the gate pulse with a frequency $1/T_f$. The output may include a circuit generating the compensation pulse. The N stages of the shift register generate N pulses in time intervals T_f/N or with a frequency N/T_f. As a rule, the even and odd numbered rows are addressed from different sides in order to double the pitch to more feasible dimensions. Very seldom, the rows are scanned from both sides if the resistance of the line is too high, but the price for double the number of drivers is, as a rule, not acceptable. The video or column driver in Figure 14.13 receives the signals for R, G and B, which are time sequentially stored on capacitors from where the operational amplifiers used as buffer amplifiers drive them as voltage sources into the columns.

In the dual scan driving in Figure 12.9(c), the columns are interrupted in the middle. Hence, we can drive the upper and lower half of the display with $N/2$ lines, each in frame time T_f with half the line address frequency that is with $N/2T_f$. This alleviates the frequency requirements for the line drivers, especially for a large number of N lines and the Pleshko–Alt limits.

A further reduction of frequencies is possible with the block parallel addressing in Figure 14.14. b words (corresponding to b pixels) are stored time-sequentially, at frequency f_1, into shift register A or B, and transferred in parallel to the latch, while the other shift register is filled. The parallel transfer from the latch with frequency $f_2 = f_1/b$ into the b blocks of latches loads the m/b cells of each latch in the time $(m/b)\cdot(1/f_2) = (m/f_1)$, which is the same time as needed for loading m words into the shift register A or B. The number of columns in the display is m; the columns are subdivided into b blocks, with m/b columns each. The storage cells $\nu = 1.2,\ldots,m/b$ in the b blocks feed their content in parallel into the column blocks. While this is happening, a second set of b blocks above the displays, which is not drawn, is being filled with data. The selection of the number b determines by how much the frequency f_1 is decreased. As we shall see later, further processing of the digital

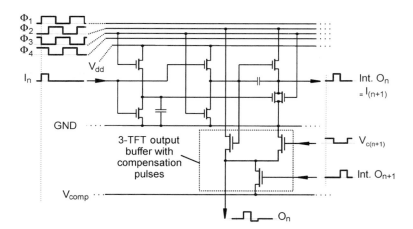

Figure 14.12 Example for one stage of a shift register for the rows of an AMLCD

Column driver OKI MSM 5280/81

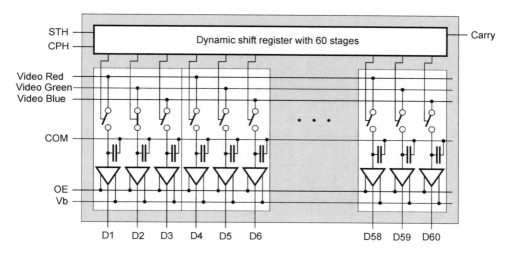

Figure 14.13 A video driver of an AMLCD

words, including D/A-conversion and amplification, can also be performed at the lower frequency f_2.

Since the cost of the addressing circuits is roughly half the cost of an LCD, halving the number of the expensive video drivers in the scheme in Figure 14.15 is most helpful (Sakamoto *et al.*, 1997). The column driver first feeds the video voltage into the pixel of the first column driven in Figure 14.15, and then into those of the second column driven. Only half the number of column lines, but twice the number of rows r_1 and r_2, are necessary. Half the number of column drivers have, time sequentially, to take care of two lines, for which they have to work at double the speed of the conventional approach. Electrically, the pixels became asymmetric because the parasitic capacitive couplings between neighbouring pixels C_R and C_L are unequal. To render them almost equal, a dummy electrode (indicated by a dashed line) has been introduced. This electrode is expected to exhibit the same shielding effect between the pixel as the column line. There are at least two pixel layouts in the panel in Figure 14.15: one with the TFT in the upper portion of the pixel and left of the column, and one with the TFT in the lower portion and right of the column. This results in different gate-source capacitances of the TFTs which, as we know, influence the voltage V_{LC}. The difference in capacitance can be balanced by an additional thin film capacitor C_b in Figure 14.15. With this new driving scheme, power consumption is also decreased by 25 percent, as only half the number of data drivers is used.

A further decrease in power consumption is possible with the power recycling shown in Figure 14.16 (Kim *et al.*, 1997). The largest power consumption occurs for a sign inversion of the voltage from pixel to pixel, because the voltage swing between two adjacent frames is largest. Between two frames the switches between two adjacent columns are closed. This allows the capacitances charged to a negative pixel voltage and those charged to a positive one to discharge, resulting in a common voltage half-way between the previous positive and negative voltages. From there, the capacitances are recharged to the inverted positive and negative voltage of the next frame by the addressing circuit. Hence, power consumption is

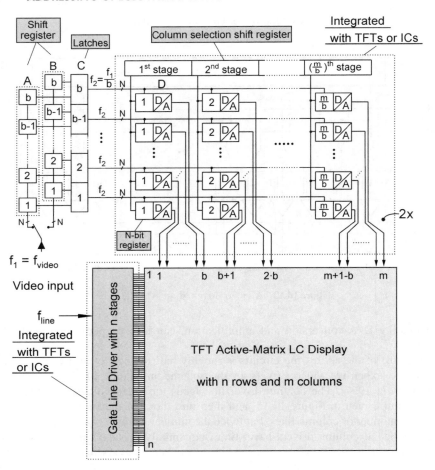

Figure 14.14 Block diagram for block parallel video drivers for an AMLCD

only half of the power consumption encountered if the addressing circuit had to provide the current for the entire recharging of the capacitances.

The maximum voltage swing V_s is with V_{max} in Equation (14.29)

$$V_s = 2\,V_{max}.$$ (14.82)

The power consumption of the conventional addressing scheme, with the voltage V_{DD} of the power supply, is

$$P_{conv} = V_{DD}\,I = V_{DD}\left(N\,C_{load}\,V_s \cdot \frac{f_r}{2}\right),$$ (14.83)

where N is the number of rows, C_{load} is the total row capacitance and f_r is the frequency of the row driver. For addressing with recycling of charge, the power consumption is

$$P_{recycl} = V_{DD}(N\,C_{load}\,V_s/2 \cdot f_r/2).$$ (14.84)

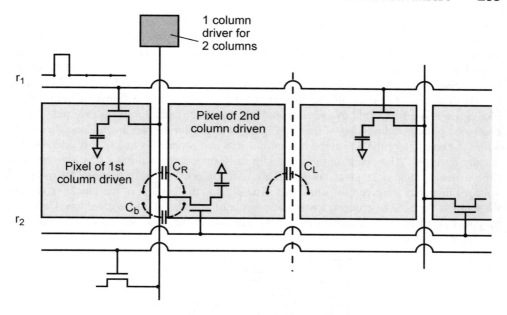

Figure 14.15   Addressing of an AMLCD with half the number of video drivers

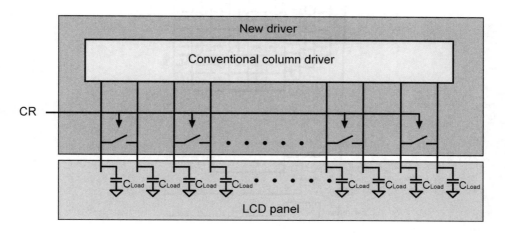

Figure 14.16   Recycling of charge by closing the switches by the control signal CR

For a 50 percent saving of power to be realized, the capacitance $C_{LC} + C_s$ of each pixel has to be discharged to a roughly zero voltage, for which the TFT has to be conductive with a zero drain-source voltage V_D.

A previous solution (Erhart and McCartney, 1997) with similar effect, used a large external capacitance on to which all the pixel capacitances were discharged. A further power reduction can be achieved by combining the charge recycling with the scheme using only half the number of video drivers.

In high resolution displays, the pixel storage capacitors can no longer be fully charged because the row address time $T_r = T_f/N$ becomes too short, especially at lower temperatures around 0°C, where the mobility of the a-Si-TFTs is low. A remedy is the multi-dot pixel inversion and the dual line addressing depicted in Figures 14.17(a) and 14.17(b) (Nishimura et al., 1998). Figure 14.17(a) shows that blocks of lines (in the example blocks of four lines), do not exhibit an inversion of the sign of the pixel voltage. From frame to frame, these blocks are shifted upwards by one line and the sign of the pixel voltage inverted. This is combined with the addressing of lines in Figure 14.17(b) during two line address times $2\,T_r$, with the exception of the line at the border of a sign inversion. This line is only addressed as usually during T_r. The pixels in the lines with the longer address times $2\,T_r$ are given more time to reach their final pixel voltage. This is made possible only because all the pixel voltages involved are given the same sign. The pixel at the borders of the blocks have only the address time T_r, in order to avoid the precharging in the opposite direction. However, they still profit from the precharging effects, because in the previous frame this pixel was

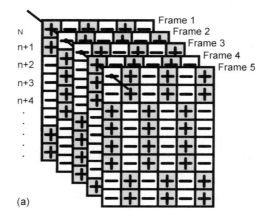

(a)

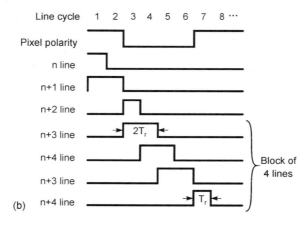

(b)

Figure 14.17 (a) Blocks of pixels with the same sign of the voltage V_{LC}; (b) addressing of a line during two row address times

already charged to the inverted polarity. The storage capacitors in a UXGA-LCD were charged only to 94 percent of the desired value when the common addressing scheme was used. The new scheme allowed charging to 99 percent of the desired value, with a beneficial effect on contrast, which reached 150 : 1, with contrast of 10 : 1 extended to 80° in all four directions. The framewise shifting of the block pattern also reduces flicker.

In cases where the format of the picture data has a smaller pixel count than the format of the LCD, the need arises to increase both the number of lines and the number of dots per line

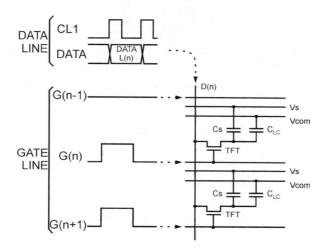

Figure 14.18 Introduction of a second line with the same information as the previous line

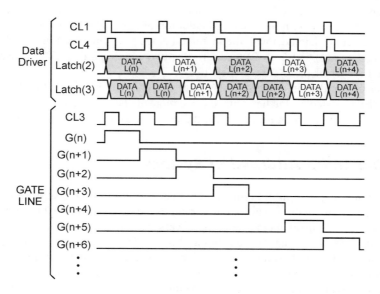

Figure 14.19 The introduction of an additional line if the capacitor C_s in Figure 14.18 is connected to the gate line

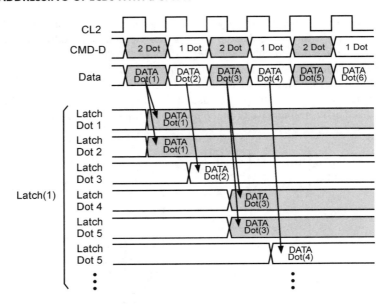

Figure 14.20 The introduction of an additional dot in an LCD

of the data before it can be written into the display. An easy to implement way of introducing two lines with the same information is shown in Figure 14.18, where two lines are scanned simultaneously while the columns feed in the same information in both lines. This works well as long as the storage capacitor C_s in Figure 14.18 has its own electrode V_s. This expansion of the number of lines does not require a line buffer, as did previous approaches. If C_s is connected to the gate line, the two falling edges of the gate pulse couple too large a parasitic voltage onto V_{LC}. In this case, the addition of a line is achieved in Figure 14.19 by writing the content of latch (2) with the increased clock frequency CL4 into latch (3), but by repeating the data $L(n)$ and $L(n+2)$ meant for the two additional lines. The gate pulses $G(\nu)$ achieve the transfer of these data to their lines.

The expansion of dots in a line starts with the original data in Figure 14.20, and transfers the dots to be repeated in latches with the location of the repeated dots underneath each other. A modified timing controller directs the dot data to their respective columns.

14.6   The Entire Addressing System

The entire addressing system in Figure 14.21 (Lauer, 1996) has the task to receive the time sequential data of the picture sources, tailor them to the needs of the LCDs by digital signal processing and then transmit them over to the drivers. The video sources are analog TV, digital TV, digital data from a computer and digital graphic data. The video adapter at the input transforms all analog data using a video decoder, a three-channel amplifier and an A/D converter into the same digital format as the other digital inputs already have. The adapter also matches the data to the format required by the LCD, such as a black and white or colour with vertical or diagonal filter stripes. The colour coding is, as a rule, at most 8-bit. The output R G B, C-Synch serves as monitor for colour and the synchronization pulse. Together with a

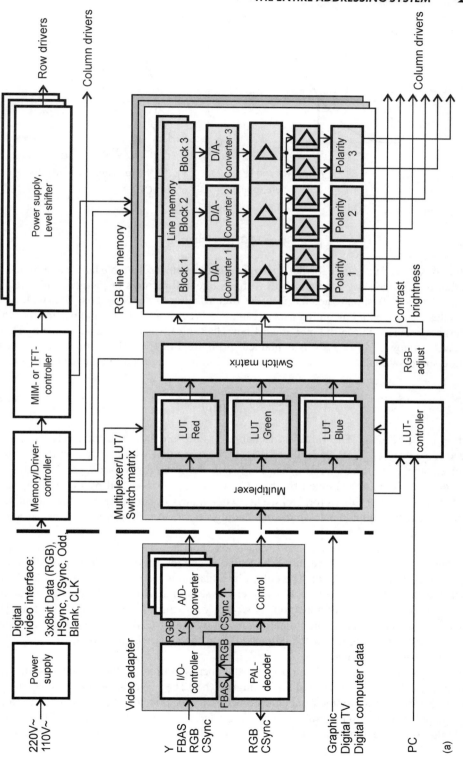

Figure 14.21 (a) The entire addressing system; (b) γ-correction

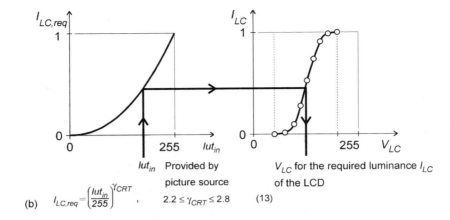

(b) $I_{LC,req} = \left(\dfrac{lut_{in}}{255}\right)^{\gamma_{CRT}}$, $2.2 \leq \gamma_{CRT} \leq 2.8$ (13)

Figure 14.21   (continued)

feature box, the adapter is enabled to generate pictures with a selectable half frame frequency, interlaced or progressive scan and with features like still pictures or picture in picture. The digital R G and B signals are fed into the digital signal processing box. The multiplexer translates the sequential incoming data stream into three parallel R-, G- and B-streams, which undergo a γ-correction in the individual Look Up Tables (LUTs) assigned to them. This γ-correction, known from CRTs, is explained in Figure 14.21(b). The luminance lut_{in} of the picture source with 256 grey shades is translated into the required luminance of the LCD

$$I_{LC\,req} = \left(\frac{lut_{in}}{256}\right)^{\gamma_{CRT}}, \tag{14.85}$$

with $\gamma_{CRT} \in [2.2, 2.8]$. The values in Equation (14.85) are stored in the LUT; I_{LCreq} is reduced to values $l_{Lcreq} \in [0, 255]$. Finally, l_{LCreq} is transposed with the measured luminance-pixel voltage curve, also stored in the LUT, into the desired pixel voltage V_{LC}. After this procedure, the colours exhibited by the LCD match those produced by the internationally defined colours co-ordinates for CRTs. The switch matrix or commutator at the output of the digital signal box interchanges lines, depending on whether progressive or interlaced addressing is used.

The RGB line latch (memory) of the last box separates the write- and read-frequencies. One line latch consecutively receives the R, the G and the B colour information of one line as prepared by the commutator. The colour information is read out in block parallel format, starting with the first data in each block, then the second data in each block, etc. To do this, we need two line latches, of which one is written while the other is read out. After the D/A conversion, all channels are amplified controlled by the 'R G B adjust' to the same brightness. The two video signals at the output differ in the sign, which is used for frame-, line- and column-wise sign inversion.

The memory driver controller supervises the operation times of the line latches, the commutation, the D/A conversion and the column drivers. The line controller supervises the gate pulses; the level shifter for the front plate compensates the offset of V_{LC}, caused by the drop of the gate pulse. The outputs of the polarity stages feed into the video drivers in Figure 14.12, while the row drivers in Figure 14.11 are connected to the power supply in the level shifter.

14.7 Layouts of Pixels with TFT Switches

The performance of a display depends strongly upon the layout of the pixels. Key features are a large enough aperture ratio A, defined in percent for a transmissive display as

$$A = \frac{\text{transmissive area of a pixel}}{\text{total area of a pixel}} \cdot 100\%, \qquad (14.86)$$

which produces high brightness and a large enough storage capacitor C_s for holding the information during the frame time, for reducing the effect of parasitic capacitive coupling in Equation (14.18) to the pixel voltage, and for shielding the pixel against detrimental couplings from pulses travelling on the columns and rows. For a reflective display, the word 'transmissive' in Equation (14.85) should be replaced by 'reflective'. The increased C_s and the shielding, as a rule, diminish the aperture ratio, as the materials used are not transparent. It is up to the designer to find an ingenious solution benefiting all criteria.

The basic pixel layout in Figure 14.22(a) (Lueder, 1998a) possesses a column made of Al or Cr-Al-Cr, which is insulated from the rows made of Cr or Mo-Ta. The rows are widened in the region of the storage capacitor C_s, which has a dielectric of SiO_2 or Si_3N_4 and an upper electrode of Al or ITO. This placement of C_s leaves more space for the ITO electrode, thus enhancing the aperture ratio to, as a rule, 65 percent. The a-Si TFT requires only about

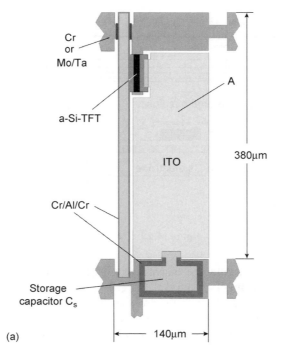

(a)

Figure 14.22 (a) Basic pixel layout of an AMLCD; (b) pixel layout with storage capacitor along the edges of the ITO-electrode; (c) cross-section along the line A−A' in Figure 14.22(b)

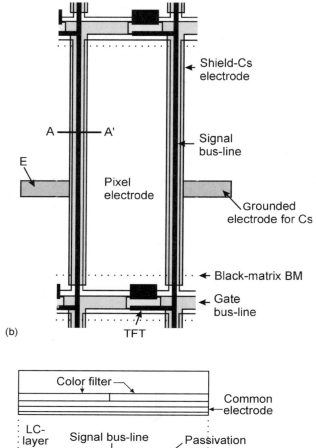

(b)

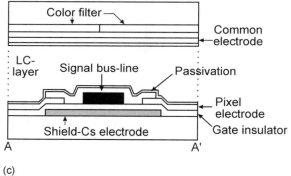

(c)

Figure 14.22 (continued)

2 percent of the area. As a rule, the black matrix is deposited on the opposite glass plate together with the colour filter.

The layout in Figure 14.22(b) with the cross-section in Figure 14.22(c) (Kitazawa *et al.*, 1994; Suzuki, 1994; Kim, 1996) places the capacitor C_s along the edges of the ITO electrode. Most importantly, it is shielded against crosstalk by a shield electrode. This electrode in Figure 14.22(c), from Mo-Ta, also serves as an additional black matrix layer. Furthermore, C_s is no longer connected to the gate-line, but has its own transparent ITO electrode E in Figure 14.22(b), not shown in the cut $A - A'$ in Figure 14.22(c). Hence C_s and, as a consequence, V_{LC} is no longer affected by the pulses on the former bottom electrode of C_s. There is, however, still the downward shift of V_{LC} stemming from the gate pulse coupled in through the gate-source capacitance C_{GS}. The black matrix BM on the active matrix plate has the beneficial effect of blocking the oblique light beam O from the backlight, shown in the cross-section of the pixel depicted in Figure 14.23. Without the BM on the active matrix plate, backlight would be transmitted in the direction shown by the dashed arrow. Further advantages of the additional BM are the shielding of the imperfectly controlled LC stripe along the edge of the ITO electrode, and the introduction of the overlap of the two BMs in the range w, allowing for small misalignments of the two glass plates (Kim, 1996).

Figure 14.24 shows schematically some layers of a pixel in which the storage capacitor C_s is completely placed underneath the ITO pixel electrode (Maier, 1997). For a transmissive pixel, all layers have to be transparent, which is satisfied by the ITO bottom and pixel electrodes and the Si_3N_4 dielectric, which is also used as gate dielectric. The electric field from the neighbouring column ends mostly on the extended bottom electrode, resulting in a shielding of the capacitor C_s. The completed pixel layout with the light shield omitted is shown in Figure 14.25. C_s possesses the large value of 200 fF on an area of 1300 μm², with a thickness of 400 nm, which helps to diminish all effects of the parasitic couplings.

The pixel layout for a reflective cell drawn in Figure 14.26(a) (Egelhaaf *et al.*, 2000) shows that the entire pixel apart from the interpixel gap is covered by an Al mirror, yielding

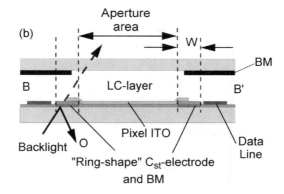

Figure 14.23 Cross-section of a pixel with an additional black matrix on the active matrix plate

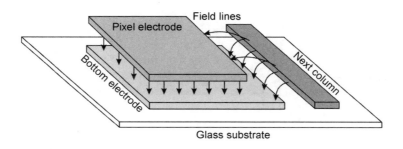

Figure 14.24 Pixel with the storage capacitor underneath the pixel electrode

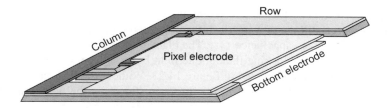

Figure 14.25 A pixel with C_s underneath the ITO electrode

a 92.6 percent aperture ratio. The TFT, the storage capacitor, the rows and columns are all underneath the mirror, as depicted in the cross-section in Figure 14.26(b). The column is shielded by a grounded electrode, preventing capacitive coupling of the video pulses onto the pixel voltage.

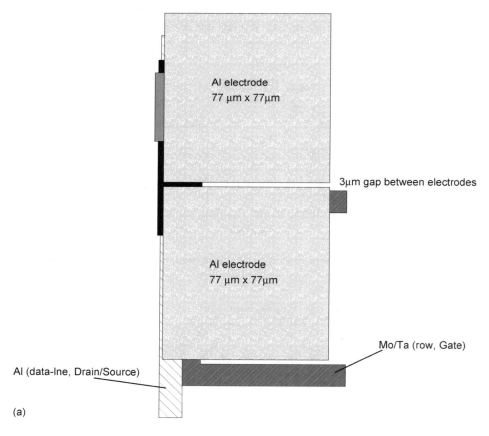

Figure 14.26 (a) The top view of a pixel in a reflective display with a mirror, which also covers the rows and columns; (b) cross-section of the pixel in Figure 14.26(a)

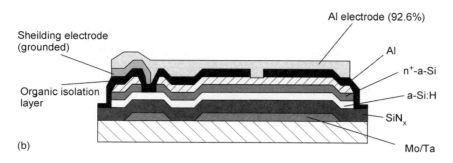

Figure 14.26   (continued)

14.8   Fabrication Processes of a-Si TFTs

A criterion for the usefulness of a TFT fabrication process is a low mask count, presently in the range of 4 to 6. The rationale behind this is that a small number of mask alignments and a small number of processing steps associated with a small mask count is synonymous with an economic process. This is, however, only true if the low mask count does not narrow the process window, and still allows for a high fabrication yield. Furthermore, a small number of masks is only acceptable if it does not degrade the performance of the display.

Figure 14.27 depicts the sequence of fabrication steps involved in a four mask fabrication of bottom gate a-Si TFTs (Glueck *et al.*, 1994; Lueder, 1994c). After a glass substrate (e.g. 0.7 mm thick) has been chemically cleaned in an ultrasonic bath, a 200 nm Cr- or Mo/Ta layer is sputtered and chemically wet etched to form the rows, the gate electrodes and the bottom electrode of the storage capacitor. The patterning shown in Figure 14.27(a) requires the first mask. The next three layers in Figure 14.27a are deposited by [Plasma Enhanced Chemical Vapour Deposition (PECVD)]. They are Si_3N_4 as gate dielectric and as the dielectric for C_s, intrinsic a-Si : H (i-a-Si : H) as semiconductor and n^+-a-Si as a layer providing an ohmic contact for drain and source. In the semiconductor, dangling bonds are neutralized by H, which enhances the electron mobility to values from $0.5\,cm^2/Vs$ to $1\,cm^2/Vs$. This mobility of the amorphous Si layer is still far below the mobility of monocrystalline Si-FETs reaching more than $500\,cm^2/Vs$. Finally, a Cr/Al/Cr- or Mo/Ta layer is sputtered and chemically wet etched in Figure 14.27(b) to form the pattern for the drain and source electrode and the top electrode of C_s. For this, the second mask is required. The pattern of the Cr/Al/Cr serves as a mask for plasma etching of n^+-a-Si in Figure 14.27(b). As plasma etching also attacks a-Si, an end point control is required by watching the edging time or by monitoring the plasma emission at 204.4 nm stemming from CF and SiF radicals. This back channel etch is the most subtle step, as it has to remove all n^+-a-Si residues while leaving an a-Si-layer with a homogeneous thickness of around only 50 nm. This small thickness off current below 1 pA. A second plasma etch is required to reduce the using a third mask in Figure 14.27(c) removes i-a-Si : H, forming the i-a-Si islands for the TFTs. The sputtering and patterning of the ITO pixel electrode by a lift-off process, or preferably by plasma etching in Figure 14.27(d), requires the fourth mask. Finally, a PECVD passivation layer of SiN_x requires a fifth mask for opening the bond pads, whereas the deposition of a metallic light shield on top of the array uses a sixth mask. The two last masks do not have to be aligned as precisely as the others, and as a rule are not counted as yield-determining masks.

(a) Sputtering of 200nm Cr and wet-etching as gate line,
storage capacitor and gate; *(1st mask)*
PECVD of 400nm SiN$_x$, 130nm i-a-Si and 50nm n$^+$-a-Si;
sputtering of 150nm Cr/Al/Cr

(b) Wet etching of Cr/Al/Cr as source line, drain/source
metallization, top electrode of storage capacitor; *(2nd mask)*
plasma-etching of n$^+$-a-Si with Cr/Al/Cr as etch mask with
end point control

(c) Plasma-etching of i-a-Si:H forming
a-Si-islands for TFT *(3rd mask)*

(d) Sputtering and patterning of 80nm ITO as pixel electrode and
column redundancy; lift-off *(4th mask)*
PECVD of 350nm SiN$_x$ as passivation;
opening of the passivation at the bond pads
not very accurate 5th mask light shield (6th mask)

Figure 14.27   The process steps for a four mask fabrication of a-Si : H TFTs

The ultimate in mask count is a two mask process, such as that depicted and explained in
Figure 14.28 (Miyata *et al.*, 1998). An interesting feature is the self-aligned generation of the
channel in Figure 14.28(c) by exposing the photoresist OFPR-800 from underneath using the
Cr gate as a shadow mask. The ITO used for the pixel electrode is also applied as drain and
source electrodes, which saves a mask. The last step is the back channel etch of n$^+$-a-Si. An
improvement of the ITO electrodes for the TFT (Ban *et al.*, 1996) consists of introducing a
doped microcrystalline Si-layer (μc–Si : H (n$^+$)) instead of n$^+$-a-Si : H, which enhances the
mobility of a TFT with ITO contacts above 0.5 cm^2/Vs. A further two-mask process (Richou
et al., 1992) does not generate a-Si islands for the TFTs, and isolates the TFTs having a
guard ring around them.

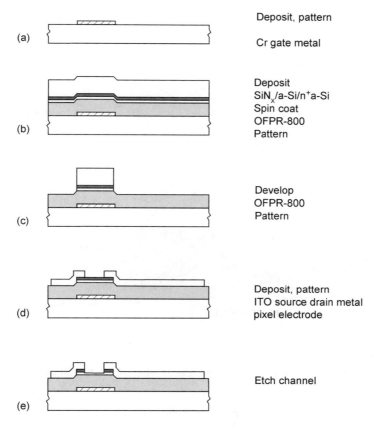

(a) — Deposit, pattern / Cr gate metal

(b) — Deposit SiN$_x$/a-Si/n$^+$a-Si / Spin coat / OFPR-800 / Pattern

(c) — Develop / OFPR-800 / Pattern

(d) — Deposit, pattern / ITO source drain metal / pixel electrode

(e) — Etch channel

Figure 14.28   A two-mask fabrication of a-Si : H TFTs

An n$^+$-doped layer is formed in the a-Si: TFT in Figure 14.29(a) by implanting P ions through an etch stop layer, using Mo-layer as a stopping layer (Maier *et al.*, 1997). After deposition and etching of the Mo source and drain electrodes, which were used before as a stopping layer, we obtain the TFT in Figure 14.29(b). This process has the advantage that back channel etch is not required, and that the i-a-Si semiconductor is, throughout the manufacture, protected by the etch stop layer. The stopping layer has the task of placing a large concentration of the dopant at the surface of the i-a-Si-layer. The process requires an implanter for large areas (Eaton Corp., 1997).

A technological breakthrough was the introduction of Cu with a specific resistance of only 2 μΩcm for conductors and gates into AMLCDs, providing a low enough resistance for rows and columns of large area displays (Fryer *et al.*, 1996). Cu had previously been virtually forbidden in Si semiconductor devices, due to its reaction with Si. Furthermore, it does not adhere well to glass, and is difficult to contact to other metals. Figure 14.30 shows a cross-section of an a-Si TFT with a copper gate (Fryer *et al.*, 1996). The lowermost layer on glass is ITO, which is used as a pixel electrode, as an adhesion layer for the copper gate and as a contact bridge to copper without exposing the Cu to other metals. The configuration in Figure 14.30 does not require back channel etching, because after deposition of i-a-Si : H by PECVD an SiN$_x$ layer was deposited and etched to only having an SiN$_x$ protection film on top of the

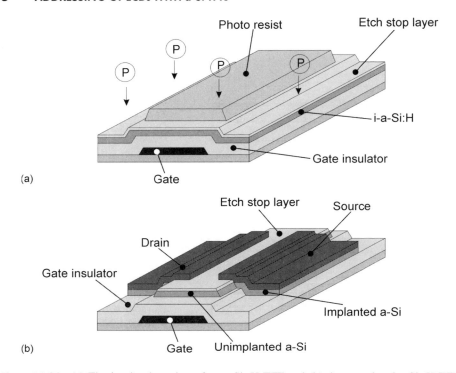

Figure 14.29 (a) The ion-implantation of an a-Si : H TFT and (b) the completed a-Si : H TFT

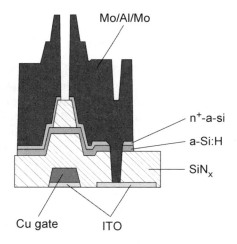

Figure 14.30 Cross-section of an a-Si:H TFT with a copper gate

channel. Then, only an n^+-a-Si : H-layer is generated, and etched off on top of the channel, as shown in Figure 14.30. This eliminates the subtle etching of n^+-a-Si : H on top of a-Si : H, and protects the a-Si : H channel by a dielectric. A via hole is necessary to connect the Mo/Al/Mo source to the ITO.

15

Addressing of LCDs with Poly-Si TFTs

Poly-Si as semiconductor provides TFTs with a much higher electron mobility than a-Si TFTs. Depending on the fabrication process, mobility lies in the range from $20\,\mathrm{cm}^2/\mathrm{Vs}$ up to $400\,\mathrm{cm}^2/\mathrm{Vs}$. As a consequence, poly-Si TFTs with the same channel geometry as a-Si TFTs allow for a considerably larger current, or if carrying the same current as a-Si TFTs, they require a smaller area within a pixel. For this reason, displays with a high pixel density and pixel sizes of $15\,\mu \times 15\,\mu$ only are best addressed by poly-Si TFTs, as they deprive the pixels of the least amount of transparent area. Applications of this kind are displays for view finders in camcorders, or the small LC light valves for projectors, discussed in Chapter 24. A further application for poly-Si TFTs, which has been envisioned for many, years, is the integration of the row and column drivers along the edges of the display substrates, thus eliminating thousands of external connections. The connections are made in the thin film technology using a mask which is in any case needed for processing the display. As poly-Si TFTs, contrary to a-Si TFTs, exhibit, a substantial hole mobility of more than $15\,\mathrm{cm}^2/\mathrm{Vs}$, they are able to realize CMOS-type devices, such as those used in shift registers, operational amplifiers or A/D converters. Interest in all these circuit applications with poly-Si TFTs was, however, diminished with the advent of the special long slim driver ICs for chip on glass technology, which requires less space than thin film solutions. A further blow was dealt to poly-Si technology by Liquid Crystals On Si (LCOS), detailed in the next chapter, where addressing is performed by monocrystalline high mobility transistors in a Si-wafer. As a result, poly-Si is now concentrating mainly on pixel switches in displays with a high pixel density, where it could also be advantageous to realize both the driving circuits and the pixel switches in the same poly-Si technology. A new application is emerging in the form of current sources for Organic LEDs (OLEDs), but only for large area displays in which the Si-wafer with the addressing circuitry underneath the OLED is too expensive.

15.1 Fabrication Steps for Top-Gate and Bottom-Gate Poly-Si TFTs

The basic structure of top- and bottom-gate TFTs was introduced in Figures 14.1(b) and 14.1(c). Poly-Si TFTs are mostly built as top-gate devices, because as we shall see, they reach a higher mobility. The fabrication process for the p- and n-channel poly-Si TFT in Figure 15.1 (Iberaki, 1999; Harrer *et al.*, 1999) will now be outlined, together with the most important features of these TFTs. Common to all processes is that they start out with an a-Si:H semiconductor, which is crystallized either thermally by high temperature processes at 600°C or by laser annealing, a low temperature process not exceeding a substrate temperature of 450°C. As the high temperature processes require expensive quartz substrates, they were virtually abandoned in favour of Low Temperature (LT) processes which can work with inexpensive soda lime glass. The fabrication to be described is of the low temperature category, with the process steps shown in Figure 15.2. On a cleaned glass substrate (Figure 15.2(a)), a SiO$_2$ or SiN$_x$ undercoat is deposited by PECVD, preventing the diffusion of sodium ions into the layers of the TFT. This blocking layer has a dominant influence on the stability of the threshold voltage V_{th}. Again by PECVD, a 50 nm a-Si precursor is deposited, and patterned into islands by dry etching. This thin a-Si layer later ensures laser crystallization continues to the very bottom of the layer, where the channel is formed, and where a high mobility is thus required. Then, a SiO$_2$ buffer is deposited (Figure 15.2). A dehydrogenation at 450°C in vacuum leaves < 1 at percent H in the a-Si film, thus avoiding hydrogen explosions and ablations during the succeeding laser crystallization. The buffer serves four purposes: passivation of the a-Si semiconductor, reducing impurities at the interface; stopping layer for the succeeding ion implantation; a coupling layer for the laser beam, reducing dose; and a gate dielectric. The subsequent crystallization with an excimer laser generates crystallized grains, enhancing mobility. Details on laser crystallization are given in Section 15.2. An advantage of the top-gate TFT is that laser annealing takes place with a-Si as the bottom layer. This avoids dealing with different thermal expansions of a set of layers, leading to cracks and ablation. For this reason, some processes (Hasser *et al.*, 1999) deposit the SiO$_2$ gate dielectric only after laser annealing. Sputtering and patterning provides the gate electrode (Figure 15.2(b)) of MoW, CrAl or poly-Si. MoW has a specific resistance of 12 $\mu\Omega$ cm. The gate electrode serves as a mask for the self-aligned ion implantation of poly-Si with B or P to form n$^+$ poly-Si, where low resistance drain and

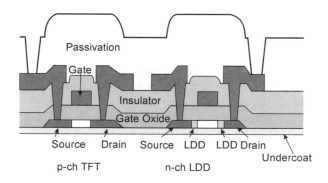

Figure 15.1 Cross-section of a p-channel and n-channel poly-Si TFT

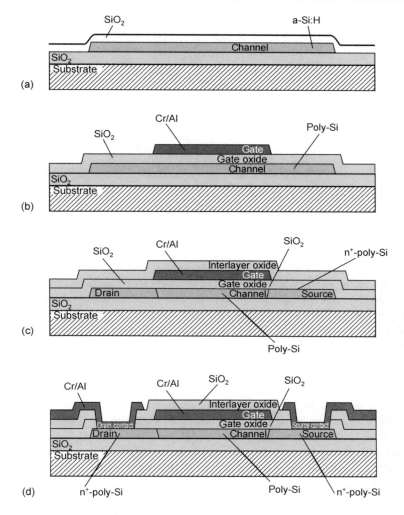

Figure 15.2 Fabrication steps of a top gate poly-Si TFT

source contacts will later be formed. This applies for both p- and n-channel TFTs. In the case where an SiO_2 stopping layer is placed on top of the poly-Si-film, the maximum dopant concentration of about $10^6/cm^3$ occurs at the interface between SiO_2 and poly-Si. The self-aligned doping ensures a minimum overlap of the gate and the drain and source electrodes. A small overlap will occur due to the scattering of the dopant ions at the edge of the gate mask. This small overlap minimizes parasitic gate-source and gate-drain capacitor, minimizing crosstalk and speeding up the dynamics of the TFT.

Doping with B is also a most effective means to adjust the threshold voltage V_{th} of n-channel and p-channel TFTs by changing the intrinsic charge in the channel. According to Equation (14.8), this charge directly influences V_{th}. Figure 15.3 (Iberaki, 1999) shows the dependence of V_{th} of an n-channel TFT and of a p-channel TFT on the B dose per cm^2. The

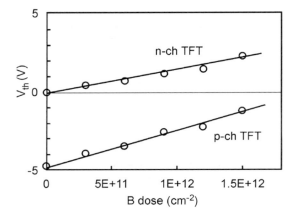

Figure 15.3 V_{th} dependant on the dose of B doping in the channels of n-channel and p-channel poly-Si-TFTs

two thresholds can be independently controlled. This allows us to correct for other influences on V_{th}, such as the grain boundaries, the gate oxide and the undercoat SiN_x or SiO_2 film. A final SiO_2 passivation layer is deposited on top of the gate (Figure 15.2(c)).

The area of the n-channel TFT next to drain and source has to be lightly doped with B in order to establish potential barriers to impede the flow of holes. The result is called LDD, standing for Lightly Doped Drain. Details on LDD are given in Section 15.3. For an n-channel TFT, LDD has the most desired effect of decreasing I_{off} for negative gate-source voltages, where holes can flow. The activation of the dopants is done by laser irradiation with a dose of about $400\,mJ/cm^2$.

The via holes for the drain and source contacts are selectively dry etched down to poly-Si, which is followed by sputtering and patterning of metals such as Al/Nd for the drain and source contacts in Figure 15.2(d). The Nd content in the Al/Nd alloy prevents the formation of hillocks during hydrogenation and final anneal. Hydrogenation is needed for the passivation of grain boundaries.

Figure 15.4 (Iberaki, 1999) shows the transfer characteristics of an n-channel TFT with the rather high mobility of $235\,cm^2/Vs$, and of a p-channel TFT with the lower mobility of $120\,cm^2/Vs$, as expected.

Figure 15.5 depicts the fabrication steps of an n-channel bottom-gate transistor (Harrer *et al.*, 1999). The process starts with sputtering and patterning of Mo/Ta gate (Figure 15.5(a)), followed by PECVD of the trilayer: SiO_2 as gate dielectric, a-Si:H as semiconductor and a thin SiO_2 capping layer, all done in one vacuum run (Figure 15.5(b)). After dehydrogenation, a photoresist mask is used for implantation with P-ions, with SiO_2 as the stopping layer (Figure 15.5(c)). Now the laser crystallization and activation of the implant is performed simultaneously, thus requiring only one laser process in the entire fabrication. After patterning the semiconductor islands and coating with a second SiO_2 layer, RIE is used to open the drain and source via holes (Figure 15.5(d)). Then the Al/Cr drain and source electrodes are sputtered. Hydrogenation in an electron cyclotron (ECR) plasma completes the process. Electron mobilities of $190\,cm^2/Vs$ can be achieved. As a rule, mobilities of a bottom gate TFT are lower, as laser crystallization of a-Si in Figure 15.5 in the presence of a set of layers with different thermal expansion coefficients must be done at a reduced dose.

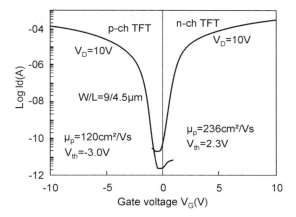

Figure 15.4 Transfer characteristics of n- and p-channel poly-Si TFTs

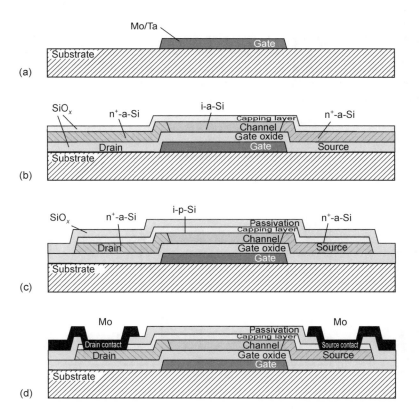

Figure 15.5 Fabrication steps of a bottom gate poly-Si TFT

15.2 Laser Crystallization by Scanning or Large Area Anneal

Laser crystallization is performed either by the conventional laser scanning (Lambda Physik, 2000), or by the recently introduced large area annealing (Prat *et al.*, 1999), otherwise known as Single area Excimer Laser Crystallization (SELC).

Laser scanning as a rule uses an XeCl excimer laser, emitting at 308 nm. The laser beam with an area of, typically, $8\,mm^2$ scans the a-Si layer. The intensity profile of the beam approaches a Gaussian. At its centre, temperatures of up to 1400°C are reached in a few ns, melting the Si (Sun *et al.*, 1994). In this region the largest crystallized grains are formed. The scan lines overlap by about 1 mm, in which region smaller grains with a lower mobility are obtained. These regions have a higher reflection coefficient for laser light, reducing the crystallization during a second exposure to laser irradiation, which results in stripes in the poly-Si with higher and lower mobilities (Tanaka *et al.*, 1993). The dose of the first laser scan is typically $270\,mJ/cm^2$ followed by a second scan of $450\,mJ/cm^2$. This provides grain diameters up to 80 nm and mobilities from $150\,cm^2/Vs$ to $200\,cm^2/Vs$, which are also high enough in the low mobility zones (Yamamoto *et al.*, 1992; Brotherton *et al.*, 1997).

The grain size depends upon the laser energy density, as demonstrated in Figure 15.6. The region with the largest grains is called the Super Lateral Growth (SLG) region. Favourable growth conditions are encountered if the seeds from which growth originates are not too densely distributed, and if the a-Si is thick enough, leaving room for lateral and perpendicular growth. If these conditions are met, grain size increases up to the peak in Figure 15.6. Beyond the peak there are a few scattered large grains separated by microcrystals. This decreases the average grain size and the associated mobility. Fabrication, however, should not take place in the SLG region, as small fluctuations in the laser dose result in a large non-uniformity of mobility. Fabrication works best in a region just before the peak (Okabe, 1996).

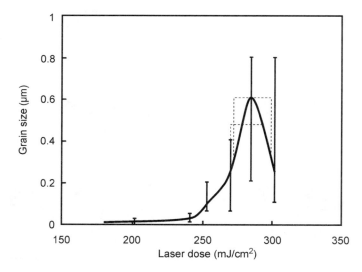

Figure 15.6 Grain size with super lateral growth region dependant on the scanning laser dose

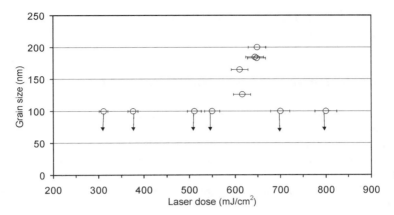

Figure 15.7 Grain size with super lateral growth region dependant on large area laser dose

An excimer laser with Xe, Ne and HCl gases was developed to provide a powerful pulse with an energy of up to 18 J and a 1.5 Hz train of pulses with an energy density up to 800 mJ/cm^2 (Prat *et al.*, 1999; Maresch *et al.*, 1999). The pulses extend over an area of 68 × 27 mm in which the energy distribution exhibits a uniformity with only ±2.4 percent fluctuations. Around the boundary of this area is a 500 µ wide stripe in which the energy decays to zero. For homogeneous discharge, the gases are pre-ionized by X-rays and ignited by voltages up to 24 kV. The shot-to-shot stability of the pulse is 2 percent in the range of 20–26 kV. The mesa-shaped pulse is able to anneal a large area of a-Si in one 210 ns shot avoiding non-uniformities along scan lines. These non-uniformities occur infrequently along the edges of the mesa.

Figure 15.7 depicts the grain size achieved versus the incident laser energy density. Grain sizes can be enhanced if several shots are applied. If two shots of 685 mJ/cm^2 followed by

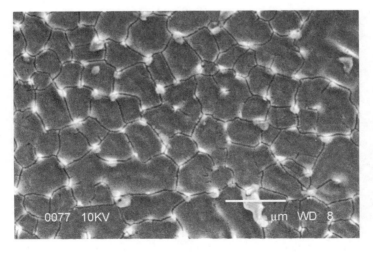

Figure 15.8 REM photograph of grain size after four large area laser shots

two shots of $640\,\mathrm{mJ/cm^2}$ are administered, the average grain size is, according to Figure 15.8, extended to 600 nm.

15.3 Lightly Doped Drains for Poly-Si TFTs

Three possibilities are examined to generate the two lightly B-doped areas in the n-channel TFT in Figure 15.1.

In the photoresist reflow method in Figure 15.9(a) a light ion implantation takes place with the photoresist masking the channel area beside the LDD stripes close to drain and source. A bake of the photoresist at 180°C for 1 hour expands the resist by reflow over the LDD areas, as depicted in Figure 15.9(b). Then the implant of the contact areas with a heavier dose is performed.

The method in Figure 15.10(a) uses a Ta mask, the outer region of which was anodized to Ta_2O_5. This mask covers the entire channel area and the heavy implantation only takes place at the contacts. After etching the Ta_2O_5 a light implantation of the now exposed LDD areas in Figure 15.10(b) takes place.

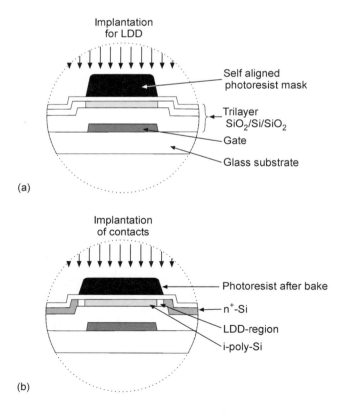

(a)

(b)

Figure 15.9 The reflow method for generating LDD and n^+ areas: (a) implantation before and (b) after reflow of photoresist

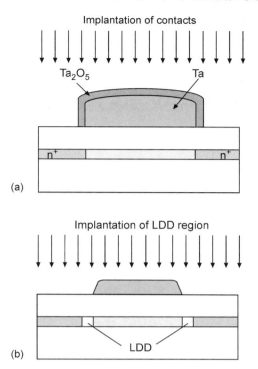

Figure 15.10 Implantation with an anodized Ta mask (a) for contacts and (b) with Ta_2O_5 etched off for LDD

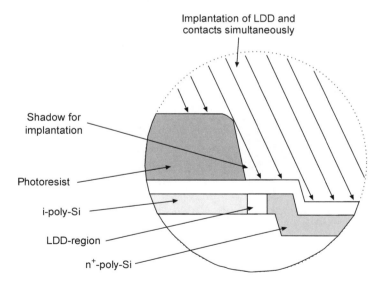

Figure 15.11 Oblique implantation for LDD and contacts simultaneous

Implantation at an oblique angle in Figure 15.11 (Harrer, to appear) uses the photomask to shadow the LDD region, reducing the implantation to a light one while simultaneously implanting the contact areas with a heavier dose.

15.4 The Kink Effect and its Suppression

An FET operating in saturation exhibits a large voltage drop in the channel region near the drain where the channel is depleted.

Charge carriers being accelerated in the high field region impact with uncharged particles, generating an avalanche of carriers. The electrons and holes produced by this impact ionization are able to flow off to ground in monocrystalline Si ICs, because the Si-wafer substrate is connected to ground. For TFTs the substrate is glass, preventing the carriers from flowing to ground, and thus increasing the drain-source current. This leads to degraded grey shades in a display. This phenomenon is called the kink effect. In an n-channel TFT, the holes are accumulated in the region below the channel, because they cannot escape through the insulating substrate this region. In a p-channel TFT the electrons accumulate.

The remedy is to extend the channel area in Figure 15.12 (Yoo *et al.*, 1999) and to counter dope this extended area p^+ in the case of an n-channel TFT. The extension works as a Lateral Body Terminal (LBT). After activation of the p^+ implantation, it serves as a collector of the holes and fixes the potential at zero. As a consequence, the increase in I_D is suppressed, the suppression increasing with increasing width, W_B of the LBT in Figure 15.12. A further beneficial effect of LBT is the decrease of the leakage current I_{off} (Yoo *et al.*, 1999).

Finally, LBT enhances the stability of the device by slowing down the decrease of mobility and the negative shift of V_{th} with increasing stress time. The mobility decreases due to the generation of defects on top of the gate dielectric, caused by lattice scattering of electrons when a high kink current is flowing. As the LBT suppresses the kink current, the decrease of mobility is also diminished. The negative shift of V_{th} is induced by hole trapping

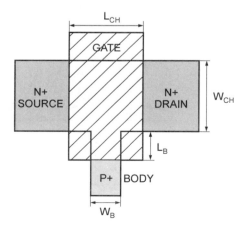

Figure 15.12 The Lateral Body Terminal (LTB) for the reduction of the kink current

in the dielectric. A poly-Si TFT with LBT collects the excess holes, and thus suppresses the hole trapping in the dielectric. This reduces the negative shift of V_{th}.

15.5 Circuits with Poly-Si TFTs

An economically attractive example for the application of poly-Si TFTs is a high resolution display which requires poly-Si pixel switches, and where the scan-line drivers and the data

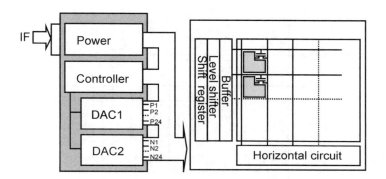

Figure 15.13 The circuit concept for a 202 dpi display with poly-Si drivers on glass and further ICs on an external circuit board

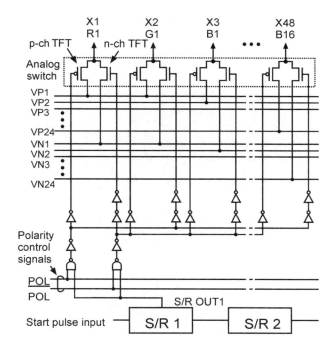

Figure 15.14 The circuit concept of a high resolution display

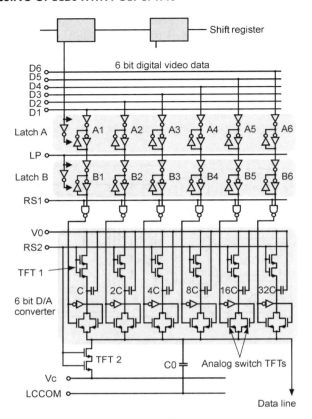

Figure 15.15 Circuit diagram of a 6-bit poly-Si data driver

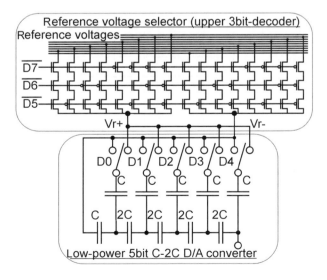

Figure 15.16 A 5-bit D/A converter with γ correction in poly-Si technology

drivers are realized simultaneously in poly-Si technology using the same masks. Figure 15.13 shows the circuit concept for such a high resolution 202 ppi 4" diagonal display with $640 \times 3 \times 480 = 0.92$ M pixels $42\,\mu \times 126\,\mu$ in size. (Manazawa *et al.*, 1996). The buffer, the level shifter and the shift register of the scan line driver and the data driver are placed along the edges, thus reducing about 2400 contacts from the display to the outside to about 100. The video sampling circuit which precedes the data driver shown in Figure 15.14 generates polarity inversion from column to column. To this aim, a p-channel poly-Si TFT samples and holds a positive video signal, whereas an n-channel TFT does the same for a negative video signal.

The 6-bit data driver in Figure 15.15 (Matsueda *et al.*, 1996) shows the data in the latches A_1 through A_6 controlled by the shift register. Timing pulses transfer these data through the latches B_1 to B_6, then through TFT1s to six MOS capacitor cells, and finally, through the analog switch TFTs to the data lines. The use of poly-Si TFTs allows for 25 MHz digital video data. The 5-bit D/A converter with γ correction in Figure 15.16 (Lee *et al.*, 1998) consists of a reference voltage selector operating with TFT switches and a C-2C DAC. The γ correction is achieved by varying the reference voltages with the poly-Si-TFT switches.

The circuits for driver ICs presented in Chapter 14 can also be implemented with poly-Si TFTs.

16

Liquid Crystal on Silicon Displays

Reflective liquid crystal displays that are formed on a silicon wafer, which contains the addressing circuits for each pixel, are called Liquid Crystal on Silicon, (abbreviated to LCOS) displays. Addressing circuits made from conventional monocrystalline Si offer all the advantages of this technology, such as electron mobilities of $800\,\text{cm}^2/\text{Vs}$, hole mobilities of $600\,\text{cm}^2/\text{Vs}$ and feature sizes in the sub-µm region. For displays, well established high yield Si technology has to be complemented by a metallic light shield on top of the Si devices, preventing light-induced formation of electron-hole pairs and doubling as a capacitor electrode. A very flat, highly reflective metallic mirror is also required for the pixel electrodes of these reflective displays. Applications for LCOS are micro-displays, also called virtual displays, and light valves for projectors, as will be discussed in this chapter and in Chapter 24.

16.1 Fabrication of LCOS with DRAM-Type Analog Addressing

The cross-section of LCOS in Figure 16.1 (Sanford *et al.*, 1998) shows in the lower half the pixel transistor and the storage capacitor and in the upper half the mirror (third level of metallization) and the light shield (second level of metallization). The circuit diagram is shown in Figure 16.2.

The transistor is an n-well on p-epi Si substrate 5V n-MOS FET device with a 450 µm gate oxide and a poly-Si gate. The storage capacitor C_G has poly-Si electrodes as well. As a DRAM (Dynamic Random Access Memory) has a similar configuration, consisting of a transistor and a capacitor as a memory, the addressing circuit is called a DRAM-type circuit. This addressing scheme is, as we know, based on pulse amplitude modulation. Hence, the

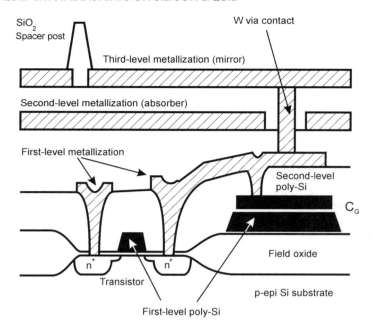

Figure 16.1   Cross-section of LCOS addressing devices

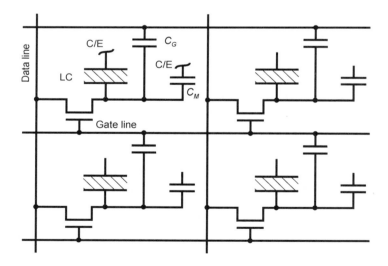

Figure 16.2   Pixel FETs and storage capacitor C_G of an LCOS array

DRAM-type circuit is associated with analog addressing. It contrasts with the SRAM-type circuit for digital addressing, which is discussed below.

The light shield (Colgan and Uda, 1998) is a TiN absorber with a reflectivity of 20 percent for blue and 65 percent for red. Light leaking into the channel between the mirror and absorber at an angle of 7° would be reflected 100 times, and would reach the opening in the absorber with an intensity reduced by a factor of 10^{-19}. This low level can no longer photoactivate the Si.

The electrodes of the first level of metallization in Figure 16.1 underlie the via hole as an additional barrier for light incident through the hole. These measures underline the importance of shielding the Si from light.

The light output of the reflective cell is maximized if the mirror covers almost the entire pixel area, rendering an aperture ratio of at least 80 percent. Further loss of light by scattering is prevented by a very flat surface of the mirror, made usually of Al. This is achieved by using a chemical-mechanical polishing technique to flatten the insulating substrate onto which the mirror and absorber layers are deposited. This layer, which is often made of SiO_2, is not shown in Figure 16.1. Hillocks in the Al mirror are suppressed by alloying the Al with Cu, and by annealing the alloy at 400°C. The combination of an absorber and mirror adds an additional capacitor C_M parallel to the LC capacitor in Figure 16.2, because the absorber is connected to the potential C/E (common electrode) of the backplate.

Finally, an etched spacer, positioned accurately in the interpixel gap, is shown in Figure 16.1. The reflective display is, as a rule, a TN display.

As the gate lines consist of poly-Si, their conductivity is much lower than for conventional Al lines; further, the row capacitance is increased by the capacitance between mirror and absorber. To counter the effect of this parasitic RC line, the rows are, as a rule, driven from both sides, decreasing the time skew by a factor of four. To enhance speed, the succeeding gate-line is precharged before the preceding one turns off. In Melcher et al. (1998), a 2048×2048 array of 4.194 million $17\,\mu \times 17\,\mu$ pixels was described. The Si-chip size was 64 mm × 64 mm. This small display size with a very high pixel density started the development of microdisplays (Melcher et al., 1998; Sato et al., 1997; Cacharelis et al., 1997).

16.2 SRAM-Type Digital Addressing of LCOS

SRAM (Static Random Access Memory) digital addressing was originally developed for Digital Micro Mirror Devices (DMDs) used for projection systems (Sompsell, 1994a, 1994b; Tew et al., 1994), and was adapted for micro-displays (Cacharelis et al., 1997; Clark, 2000). Directly underneath each pixel of a micro-display, an SRAM cell is located in the Si wafer, and is loaded time sequentially with the bits of a digital word representing the grey shade. The bits with the values 1 or 0 have to be translated into a grey shade of the pixel. This is achieved by Pulse Width Modulation (PWM). The number G of grey shades associated with an m bit word is

$$G = 2^m - 1, \tag{16.1}$$

which is $G=31$ for $m=5$ or $G=255$ for $m=8$. These grey shades are generated by the width of a voltage pulse across a pixel during the row address time

$$T_r = T_f/N. \tag{16.2}$$

The brightest pixel is generated if the pulse lasts the time T_r and the darkest grey shade if the pulse lasts the time

$$t_s = T_r/G, \tag{16.3}$$

corresponding to the Least Significant Bit (LSB); t_s is called the slot time. For a picture frequency of $f_p = 60$ Hz the frame time $T_f = 1/f_p = 16.66$ ms, and for a 1280×1024 pixel display (SXGA) the row address time is $T_r = 18.66$ µs; for 8-bit grey shades this leads to a slot time of $t_s = 73$ ns. The addressing of a pixel in the selected row is shown in Figure 16.3 (Clark *et al.*, 2000) for 5-bit grey scales as an example. The time slots are indicated by dots. The Most Significant Bit (MSB) has $2^4 = 16$ time slots, in which the pixel is bright if the SRAM exhibits a 1 and black for a 0. The addressing circuit has to hold the voltage pulse corresponding to the MSB for a pulse width of 16 time slots. The SRAM feeds in the next bit below the MSB, which determines the corresponding pulse width, which in the example is $2^3 = 8$ time slots. This continues in Figure 16.3 down to the LSB $2^0 = 1$, with 1 time slot indicated by 0 on top. In this way, the display is fully digitally addressed line after line. The result is the analog value in the form of the width of the voltage pulse, resulting in the grey shade corresponding to the m bit word fed in.

As a rule, two SRAM cells per pixel are required, one establishing the voltage across the pixel, while the other is loaded with the next bits. Around the 1 bit LSB time slot in Figure 16.3, the numbers 1 and 2 underneath the slots indicate which of the two SRAMs is connected to the pixel. This demonstrates that the fastest reprogramming of an SRAM has to be achieved within two slots. The data transfer of 1 bit in the time $2t_s$ results in the bit rate per pixel of

$$BR_p = \frac{1}{2t_s},$$ (16.4)

and for all the M pixels in a line to be addressed simultaneously, we obtain

$$BR = \frac{M}{2t_s},$$ (16.5)

or with Equations (16.1), (16.2) and (16.3),

$$BR = \frac{MN(2^m - 1)}{2T_f}.$$ (16.6)

For the example given above, Equation (16.6) yields for $m = 8$ $BR = 1.0$ Gb/s. As a rate of 100 Mbit/s can presently be handled the rate for a 1280×1024 display with 8-bit grey shades is far too high. Equation (16.6) reveals that for $m = 8$ and $BR = 100$ Mbit/s, one can handle $MN = 13$, an uninteresting display. Therefore, we have to consider a means to reduce the bit rate, or in other words, the bandwidth. As each pixel requires two SRAMs, one needs a total of two MN SRAMs. This number has also to be reduced in order to enhance fabrication yield.

Figure 16.3 The time slots for a pixel with 5-bit grey shades. Numbers ν on top indicate weight 2^ν of a bit; numbers s below indicate if SRAM number 1 or number 2 feeds in the information

A variety of remedies to decrease bandwidth and, in some cases, also the memory count have been proposed (Clark *et al.*, 2000). An easy solution is to cut the picture frequency to 30 Hz, thus increasing t_s. If 60 Hz have to be kept in order to be compatible with CRT frame rates, the following techniques are helpful.

Deleting the LSB in a given *m*-bit word for the grey shades while keeping the weights of the bits decreases the number of bits to $m - 1$, and therefore also decreases the bit rate in Equation (16.6). The value of the grey shade is not changed as long as the deleted LSB is zero. If it is 1, the value of the *m*-bit word is both rounded up by adding 1 to the LSB of the $(m - 1)$-bit words, and in the next frame rounded down by putting it to zero and deleting it. In this way, the average of the $(m - 1)$-bit words in two frames represent the correct grey shade of the *m*-bit word. As an example, we consider the 4-bit word $1001 \hateq 9$, which is represented in the first frame by the 3-bit word with the weights of the 4-bit word, namely by $101 \hateq 10$ and by $100 \hateq 8$ in the second frame with the average 9, as desired. This procedure is called a *frame-to-frame dither*. It can be extended to the deletion of additional less significant bits, thus generating a larger reduction in Bit Rate (BR) as desired. If one bit is deleted, the bit rate in Equation (16.6) is roughly halved.

As known from Pulse Amplitude Modulated (PAM) LCDs with an analog addressing scheme, block parallel addressing in Chapter 14 and in Figure 14.14 is a powerful means of decreasing the required bandwidth. A display with *M* columns is vertically divided into *B* blocks or banks with M/B columns each. As explained in Chapter 14, the simultaneous or parallel addressing of the *B* blocks reduces the addressing frequency or the required bandwidth by a factor of *B*. The bit rate BR_b for block addressing is

$$BR_b = \frac{BR}{B} = \frac{MN(2^m - 1)}{2BT_f}. \tag{16.7}$$

Figure 16.4 shows the digital block parallel addressing for $m = 5$ bit words and $B = 4$ blocks. Each line in Figure 16.4 represents a block with $2^5 - 1 = 31$ time slots. The numbers 4, 3, 2, 1 and 0 indicate when a new bit with a different weight becomes active. At those times the information of the individual SRAMs of the pixels is needed. As the SRAMs are needed simultaneously, they cannot be shared among the pixels.

The sequence of the weighted bits is arbitrary, as only the sum of their time slots determines the width of the voltage pulse, and hence the grey shade. The addressing scheme in Figure 16.5 (Tew *et al.*, 1994) makes use of this observation, and arranges the onsets of a new weight for all blocks in such a way that they do not occur at the same time. As a result, only one onset takes place at any given time. This cuts the bandwidth required in Figure 16.5 again by a factor of four from the previous scheme in Figure 16.4 and in Equation (16.7). In

```
1st block:  4 * * * * * * * * * * * * * * 3 * * * * * * * 2 * * * 1 * 0
2nd block:  4 * * * * * * * * * * * * * * 3 * * * * * * * 2 * * * 1 * 0
3rd block:  4 * * * * * * * * * * * * * * 3 * * * * * * * 2 * * * 1 * 0
4th block:  4 * * * * * * * * * * * * * * 3 * * * * * * * 2 * * * 1 * 0
```

Figure 16.4 Block parallel addressing with four blocks and the time slots for 5-bit grey shades in each block

1st block: 4 * * * * * * * 3 * * * * * * * 0 1 * 2 * * * 4 * * * * * *
2nd block: 4 * * * * * * * * 2 * * * 0 3 * * * * * * * 1 * 4 * * * * * *
3rd block: 4 * * * * * * * * * 1 * 3 * * * * * * * 0 2 * * * 4 * * * * *
4th block: 0 1 * 2 * * * 4 * * * * * * * * * * 3 * * * * * * * 4 * * * *

Figure 16.5 Digital block addressing in which SRAM outputs are required at different times

general, the reduction in the case of B blocks is by a factor of B leadings to a bit rate for the irregular scheme in Figure 16.5 of

$$BR_{bi} = BR_b/B = \frac{MN(2^m - 1)}{2B^2 T_f}.$$ (16.8)

As all the SRAMs are updating their pixels at different times, they may be shared among the pixels. This lowers the SRAM count, and thus enhances the fabrication yield.

So far we have grouped adjacent rows into blocks. The grouping can, however, be done in many different ways. We may assign, for example, each fourth row to one block, or we may select even more irregularly numbered rows for one block. These interleaved assignments to blocks generate fewer artifacts.

The bandwidth can be decreased even further by cutting a pixel into two or more subpixels. We investigate this possibility for $m=6$ bit words for grey shades, and by the two subpixels in Figure 16.6 with areas exhibiting the ratio $8:1$. The weights of the 6-bit word are

$$\underbrace{2^5 \, 2^4 \, 2^3}_{56} \quad \underbrace{2^2 \, 2^1 \, 2^0}_{7}.$$

The first part of the word with the total weight of 56 addresses the pixel with area 8, whereas the second part with a total weight 7 addresses the pixel with area 1. The ratio of the weights $56:7=8:1$ is the same as the ratio of the areas. The three bits with the lower weight require seven time slots to represent the non-zero values $1, 2, , 6, 7$. Those seven values cause area 1 of the smaller subpixel to be bright, thus generating seven degrees of brightness from 1 to 7. The three bits with the higher weight again require seven time slots to represent their non-zero values 8, 16, 24, 32, 40, 48 and 56. The bits, however, by which they address the larger part of

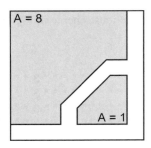

Figure 16.6 Addressing of two subpixels with area ratio 8:1

the pixel, are only 1s and 0. The larger area is multiplying them by a factor of eight, yielding the series of grey shades from 8, 16 up to 56. The values in between are realized by the smaller subpixel. The decreased bit rat BR_{sbi} for subpixelation applied together with block parallel addressing and the irregular scheme is derived from Equation (16.8) by replacing m by $m/2$ and doubling the expression in Equation (16.8) due to the two bit streams to the subpixels. This yields

$$BR_{sbi} = \frac{MN(2^{m/2} - 1)}{B^2 T_f}. \tag{16.9}$$

This represents a substantial reduction of the bit rate in Equation (16.8) by a factor $2^{m/2} - 1$.

If an m-bit word is divided into two parts with m_1 and m_2 bit, respectively, where

$$m_1 + m_2 = m, \tag{16.10}$$

we obtain from Equation (16.8) the bit rate BR_{2part} as the sum of the bit rates associated with m_1 and m_2 as

$$BR_{2part} = \frac{MN}{2B^2 T_f}(2^{m_1} + 2^{m_2} - 2). \tag{16.11}$$

The very strong effect of the subpixellation can of course also directly be applied to Equations (16.7) or (16.6). Another explanation for the decreased bandwidth is the increase of the slot time t_s in Equations (16.1) and (16.3) with decreasing m.

The time sharing of memory among blocks and pixels may lead to fixed pattern noise, stemming from bit weighting appearing as artifacts. These effects are associated with rapid eye movements, resulting in strobing effects, where large bit weight changes are perceived as spatial intensity differences. An irregular pattern of sharing memory is, as a rule, preferable.

Figure 16.7 depicts a circuit diagram for an SRAM cell with a shadow latch (second latch) and a column driver.

The addressing of TN displays requires a periodic alteration in the sign of the pixel voltage. If a bistable display such as an FLC is digitally SRAM addressed, this alteration

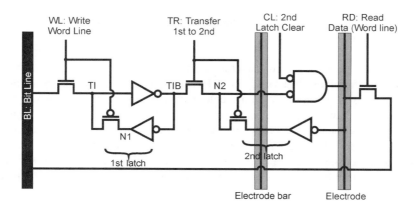

Figure 16.7   An SRAM cell

in sign is not feasible as the on- and off-states are each reached with a given sign. AC-free operation may be achieved (Clark *et al.*, 2000) by adding a frame with inverted contrast (which is made invisible by turning off the illumination) by an electro-optic efficiency doubler, or by applying a modified dc-free Harada addressing scheme.

16.3 *Microdisplays Using LCOS Technology*

A microdisplay consists of a finger-nail sized reflective LCOS as the picture source, and illumination and magnifying optics (Schott *et al.*, 1999). Figure 16.8(a) and 16.8(b) depict such systems with a simple magnifier as is used for magnifying glasses, and with a folded magnifier derived from a microscope. The latter is required for magnifications greater than 10,

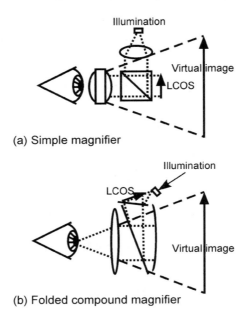

(a) Simple magnifier

(b) Folded compound magnifier

Figure 16.8 The components of a virtual display, (a) with the optic of a magnifying glass, and (b) with the optic derived from a microscope

which is needed to generate a virtual picture with a diagonal of 10" at a distance of 18". A typical value for the picture frequency is 85 Hz. Resolution required for the display is decreased by the use of a frame sequential colour presentation, with a frequency of 255 Hz reflecting in a frame time of 3.9 ms per colour. This small value leads to an input bandwidth of 300 MHz. This requirement is cut down by block parallel addressing with, for example, four blocks. As a rule, SRAM-type addressing is chosen for microdisplays due to its lower voltage and lower power consumption than DRAM addressing, if the measures reducing the SRAM cell count discussed in Section 16.2 are realized. SRAM circuits require more chip area than the DRAM approach. To render it feasible, only 3 bit grey shades have been realized so far.

Projectors, the second application for LCOS, will be discussed in Chapter 24.

17

Addressing of Liquid Crystal Displays with Metal-Insulator-Metal Pixel Switches

Metal-Insulator-Metal (MIM) devices are two terminal components with the structure of a capacitor as shown in Figure 17.1. The importance of MIMs as pixel switches has, for displays with glass substrates has declined during the last ten years in favour of TFTs. Due to their low fabrication temperature, however, they became, attractive for displays with plastic substrates. Therefore, the essentials of MIMs will be discussed.

The dielectric of a MIM, usually Ta_2O_5, differs from a conventional insulator in exhibiting a large number of defects and charge donors. The donor can release a trapped electron under the influence of an external field. During the release, the field has to overcome a coulombic potential barrier. After the electron has left, the donor is left behind with a positive charge. The defects, on the other hand, are neutral as long as they did not trap an electron, and hence become negatively charged when an electron is trapped. In a defect there is no coulombic potential barrier which the field has to overcome when an electron is released. Both effects lead to the electron current increasing exponentially with the applied electric field. The relationship between the voltage V_{mim} across the MIM and the current I_{mim} through the device can be derived by investigating the current transport mechanisms in such dielectrics (Simmons, 1967). The result is

$$I_{mim} = \sigma V_{mim} e^{\beta \sqrt{|V_{mim}|}} \qquad (17.1)$$

with the conductance

$$\sigma = e\mu \frac{A}{d} \exp(-\phi_{st}/kT)$$

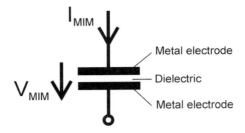

Figure 17.1   The capacitor structure of a MIM

and the gain factor

$$\beta = \left(\frac{e^3}{\pi n^2 \varepsilon_0 \varepsilon_r} \right)^{1/2} \frac{1}{\sqrt{dkT}} \tag{17.2}$$

where $e=$charge of electron; $A=$area of MIM; $d=$thickness of dielectric; $\mu=$electron mobility; $k=$Boltzmann's constant; ε_0 and $\varepsilon_r=$absolute and relative dielectric constant; $\phi_{st}=$ energy gap between trap levels and lower level of conduction band; $n=$density of charge carriers; and $T=$temperature in K.

β in Equation (17.2) applies if the current through the dielectric stems from donor-type traps. This effect is called the normal Poole–Frenkel effect. For a current originating from the defects, β in Equation (17.2) has to be replaced by $\beta_a = \beta/2$. The underlying effect is called the abnormal Poole–Frenkel effect. Finally, a third effect affects current flow (Shannon et al., 1993). The surface states between the dielectric and electrodes generate another potential barrier. At higher and increasing voltages, an increasing number of electrons is able to tunnel through this barrier, resulting in a rise in current that is steeper than for the Poole–Frenkel effects. In MIMs with Ta_2O_5 as the dielectric, the normal and abnormal Poole–Frenkel effect is dominant (Togashi, 1992; Hochholzer et al., 1994) whereas for PECVD SiN_x dielectrics, tunnelling plays an important role (Shannon et al., 1993).

Equation (17.1) is plotted in Figure 17.2. For an inverted electric field or an inverted voltage, all the effects are the same besides an inversion of the sign of the current. This fact is reflected in the dependance on $|V_{mim}|$ in the exponent in Equation (17.1) and which is shown in Figure 17.2 as a curve which is symmetrical about the origin. For positive voltages, the characteristic in Figure 17.2 is the same as for a diode. This is obviously not true for negative voltages. Rather, MIMs behave like a kind of a double diode. In Figure 17.3 the relation reduced by $I_0 = 1A$

$$\log I_{mim}/I_0 = \beta\sqrt{|V_{mim}|} + \log \frac{\sigma V_{mim}}{I_0} \tag{17.3}$$

is plotted versus $\sqrt{|V_{mim}|}$ exhibiting an almost linear dependence on $\sqrt{|V_{mim}|}$. A threshold $V_{th\,mim}$is arbitrarily defined to occur at $I_{mim} = 10^{-13}A$. Due to their nature as double diodes, MIMs operating as rectifiers for both polarities of the video voltage V_d in Figure 17.4 are able to serve as pixel switches. The narrow lines in Figure 17.4 indicate the components

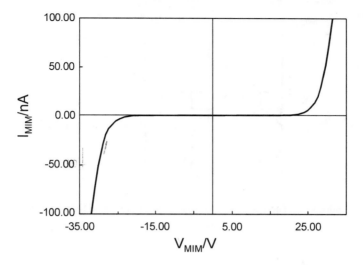

Figure 17.2 The $I_{MIM} - V_{MIM}$ characteristics

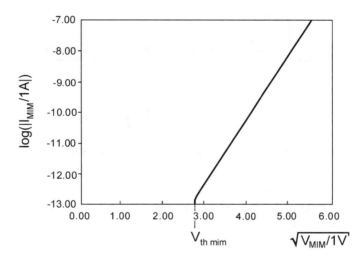

Figure 17.3 The log $I_{MIM}/I_0 - \sqrt{|V_{MIM}|}$ characteristics

placed on the glass substrate carrying the MIMs, whereas the grounded columns drawn with bold lines are located on the other glass substrate. After selecting a line by applying the voltage V_r, the voltage across the pixel in Figure 17.4 becomes

$$V_{LC} = V_x - V_d - V_{mim}. \qquad (17.4)$$

The line voltage is plotted in Figure 17.5, where V_s stands for the selection voltage and V_h for the holding voltage. Both alternate in sign together with V_d, as in TFT addressing. The video voltage consists of rectangular pulses, the height of which determines the grey shade,

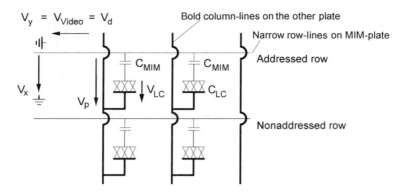

Figure 17.4   The addressing of pixels in one line at a time by MIMs

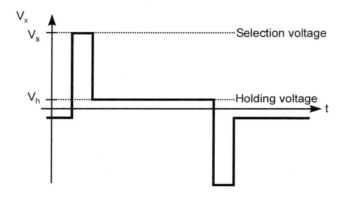

Figure 17.5   The line selection voltage for MIM addressing

again as for TFTs. The selection voltage V_s is chosen to lie at the mid point of the transmission curve in Figure 17.6 and is denoted by V'_s. The value V_s according to Figures 17.6 and 17.3 is

$$V_s = V_{\text{th mim}} + V'_s = V_{\text{th mim}} + V_{\text{th}} + \frac{1}{2}(V_{\text{sat}} - V_{\text{th}}) = V_{\text{th mim}} + \frac{1}{2}(V_{\text{sat}} + V_{\text{th}}). \qquad (17.5)$$

This choice ensures that the data voltage only has to cover the voltage range $\pm 1/2(V_{\text{sat}} - V_{\text{th}})$ in order to use the full linear part of the T-V curve. This is the same voltage range as for TFT addressing, allowing the use of the same data drivers as for TFTs. The holding voltage V_h in Figure 17.5 is chosen to be

$$V_h = \pm \frac{1}{2}(V_{\text{sat}} + V_{\text{th}}). \qquad (17.6)$$

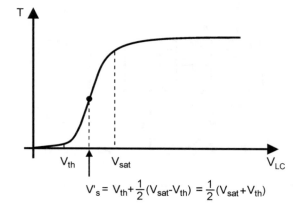

$$V'_s = V_{th} + \frac{1}{2}(V_{sat} - V_{th}) = \frac{1}{2}(V_{sat} + V_{th})$$

Figure 17.6 The transmission versus voltage curve of an LCD and the voltages V_{th}, V'_s and V_{sat}

This choice minimizes the voltage appearing across the MIM during the non-addressed periods of a line, causing a MIM current, and hence a damaging crosstalk (Togashi, 1992).

The addressing scheme discussed so far is called the four level scheme, derived from $\pm V_s$ and $\pm V_h$ in Figure 17.5 (Togashi, 1992). Two conditions have to be met for satisfactory operation. At the beginning of the line selection, the voltage across the MIM is, according to Figure 17.4,

$$V_{mim} = V_p \frac{C_{LC}}{C_{mim} + C_{LC}},\qquad(17.7)$$

where V_p stands for the pixel voltage in Figure 17.4. V_{mim} should be as large as possible to ensure a large current to charge C_{LC}. This requires

$$C_{mim} \ll C_{LC}.\qquad(17.8)$$

In non-addressed pixels the parasitic voltage

$$V_{LC} = V_p \frac{C_{mim}}{C_{mim} + C_{LC}}\qquad(17.9)$$

is crosstalk, and is also minimized by constraint (17.8). As a rule, C_{mim} is chosen as $C_{mim} = (0.1, \ldots, 0, 3)C_{LC}$.

As a consequence of constraint (17.8), small pixel sizes with small values of C_{LC} require unrealizably small values for C_{mim}. Thus, MIM addressing is not feasible for displays with small pixel sizes, such as occur in high resolution displays of HDTV or even XGA.

As $V_{th\,mim}$ in Figure 17.3 is rather large, V_s in Equation (17.5) will also become large, reaching values of ± 30 V. This requires more costly line drivers than for TFTs.

Further limitations for MIMs are parasitic capacitive couplings between the pixels, as outlined in Figure 17.7. Remedies are (as for TFTs) reversal of the sign of the data voltage line by line, column by column and frame by frame, and a large gap of up to 7 μm between the

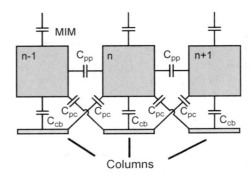

Figure 17.7 Capacitive couplings in a MIM-driven display

pixel electrodes. The latter, however, shrinks the aperture ratio. Finally, the small currents through a MIM at low voltages as visible in Figure 17.2 prevent the MIM addressed pixel from reaching its desired voltage in the row-address time T_r. This can also be demonstrated by numerically solving the highly nonlinear differential equation for the dynamics of a MIM (Möbus, 1990). A remedy would be a predistortion of the data voltage (Fuhrmann *et al.*, 1995), which would require a costly processor for the data dependent calculations. On the other hand, the current through a TFT at low drain-source voltages assumes, according to Figure 14.2(a), values that are still large enough to fully charge the storage capacitor within T_r.

The shortcomings of MIMs mentioned so far render their application for high resolution and high performance displays uncompetitive with TFT driven displays.

For low resolution displays with a limited number of about 32–64 grey shades, MIMs are still a viable alternative, mainly due to their inexpensive manufacture and their low process-ing temperatures and are suitable for displays with plastic substrates.

Figure 17.8 depicts and explains a two-mask fabrication for MIMs, the row and pixel electrode, and a self-aligned top ITO electrode of the MIM. This is the same ITO as for the pixel electrode, which saves one mask. The upper portion of the sputtered Ta bottom electrode is anodically oxidized to Ta_2O_5 in an electrolyte of 0.01% citric acid. The top electrode is patterned by exposing the image reversal photo resist from the back side to UV light using the Ta bottom electrode as a shadow mask. Light is scattered around the Ta edge, and penetrates into the resist. The depth of penetration (e.g. 0.3 µm) is controlled by the exposure time. Thus, a very small overlap of the electrodes of 0.3 µm ± 10%, and hence, as desired by constraint (17.8), a very small MIM capacitance is generated. The process is very economic as it requires only two masks, only sputtering instead of the more costly CVD, and only low temperature processes associated with sputtering are used (Hochholzer *et al.*, 1994).

Figure 17.9 shows the measured $I_{mim} - V_{mim}$ characteristics of Ta_2O_5 MIMs with various top electrodes. The curves for the third quadrant are plotted in the first quadrant as dotted lines. This reveals asymmetries of the MIMs which are very pronounced for ITO electrodes, and less so for Al and Ta. For Al and Ta top electrodes a third mask is required. The MIM character-istics shift with time. Shifts and asymmetries are successfully compensated by using a pair of MIMs connected back-to-back in Figure 17.10 instead of a single MIM. As the voltage at the two MIMs is applied in one case in the direction of the fabrication sequence of the layers, and in the other case in the opposite direction, the shifts and the asymmetries compensate each other. If need arises, the compensation may further be improved by selecting different positive and negative values of the holding voltage V_h.

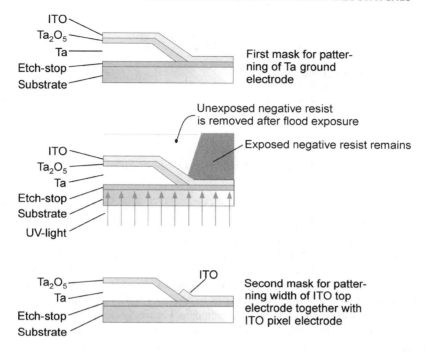

Figure 17.8 A two mask fabrication of Ta_2O_5 MIMs with self-aligned ITO top electrode

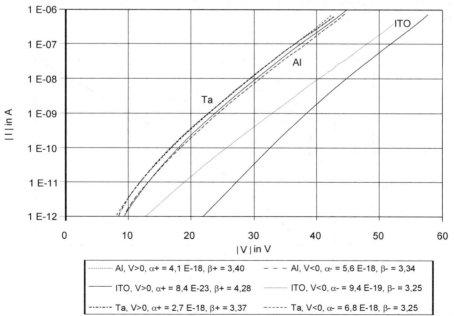

Figure 17.9 $I_{MIM} - V_{MIM}$ characteristic for MIMs with ITO, Ta and Al as top electrode (as indicated in the legend)

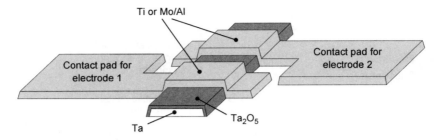

Figure 17.10 A back-to-back pair of MIMs

The threshold voltage $V_{\text{th mim}}$ can be shifted by changing the thickness of the Ta$_2$O$_5$ layer, which is achieved with different anodization voltages, extending from 30–80 V.

A steeper rise in I_{min} and smaller threshold voltages are observed for MIMs with PECVD SiN$_x$ as dielectric mainly due to the tunneling of electrons through the pronounced potential barrier at the interface between dielectric and electrode. This is demonstrated in Figure 17.11. These advantages are counterbalanced by use of the more costly PECVD process.

The pixel layout with a MIM is depicted in Figure 17.12. The columns, in contrast to the TFTs, are on the back plane, which is depicted in Figure 17.4, so no area is lost for the columns and gaps around them. This enhances the aperture ratio to an easily achievable 80 percent, rendering MIM displays the brightest AMLCDs. A further advantage is the insensitivity of MIMs to photoactivation.

Unfortunately, MIMs do not allow for the addition of a storage capacitor C_s between the upper ITO electrode of the pixel in Figure 17.4 and the succeeding line, as was possible for TFT displays. The reason is that the lines are never grounded, but lie at $\pm V_s$ or at $\pm V_h$. The lack of C_s may cause flicker.

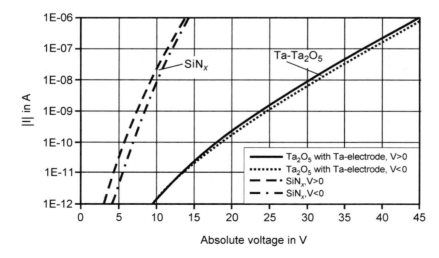

Figure 17.11 Characteristics of MIMs with SiN$_x$ as a dielectric in comparison to those with a Ta$_2$O$_5$ dielectric

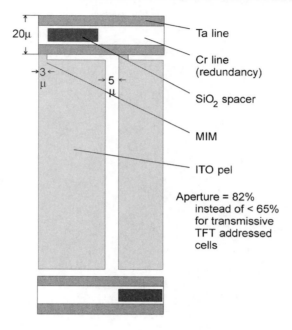

Figure 17.12 The pixel layout of a MIM display

The main advantages of MIMs, the high aperture ratio, the low fabrication cost and the low process temperatures, are counterbalanced by the expensive higher voltage line drivers and by all the above-mentioned shortcomings limiting MIMs to applications with a low pixel density and a small number of grey shades. The low process temperature associated with sputtering renders them appealing for displays with plastic substrates.

18

Addressing of LCDs with Two-Terminal Devices and Optical, Plasma, Laser and e-beam Techniques

In the early days of active matrix addressing, semiconductor diodes and nonlinear resistors called varistors were investigated, and as far as the diodes are concerned were also commercially used. As their importance has greatly declined, they are discussed only very breifly.

The antiparallel diodes in the pixels in Figure 18.1, also called a *diode ring* (Togashi *et al.*, 1984), possess the same characteristic as the MIMs shown in Figure 17.2 and are hence able to charge the pixel capacitor with both polarities. The fabrication of a thin film diode with a-Si as the semiconductor is almost as expensive as the fabrication of a TFT. As diodes suffer from similar shortcomings as MIMs, they have been virtually abandoned in favour of TFTs.

The varistor also exhibits current voltage characteristics similar to a MIM (Castleberry, 1979). As similar restrictions also apply, and as the substrate carrying the varistors is not transparent, varistor displays were not commercialized.

So far we have discussed only electronic addressing schemes for LCDs. In optical addressing the input is a still or a moving picture, usually displayed on a high resolution CRT. The image converter shown in Figure 18.2 (Bleha *et al.*, 1978) transfers this image with only a short delay of around 10 ms, into a copy on the output face. It can serve several purposes: it may operate as a light amplifier by increasing luminance of the picture, it may change the picture format; it may reproduce the picture using different wavelengths of light than the original, or it may convert an input with incoherent light into a coherent output. Applications use the image converter as a light valve for projectors, or as a source of information for optical signal processing.

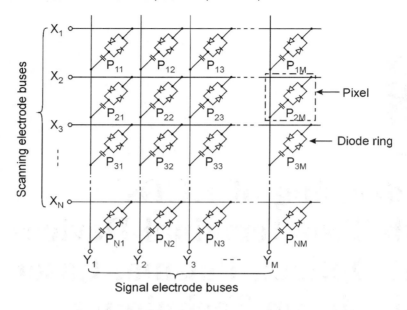

Figure 18.1 Liquid crystal pixels addressed by a diode ring

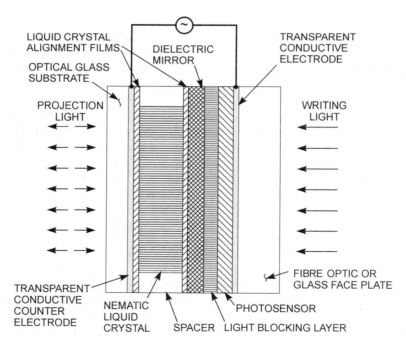

Figure 18.2 The optical image converter with an optical receiver on the right and an optical output LCD on the left

The image converter in Figure 18.2 consists of two glass plates and a variety of thin film layers sandwiched between them. An ac-voltage with amplitude 10 Vrms and a frequency of around 2–10 KHz is applied across the two transparent ITO electrodes. The layers on the right-hand side up to the light blocking layer, made of CdTe, represent the optical receiver, whereas the left-hand side up to the dielectric mirror, which is made of SiO_2 operates as a reflective LCD which displays the information received.

The receiver contains an a-Si photosensor which lowers its impedance in proportion to the intensity of the incident light. As a consequence, a larger fraction of the external ac voltage now appears across the liquid crystal cell, and is used for controlling the grey shade provided by the cell.

In the LCD the directors of the nematic molecules are aligned parallel to the plane of the orientation layers and to each other. They are twisted by 45° from the top to the bottom layer as in an MTN-cell. In Figure 18.3 light, for example from a Xenon lamp is linearly polarized, and shines obliquely onto the reflective display. The direction of the polarization is parallel to the directors of the top layer. The reflected light passes through the analyser in Figure 18.3. In the field-off state the polarization follows the twist of the LC helix to the dielectric mirror, where 98 percent of it is reflected and twisted back to the original direction at the input. In the crossed analyser in Figure 18.3, this light is blocked.

In the fully on state, the LC molecules align because $\Delta\varepsilon > 0$ parallel to the field. The incident and reflected light do not experience birefringence. Therefore, the reflected light is again blocked in the analyser. Between the two blocking states, at zero voltage and at maximum voltage there must be a state with maximum transmission through the crossed analyser. Calculations of the voltage dependent director field (Berreman, 1973) reveal a maximum transmission around 4 Vrms across the pixel for a layer thickness of 2.2 μm. An increasing pixel voltage starting from zero volts caused by an increasing luminance of the picture on the CRT ensures an increased transmission through the analyser.

A closer look at the equivalent circuit in Figure 18.4 for the image converter demonstrates how the pixel voltages are generated. The photosensor and light blocking layer form the heterojunction diode in Figure 18.4 with a capacitor C_3 and a forward resistor R_3 in parallel, dependent on the illumination from the CRT. With zero illumination $R_3 = \infty$; $R_3 + R_4$ is the illumination-dependent back resistor of the diode and C_2 the capacitor of the mirror, while R_1 parallel to C_1 stands for the liquid crystal. The circuit represents a rectifier in which

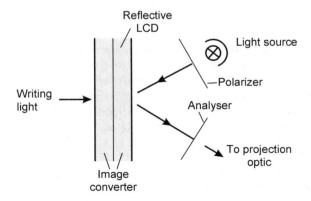

Figure 18.3 The light source and reflected light at the output of the image converter

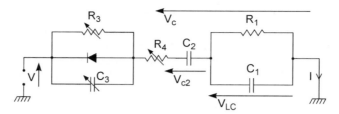

Figure 18.4   Equivalent circuit of the image converter in Figure 18.2

$R_1 \gg R_4$, R_3 is set to $R_1 = \infty$. Without illumination ($R_3 = \infty$) the diode has a lossless capacitive load C_2 in series with C_1 and C_3 in parallel, which is loaded to $V_c = V_{max}$ where V_{max} is the peak of the voltage V as depicted in Figure 18.4. V_{c_2} is part of this voltage. The other part is V_{LC}; in the unloaded case it has to be below V_{th} of the LCD. With increasing illumination the resistive load R_3 of the diode increases and V_{c_2} in Figure 18.4 decreases, with the consequence that $V_{LC} = V - V_{c_2}$ also increases. This effect controls the transmission of the LCD in proportion to the input illumination. The LCD and photosensor do not need to be pixellated, as the pattern of the luminance of the CRT is transferred without sampling to the photosensor, and from there to the director field of the LCD. This enhances the picture quality by avoiding all degrading effects of the spatial rectangular sampling by a pixel such as aliasing or not blending information with neighbouring pixels.

The application of this image converter as a light valve for a high quality projector will be discussed in Chapter 24.

The transfer of AMLCD technology to large area displays such as large area TV screens becomes difficult, because of the requirements to manufacture a large number of TFTs defect-free and with high uniformity over a large area. A remedy for this problem of manufacturing yield might be tiling of the displays which, however, suffers from the necessity of hiding the seam and of precisely aligning the tiles. Another possible solution is Plasma-Addressed Liquid Crystal (PALC) displays (Buzak, 1990; Kakizati *et al.*, 1996).

A plasma channel represents an excellent switch with a high switching ratio $R_{off}/R_{on} \geq 10^{12}$, whereas semiconductor switches struggle to reach 10^8.

The PALC display in Figures 18.5 consists of a regular pixellized LCD separated by a $50\,\mu\text{m}$ glass sheet, a micro-sheet, from the plasma channel and the backlight. The channels have silk screen printed spacers as well as anodes and cathodes. They run between the spacer ribs underneath the rows of the LCDs, are filled with a gas at a pressure of $10^4\,\text{Pa}$, and contain the plasma as soon as they are ignited within the short time of $1\,\mu\text{s}$ by a $350\,\text{V}$ pulse. For a 43" display, the plasma channels consume a total of only 6 W.

The rows of the display exhibit a pitch of $700\,\mu\text{m}$ or more; it is difficult for this technology to realize a smaller pitch. In order for the $50\,\mu$ glass sheet to seal the plasma channels hermeticly, the upper surfaces of the ribs must be polished. The front plate of the LCD carries the colour filter and, orthogonal to the plasma channels, the stripe $10\,\Omega/59\,\text{ITO}$ electrodes which feed in the video signals.

The operation of the plasma switches is explained in Figures 18.6(a) and 18.6(b) which represent cross-sections of the plasma channels. After ignition, the channel is filled with charge carriers (Figure 18.6(a)) which finally establish a low resistance connection between the surface of the $50\,\mu$ sheet and the potentials between anode and cathode. As soon as the cathode is grounded (Figure 18.6(b)), the separation sheet also assumes ground potential. The video

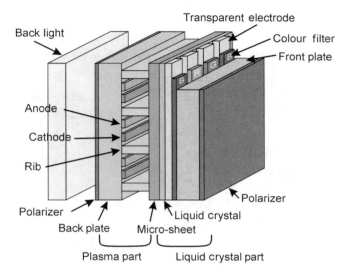

Figure 18.5 A Plasma Addressed Liquid Crystal (PALC) display

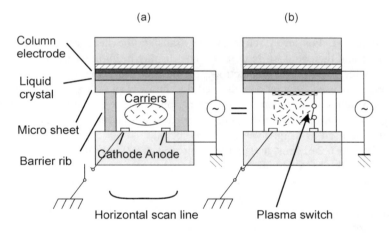

Figure 18.6 Operation of the plasma switch (a) after ignition of the plasma and (b) after the subsequent grounding of the cathode

pulses on the ITO stripes apply a voltage between pixel electrode and the grounded lines. This voltage must be high enough, of the order of 60 V, as it also has to provide the voltage drop across the separation sheet to ground. The backlight is fed in through the polarizer in the back and the plasma channel. As the anode and cathode in this channel are not transparent at present, the aperture ratio of the pixel is limited to 40 percent. A luminance of $500\,\text{cd}/\text{m}^2$ and a contrast of 100 : 1 was achieved with a 42" diagonal LCD (Buzak, 1999). PALC is competing with Plasma Display Panels (PDPs) in the market for large area flat panel displays.

A focused laser beam in Figure 18.7 (Shields and Bleha, 1990) heats up a dot of a liquid crystal display causing a phase transition into the isotropic state. The LCD is obviously

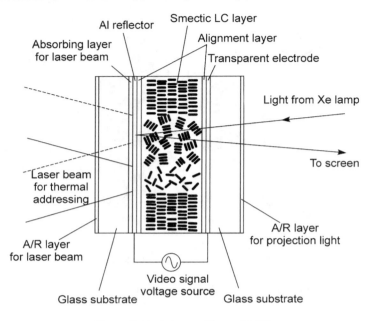

Figure 18.7 A laser-addressed LCD

thermally addressed. For an effective energy transfer, the laser beam first hits an anti-reflection layer and then an absorbing layer. The display in Figure 19.1 is reflective, having an Al-mirror, which also serves an electrode, on the back plate. In our example, an LC in the smectic A phase is used. In a laser-addressed dot the isotropic state cools down to the light scattering focal conic state when no voltage is applied. If a voltage is applied, the liquid crystal aligns homeotropically, thus generating the transparent state. The laser addressed LCD can also be used as a light valve in a projector.

The sensitivity is $0.2\,\text{pJ}/\mu\text{m}^2$ and can be improved to $0.04\,\text{pJ}/\mu\text{m}^2$ by adding an infra-red absorptive dye to the LC material.

An image written in by the laser remains stable for several weeks. As with all thermal effects, the writing speed is slow. Since the laser beam can be focused into areas of several μm in diameter, dot size is small and the system provides a high resolution display.

The e-beam of a CRT in Figure 18.8 (Shields and Bleha, 1990) operating as a writing gun charges the dielectric target electrode, thus generating an electric field with a location-

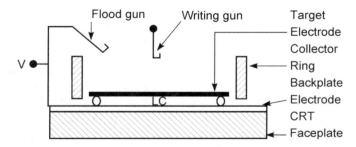

Figure 18.8 An electron-beam addressed LCD

dependent strength distributed over the adjacent liquid crystal display. The strength of the field corresponds to the grey level of the image. Obviously, the effect is the same as in a conventional field addressed LCD. The charge can be erased by electrons from the flood gun. A response time $< 100\,\text{ms}$, an optical efficiency of $11\,\text{lm/W}$ and a contrast of $85:1$ were reported. Also, this system may be applied as a light valve for projectors.

19

Colour Filters and Cell Assembly

Colours are produced either by absorptive or by reflective colour filters at each pixels. Absorptive filters used to be made from polymers, but are now made from more stable pigment dispersed materials. Reflective colour filters are dichroic, consisting either of a large number of dielectric layers or of chiral nematic LCs. Dichroic means that they have different colours when viewed from different directions.

The generation of colour can be achieved in an additive or subtractive mode. Additive colour generation superimposes the light emerging from three adjacent colour pixels. These pixels transmit the primary colours red, green and blue, and absorb the remaining spectral components of white light. They are, as a rule arranged as a triangle in Figure 19.1(a), in stripes in Figure 19.1(b), or in a diagonal mosaic in Figure 19.1(c). The triangle and diagonal mosaic are suitable for moving pictures, whereas the stripes suit graphic information.

Subtractive colour generation is, as a rule, based on reflective layers realized by a large number of dielectric layers, or by chiral nematic LCs in a helical structure. They reflect the complementary colours yellow (red + green), cyan (blue + green) or magenta (red + blue). In each case, one primary colour passes the layer, and is hence subtracted from the reflected portion of the spectrum. For yellow, blue is subtracted, for cyan, red and for magenta, green. Another subtractive colour generation is based on the spectral absorption of a band of wavelengths, as in guest-host displays.

For the most important schemes of colour generation, a somewhat more detailed description is given below.

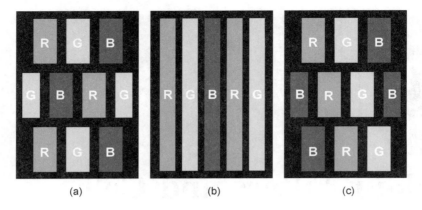

(a)   (b)   (c)

Figure 19.1   Arrangement of colour pixels (a) in triangles, (b) in stripes, and (c) on diagnols

19.1 Additive Colours Generated by Absorptive Photosensitive Pigmented Colour Filters

The goal is the fabrication of the colour filter in Figure 19.2 on the front-plate of an LCD. The colour pixels for red, green and blue are separated by a black matrix and protected by an overcoat layer. The top layers are the ITO electrode and the orientation layer. The black matrix prevents uncontrolled light from leaking between the pixels, which enhances the black state. For the same reason the black matrix material should exhibit a low reflectance not exceeding 4 percent. The overcoat layer also has to equalize the different heights of the colour filters.

The TFTs or MIMs for AM addressing are placed on the back glass plate. For TFTs the ITO on the colour plate is unstructured, whereas for MIMs it is structured in stripes.

The colour material contains photoinitiators providing the photosensitivity, pigments for the colours, solvents and an acrylic ester binder which is photo-stable, temperature stable up to 260°C, mechanically hard and has good adhesion.

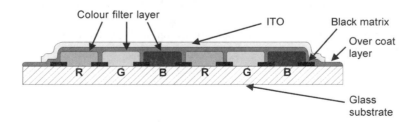

Figure 19.2   Cross-section of a pixelized colour filter

The fabrication steps for the colour filter are depicted in Figure 19.3 (Tani and Sugiura, 1995). The first step is the deposition and patterning of the black matrix, consisting of photopolymer or etchable polymer with dispersed carbon, dispersed graphite, patterned by lift-off, all with a reflectance < 4 percent. The fewest processing steps are required by the photopolymer. After deposition of the pigment dispersed colour material by spin coating and a pre-bake at 85°C to evaporate the solvents, the colour material is exposed to UV light. Only around $10 \, mJ/cm^2$ are needed if the i-line with a wavelength of 365 nm is used. Newer pigment dispersed filters are also more sensitive to the q-line at 436 nm, which is widely applied for photoresist. The development of the pattern and the post-bake at 220°C complete the cycle for one colour. The photomask is shifted to the next pixel, and the process is repeated for the next colour and again for the third colour. The use of photo-sensitive colour material saves etching steps. The thickness of all three colour layers is around $1.1 \, \mu$ to $1.22 \, \mu m$.

Figure 19.4 shows the transmittance versus wavelength, and Figure 19.5 shows the chromaticity of the three colours. The transmittance of blue, and even more of green, suffers from a strong absorbance of the material and from the fact that the slope of the absorption envelope is not steep enough.

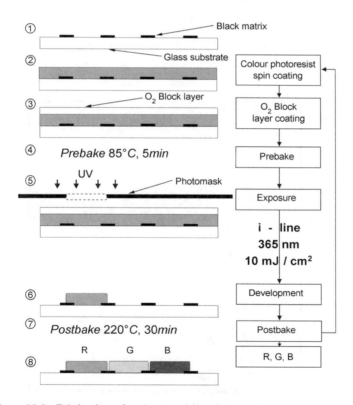

Figure 19.3 Fabrication of a photosensitive pigment dispersed colour filter

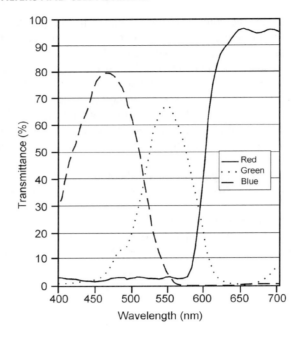

Figure 19.4 Transmittance of pigment dispersed colour filters. Thickness: $R = 1.1\,\mu$, $G = 1.14\,\mu$, $B = 1.22\,\mu$

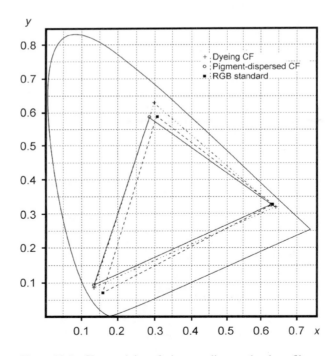

Figure 19.5 Chromaticity of pigment dispersed colour filters

19.2 *Additive and Subtractive Colours Generated by Reflective Dichroic Colour Filters*

A reflective colour filter in Figure 19.6 can be constructed by a sequence of about 20 layers out of different dielectrics (e.g. SiO_2 and TiO_2) with varying thicknesses and refraction indices. The effect is based on interference, and not on spectral absorptance. The filters are designed for incident light and viewing along the layer normal to form an optical bandpass for the reflected light, as well as for the transmitted light. Hence, the incoming white light is split into two colours, a reflected one and a transmitted one from which the denotation dichroic is derived. Depending on the design in Figure 19.6 (Sperger *et al.*, 19??) yellow is reflected, while blue is transitted, or cyan is reflected with red transmitted, or magenta is reflected and green transmitted. The spectral transmittance of the three primary colours is depicted in Figure 19.7. They exhibit the otherwise unmatched high transmittance of around 90 percent

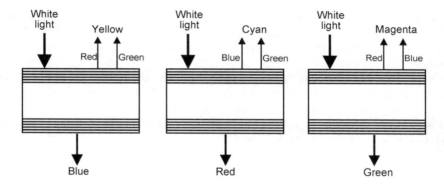

Figure 19.6 Cross-section of dichroic filters for various colours

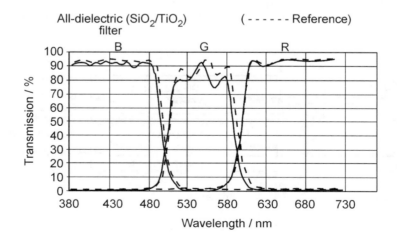

Figure 19.7 Transmittance of dichroic colour filters for R, G and B

for each colour. The steep flanks of the transmission bands guarantee a high colour purity. Only about 1 percent of the light is absorbed by the colour filter, yielding a much lower loss of light than in the absorptive colour filters. The dichroic filter can be used in an additive colour scheme in the transmissive mode and in a subtractive scheme in the reflective mode. In both modes it is strongly dependent on the viewing angle, because of the different thicknesses of the layers when viewed off the layer normal. As a consequence, these dichroic filters are most suitable for projection systems, where light always passes the filters in a normal direction. However, due to the large number of layers and the precise fabrication of the layer thicknesses, these filters are expensive but exhibit a very low loss of light and a high colour purity.

19.3 Colour Generation by Three Stacked Displays

The stacked three-layer reflective PC-GH-display in Figure 7.18 generates the primary colours R, G and B by the absorption of the dyes in the three layers. Each of those layers allows a complementary colour to pass. Further details are given towards the end of Chapter 7.

Also, the three stack reflective chiral nematic display in the right stack in Figure 8.15 is able to generate the three primary colours directly by reflecting them in one layer of the stack. The complimentary colour passes this layer. The optical effect is interference associated with Bragg reflection, and not absorption as in the first case. A detailed description was presented towards the end of Section 8.2.

19.4 Cell Assembly

The components of an LC cell are shown in the cross-section of a cell in Figure 19.8, whereas Figure 19.9 lists the sequence of the setps for cell assembly. Figure 19.10 depicts the three most common bonding techniques for the external ICs to the LC cell.

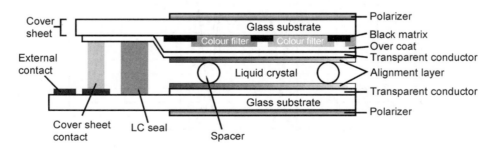

Figure 19.8 Cross-section of an assembled LC cell

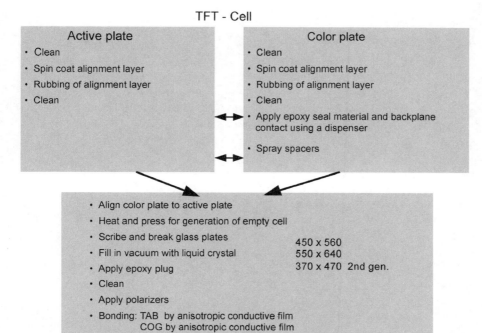

Figure 19.9   Steps for cell assembly

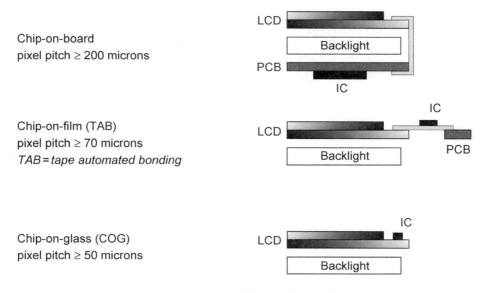

Figure 19.10   Bonding of driver ICs to the LC cell

20

Projectors with Liquid Crystal Light Valves

Liquid crystal displays can be used as the image source in a projector. Lenses project this image onto a screen. An LCD in such a projector is called a *light valve*, or more to the point, a Liquid Crystal Light Valve (LCLV). This chapter focuses on the various forms of LCLVs. A description of the elaborate optical components in a projector can be found in Stupp and Brenneshotz (1999).

A colour projector may use one, two or three LCDs as light valves; they may be transmissive or reflective and may exhibit a variety of implementations such as untwisted, especially vertically aligned, or twisted nematic LCDs, PDLCDs, FLCDs, or most promising, LCOS devices. They are, as a rule, active matrix addressed or optically addressed. The temporal handling of the information in a colour image is performed in parallel for all three colours or in a sequential form within a frame time, otherwise known as *field sequential colour scheme*. The overall system either can be a front projector, which projects the image in an open beam to a screen or a rear projector, which projects the image onto the back of a screen. The image is viewed from the front side. This system is fully encased and has no open beam.

We now examine the most important architectures of projectors that use LCLVs.

20.1 Single Transmissive Light Valve Systems

20.1.1 The basic single light valve system

The single light valve system in Figure 20.1 (Brenesholtz, 1997) consists of a light source, an integrator which generates a homogeneous illumination over the rectangular area required for LCDs, a transmissive colour LCD with the colour filter shown on the left, and projection lenses, which create the image on the screen. This is the simplest system, in which the path length from

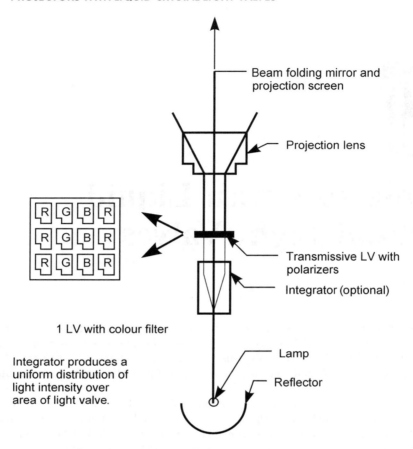

Figure 20.1   Single light valve colour projector

each differently coloured pixel to the screen is identical. For the luminance on screen, the aperture ratio of the pixels is a crucial quantity. An LCD light valve driven by MIMs (Schneider, 1997) exhibits a large aperture of 80 percent, and hence provides one of the brightest projectors, but suffers from the limited number of grey shades of MIM displays. A single light valve with colour requires a high pixel density as all color pixels are placed on one display.

20.1.2   The field sequential colour projector

Limitations in pixel density can be overcome by the field sequential colour projector in Figure 20.2 (Stupp and Brennesholtz, 1999; Lauer *et al.*, 1990) where a single light valve is illuminated sequentially with the three primary colours to produce sequential images which the eye integrates into a single coloured image. For achieving this, the addressing of the LCD has to work at triple speed. The display needs either only a third of the pixels as before, as they are sequentially used three times, or it can generate an image three times the

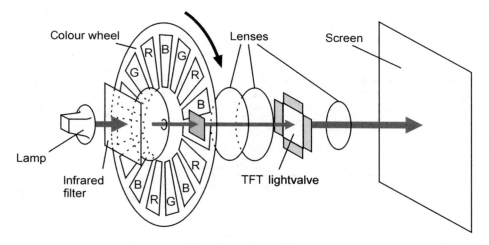

Figure 20.2   A single light valve field sequential colour projector

resolution. The temporal sequence of the colours red, green and blue is generated by a rotating colour filter fed by white light. The colour wheel can be replaced by three pulsed coloured lamps, such as three fast switching LEDs for red, green and blue. A further solution could be an electronically switchable colour filter the realized, for instance, by the stacked colour filter described in Chapter 19 and also in Brennesholtz (1997).

20.1.3 A single panel scrolling projector

The single panel scrolling projector in Figure 20.3 (Stupp and Brennesholtz, 1999) subdivides an LCD into three separate colour bands which are projected simultaneously. The

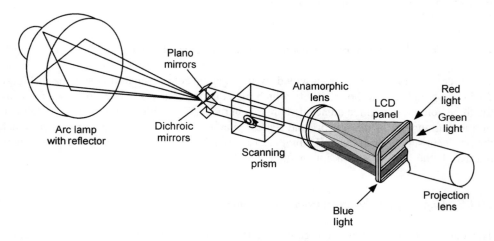

Figure 20.3   Single light valve scrolling projector

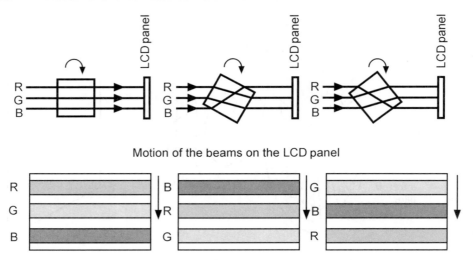

Figure 20.4 Generation of scrolling colour bands by a rotating prism

three colours generated in dichroic mirrors are fed into a rotating prism, which causes them to repeatedly scroll down the panel as illustrated in Figure 20.4. An anamorphic lens expands the three colours into bands as wide as the lines in the LCD. The LCD modulates the throughput of the three colour bands according to the pixel information.

The addressing differs from the regular line at a time scheme used for LCDs. For each colour, several lines are modulating the light simultaneously. This requires that image information for separate colours in separate lines has to be fed in simultaneously, as is further detailed in Stupp and Brennesholtz (1999). Although the scrolling colour system avoids the loss of light associated with the presence of only one colour at a time in the field sequential colour approach, it requires powerful lamps as only part of the panel transmits light of one colour at any instant in time.

20.1.4 Single light valve projector with angular colour separation

The three colours may also be provided by the angular colour separation scheme in Figure 20.5 (Stupp and Brennesholtz, 1999), which uses dichroic mirrors and microlenses focusing the three colours onto three adjacent subpixels in the LCD. Problems are the precise fabrication and alignment of the mirrors and lenses, guaranteeing that the colours precisely reach the pertinent subpixels in the LCD. In this case, the LCD again requires a tripled pixel count, but avoids the light loss of a colour filter.

20.1.5 Single light valve projectors with a colour grating

The colour grating system in Figure 20.6 (Schmidt, 2000) is based on the wavelength dependent diffraction of light by a surface grating. The diffracted colours R, G and B are

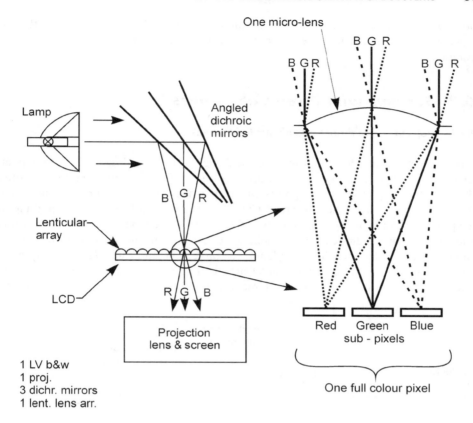

1 LV b&w
1 proj.
3 dichr. mirrors
1 lent. lens arr.

Figure 20.5 Single light valve angular colour separation projector

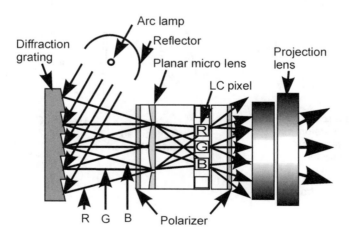

Figure 20.6 A single light valve projector with colour grating

guided by microlenses to the pertinent colour pixels of the LCLV, from where they are projected to the screen. This system is still under development, aiming for a perfect colour separation.

20.2 Systems with Three Light Valves

20.2.1 Projectors with three transmissive light valves

In the projector with three transmissive LCD light valves in Figure 20.7 (Stupp and Brennesholtz, 1999), each light valve modulates only the image pertaining to one colour. This alleviates the problem of a high pixel density as the adjacent colour pixels in a single light valve are now distributed to three LCDs. The white light of the lamp is split into the three primary colours by dichroic mirrors. After having passed the LCDs, the three colours are recombined in a dichroic colour combining cube or by separate dichroic mirrors and are then projected onto a screen. This one lamp, one projection lens, three light valve system is the most commonly used. It exhibits the advantage of allowing for a high resolution image, as the colour pixels are distributed over three displays. The penalties are the requirement to

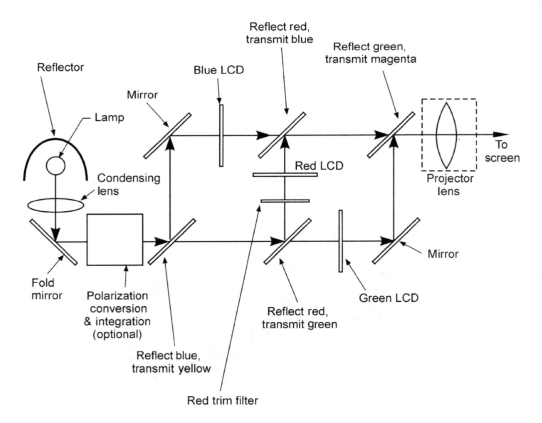

Figure 20.7 Three light valve projector with three equal optic paths

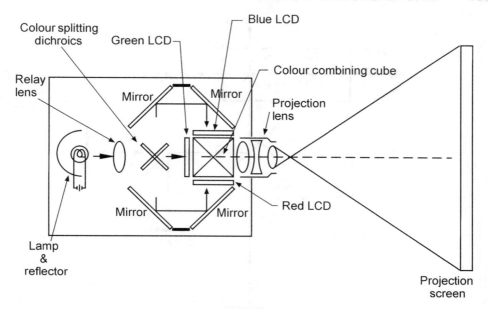

Figure 20.8 Three light valve projector with unequal optic paths

have three pertinent colour pixels converge into one dot on the screen and higher cost than for a single light valve system. The projector in Figure 20.7 has equal optical path lengths for each colour providing equal illumination and magnification of each LC panel. A more compact system is the unequal optical path three light valve projector in Figure 20.8 (Stupp and Brennesholtz, 1999) In this projector, a colour shift may occur over the area of the light valve, because the angle of incidence of light on the left side of the LC panel differs from that angle on the right side of the panel. As a consequence, angle dependent variations in the optical properties of dichroic filters show up.

20.2.2 *Projectors with three reflective light valves*

A simplified example without relay and field lenses for a projector with three reflective LCDs is depicted in Figure 20.9 (Stupp and Brennesholtz, 1999). The polarizing beam splitter provides polarized light for the two dichroic mirrors, which pass red, green and blue light on to the LCDs. The reflected light travels in the reverse direction back through the mirrors and the beam splitter to the screen. Reflective LCDs offer the advantage of an aperture ratio greater than 90 percent, as the addressing circuit is hidden underneath the reflective pixel electrode.

LCDs used in the reflective mode are commonly TN cells, but also HAN cells (Glueck *et al.*, 1992) and VA cells (DAP cells) (De Suet *et al.*, 1999). The latter cells offer an excellent black state, and hence good contrast ratios up to 500 : 1 when viewed perpendicularly. As all the LC light valves operate only with light beams perpendicular to the surface of the substrates and the screen, viewing angle properties for off-axis angles at

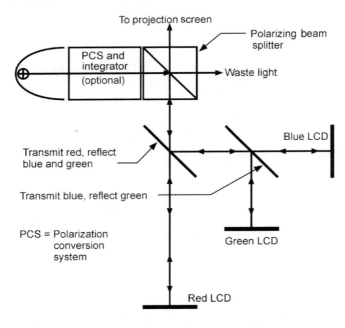

Figure 20.9 Projector with three reflective LC light valves

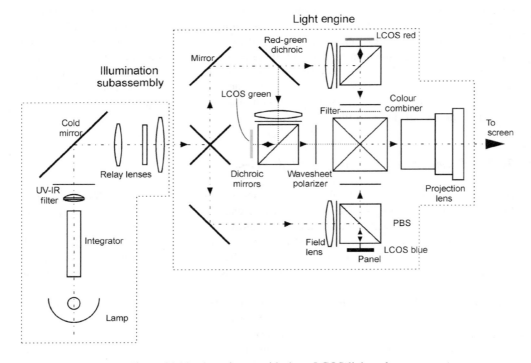

Figure 20.10 A projector with three LCOS light valves

the LCDs are not important. Off-axis viewing of the screen has to be taken care of by the reflective properties of the screen.

20.2.3 Projectors with three LCOS light valves

Due to their small size, LCOS devices described in Chapter 16 offer the opportunity to build a very compact projector as realized by the system in Figure 20.10 with three reflective LCOS light valves (Melcher *et al.*, 1998). The cold mirror allows IR to pass, and reflects visible light into the projector. The dichroic mirrors and the colour combiner are already known from other systems. The Polarizing Beam Splitters (PBS) provide the linearly polarized light for the LCD light valves.

20.3 Projectors with Two LC Light Valves

A compromise between the costly, high performance, three light valve systems and the cheaper and lower resolution one LC panel system is the two LC light valve projector in Figure 20.11 (Stupp and Brennesholtz, 1999). One light valve transmitts only one colour, the monochrome LCD, for which the most deficient colour in the light source is chosen. The

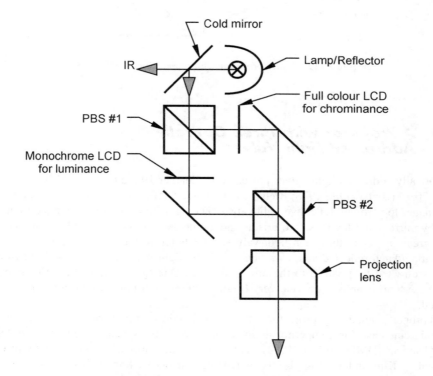

Figure 20.11   Projector with two light valves

second light valve modulates the two other colours. The second PBS recombines the colours before they are projected onto the screen.

20.4   A Rear Projector with One or Three Light Valves

Figure 20.12 depicts a rear projector with one or three light valves (Lueder, 1994a). For simplification, only one light valve is shown. To reduce the depth of the system, the modulated beam transmitted by a light valve is folded by one or several mirrors before being imaged onto, a rear projection screen. The light rays are converged onto the screen with a Fresnel lens to maximize the light flux within the preferred viewing cone. If the image is reversed left to right by the mirrors, it is corrected electronically by reversing the image written on the light valve.

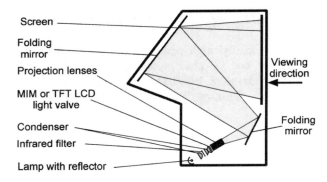

Figure 20.12   A rear projector with one or three light valves

20.5   A Projector with Three Optically Addressed Light Valves

The optically addressed light valves presented in Chapter 19 are the core of the projector in Figure 20.13 (Bleha *et al.*, 1978; Bleha, 1995). Three CRTs for red, green and blue transmit their images to pertinent reflective light valves. The light from the lamp is guided to the light valve by mirrors, a relay lens, two prisms and a blue and a red dichroic filter. The red filter, where green passes to the green light valves, reflects red to the red light valve.

The modulated reflected and, as a rule, amplified light with respect to the light from the CRTs is directed by prism 2 to the projection lens. Due to the amplification of light, this complex system is able to provide very bright pictures on a huge screen greater than 10 m diagonal.

This property renders the projector with optically addressed light valves suitable for use in electronic cinemas. The competitor in this application is the system based on micro Digital Mirror Devices (DMDs) (Gove, 1994). As this approach works with electronically or piezo-electrically (Kim and Hwang, 1999) switchable mirrors as light valves rather than with LCDs, it is not examined here.

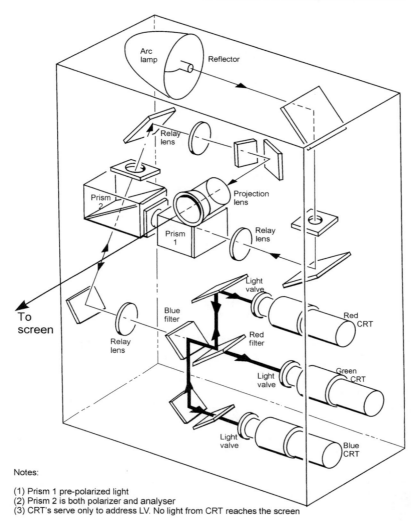

Notes:

(1) Prism 1 pre-polarized light
(2) Prism 2 is both polarizer and analyser
(3) CRT's serve only to address LV. No light from CRT reaches the screen

Figure 20.13 A projector with three optically addressed LC light valves

A further but not yet mature enough candidate for electronic cinema is the laser projector (Natamura *et al.*, 1997), which still suffers from too low a light output of the semiconductor lasers.

In all light valve based projectors, barely 10 percent of the luminance of the lamp reaches the screen (Lueder, 1994c). The reasons are the large light losses caused by collimation, by the more than 50 percent loss in the polarizers, the loss due to an aperture ratio being 95 percent for reflective displays but dropping to as low as to 40 percent for transmissive displays, and by the loss in the colour filters being only 10 percent for dichroic filters but more than 50 percent for the absorptive ones.

21

Liquid Crystal Displays with Plastic Substrates

21.1 Advantages of Plastic Substrates

Plastic substrates for the cells of LCDs exhibit only 1/6 of the weight of glass substrates. They are virtually unbreakable and their flexibility allows the designer to do elegant styling, to bend it so as to suppress reflections or for placing it on a curved surface. They can be as thin as 100–200 µm, which suppresses shadow images in reflective displays (Lueder, 1999).

Reflective displays suffer from those shadow images which are visible with a reduced luminance next to the primary image. The generation of this shadow is outlined in Figure 21.1. A full explanation can be found in Berman and Oh (1980). We focus on ray 3 at the edge of an image passing only once through the addressed pixel and being absorbed in the analyser. However, diffuse incoming rays around ray 3 have a different retardation, are reflected at the surface R of the reflector, and their plane of polarization is rotated by 90° while passing the unaddressed pixel. Hence, they pass the top analyser, generating a grey shadow next to real edge ray 2. Thinner substrates shift the lower surface R of reflection closer to the LC layer, thus shifting the edge of the shadow closer to the edge of the regular image. For plastic substrates only 100–200 µm thick instead of glass with 0.7–1.2 mm, the shadow is so close to the real edge that it is no longer visible.

These advantages render displays with plastic substrates attractive for portable systems such as mobile phones, pagers, smart cards, Personal Digital Assistants (PDAs) and portable computers. However, as production volume is still low, the costs for plastic substrates are high.

21.2 Plastic Substrates and their Properties

Plastics applicable for displays are polycarbonate (PC), polyarylate (PAR), polyimide (PI), polyethylenterephtalate (PET), polyestersulfone (PES) and polyolefin. The list of

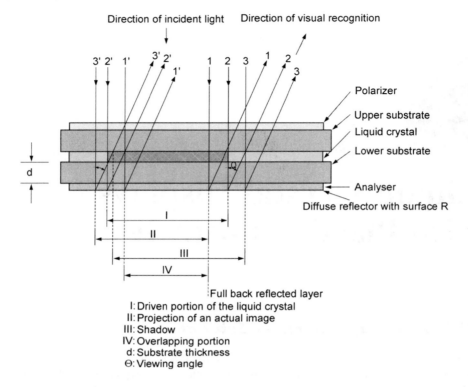

Figure 21.1 Explanation of the shadow figure in reflective displays

requirements of plastics in Table 21.1 for display applications is mostly self-explanatory. Low retardation is required to keep the influence of the substrate on the polarized state of light in the display within acceptable limits. Most crucial are the limits for the permeation of

Table 21.1 Requirements of plastic substrates for LCDs

Temperature stability	as high as possible
Temperature coefficient	$\leq 50\,\mathrm{ppm/K}$
Irreversible shrinking	$< 1/20$ of pixel pitch
Surface roughness	$\leq 10\,\mathrm{nm}$
Transparency	$\geq 90\%$
Retardation $d\Delta n$	$< 15\,\mathrm{nm}$
Permeation of H_2O	$< 0.15\,\mathrm{g/m^2}$ 24 h at 40°C and 90% relative humidity
Permeation of O_2	$< 0.1\,\mathrm{cm^3/m^2}$ 24h bar
Chemical resistance against etchants, solvents and other chemicals used in display manufacture	

Table 21.2 Properties of three plastic substrates

Material	PC	PAR	PES
Glass transition temperature in °C	195	215	225
Max. process temperature in °C	130	160	180
Flatness in *nm*	< 10	< 10	< 10
Waviness in *nm*	< 100	< 100	< 100
Thermal expansion in *ppm/K*	$60, \dots, 70$	$55, \dots, 60$	$45, \dots, 55$
Irreversible shrinking after annealing in *ppm*	10^2 ppm after 1h at 130°C	10^2 ppm after 1h at 150°C	10^2 ppm after 1h at 200°C
Modulus of elasticity in *kN/mm²*	$2, \dots, 2.5$	$2.2, \dots, 2.6$	$2.4, \dots, 3$
Transparency in %	90	90	87
Retardation $d\Delta n$ in *nm*	< 15	< 15	10
Permeability for O_2 in cc/m^2 24 h atm	0.1	0.1	0.1
Permeability for H_2O in g/m^2 24 h	15	15	
Chemical resistance against:			
Etchant, developer, photoresist	yes	yes	yes
Aceton/butylacetate	no	yes with coating	no/yes
Alcohol	yes	yes	yes

humidity and oxygen into the cell, which prevent the LC material from deteriorating. A high temperature stability is desirable, as most processing temperatures for thin film components are around 250°C.

The properties for the three most suitable plastics PC, PAR and PES improve in this sequence, and are listed in Table 21.2.

A look at Tables 21.1 and 21.2 reveals that we mainly have to deal with three problems. The first is minimizing the permeation of humidity and O_2 by depositing barrier layers on the plastics. The thermal expansion of the plastics lie between 45 ppm/K for PES and 70 ppm/K for PC, whereas the thermal expansion of a-Si, poly-Si, SiN_x and SiO_2 is only around 5 ppm/K. This will generate thermal tensions, leading to cracks and the delamination of layers. A solution to this problem is essential for the use of plastic substrates. The third task is to lower the process temperature for MIMs, TFTs and colour filters, for instance to 180°C for PES, while still achieving acceptable properties of these components.

Mainly because of its tolerance of high process temperatures and its low thermal expansion, PES is a preferable plastic.

21.3 Barrier Layers for Plastic Substrates

A good barrier layer for humidity and O_2 known from food packaging is SiO_x, $x\varepsilon$ [1.4, 1.8]. The x-value indicates the presence of unsaturated O-bonds, which may diminish the flow of O_2 through this layer by trapping O_2. An evaporated 60 nm SiO_x layer reduces the permeation of water and O_2, respectively, to 0.1 g/m^2 24h at 90 percent relative humidity (RH) and 3.95 cm^3/m^2 24h bar. A further reduction is achieved by depositing, on the other side of the plastic, a 50 nm layer of Ta_2O_5, reducing the permeation of O_2 further to 0.4 cm^3/m^2

24h atm (Baeuerle, 2000). The combination of 30 nm SiO_x and 5 μ Ormocere provides a sufficiently low permeation for water of $0.04 \, g/m^2$ 24 h at 90 percent RH. and for O_2 of 0.1 cm^3/m^2 24h bar). (Baeuerle, 2000). These permeations are found to be sufficiently low for maintaining a proper operation of the most sensitive FLCD cells. Even the combination of SiO_x and Ta_2O_5 barriers did not give rise to malfunctions of FLCDs after six months, as such a device is not permanently operated at 90 percent RH. The Ta_2O_5 film is deposited by reactive sputtering in an O_2 atmosphere from a Ta target. It may still exhibit oxygen deficiencies, lowering oxygen permeation. One of the best barrier layers is a metallic film, such as an Al mirror or an ITO electrode, but it cannot be placed everywhere over the substrate. Another good barrier is a 50 μm glass sheet. As an unsupported 50 μm glass sheet is prone to break, it is advisable to glue it onto a flexible plastic foil.

21.4 Thermo-Mechanical Problems with Plastics

If the thermal expansion coefficient α_s of a substrate differs from the thermal expansion α_l of a layer on the substrate, then the tension σ_{tot} in the layer, including its intrinsic tension σ_i at room temperature T_R, changes for a process temperature T_P (Gose, 1990; Entenberg et al., 1987) to

$$\sigma_{tot} = \sigma_i + (\alpha_l - \alpha_s)(T_P - T_R)E_l/(1 - \nu_l) \tag{21.1}$$

where E_l stands for the modulus of elasticity and ν_l for the Poisson ratio of the layer. As a rule, we have $\alpha_l < \alpha_s$ and $T_R < T_P$, resulting in a decrease in σ_{tot}, indicating an additional compressive stress. This stress may lead to cracks in the layer and to a curvature of the substrate with the radius R (Entenberg et al., 1987; Baeuerle, 2000)

$$R = \frac{E_s d_s^2/(1 - \nu_s)}{6d_l \sigma_{tot}} \tag{21.2}$$

valid for $E_s d_s \gg E_l d_l$.

In Equation (21.2), E_s is the modulus of elasticity, d_s the thickness and ν_s the Poisson ratio of the substrate, d_l the thickness of the layer, and σ_{tot} is taken from Equation (21.1). For $E_s d_s \gg E_l d_l$, the radius is so large that the substrate remains planar. For decreasing d_s the substrate starts bending, hence the substrates should not be too thin and the layers not too thick. An analytical investigation will reveal numbers for the desirable thicknesses d_s and d_l (Polach, 2000; Polach et al., 2000) based on the assumption that plastic substrates with layers are pressed flat during all processing steps. The layers for an a-Si : H or a poly-Si TFT consist of PECVD-generated SiN_x as gate dielectric, i-a-Si : H for the channel and n^+a-Si : H for the contact areas, with the moduli of elasticity being $E_{SiN_x} = E_{i-a-Si} = E_{n^+a-Si} = E_{a-Si}$ and the layer thicknesses being d_{SiN_x} and d_{a-Si}, the latter for all three a-Si-layers together. Doping of i-a-Si does not change the elasticity. The total modulus of elasticity E_t of both layers is (Timoshenko and Goodier, 1970)

$$E_t = E_{SiN_x} \frac{d_{SiN_x}}{d_{SiN_x} + d_{a-Si}} + E_{a-Si} \frac{d_{a-Si}}{d_{SiN_x} + d_{a-Si}} \tag{21.3}$$

The different thermal expansion coefficients $\alpha_{Si} = 5 \, ppm/K$ for all Si-containing layers and $\alpha_s = 44 \, ppm/K$ for a PES substrate lead to distortions of the substrate and stress at the interface between the substrates and the layers. The substrate without layers has at room temperature T_R the length L_{1x}. After a layer has been deposited, at process temperature $T_P > T_R$ onto a plastic substrate the combination of layer and plastic exhibits the length L_{0x} and the following electro-mechanical behaviour: at temperature T_P there is neither a distortion nor a stress in the combination; after cooling down by $\Delta T = T_R - T_P < 0$, the two materials still have the same length L_{Rx} as the layer firmly adheres to the substrate; however, because of the different expansion coefficients $\alpha_s \neq \alpha_l$, a distortion ε_ν and a tangential stress σ occurs at the interface of the two materials. The two quantities ε_ν and σ will now be determined.

The amorphous layers and amorphous plastic substrates are isotropic, therefore the distorted shape of a rectangular substrate in Figure 21.2 is again rectangular. The distortion in the x-direction at room temperature after cooling down is

$$\varepsilon_{\nu x} = \frac{\Delta L_x}{L_{1x}} \tag{21.4}$$

with

$$\Delta L_x = L_{Rx} - L_{1x}. \tag{21.5}$$

ΔL_x shown in Figure 21.2 indicates by how much the substrate does not return to its original length L_{1x} due to the restraining force of the layer. Due to the isotropy, the distortions are the same in both directions, as stated in Equation (21.6). For this reason, we shall only investigate one direction, and denote it by ε_ν.

$$\varepsilon_{\nu y} = \frac{\Delta L_y}{L_{1y}} = \varepsilon_{\nu x} = \varepsilon_\nu. \tag{21.6}$$

The investigation in Polach (2000) and Polach et al. (2000) is based on elasto-mechanics and Mohr's circle of stress (Timoshenko and Goodier, 1970). The result is

$$\varepsilon_\nu = \frac{\alpha_{\mathrm{eff}} - \alpha_s}{(1/\Delta T) + \alpha_s}, \tag{21.7}$$

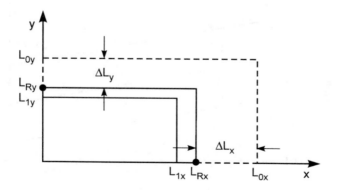

Figure 21.2 Distortion of a rectangular isotropic body

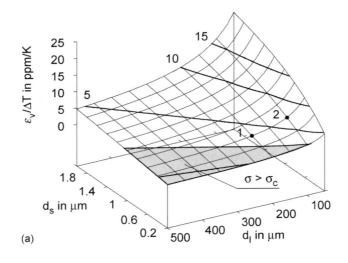

(a)

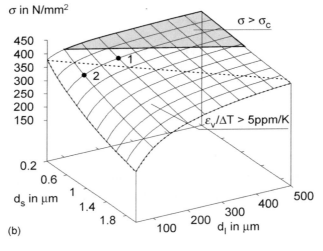

(b)

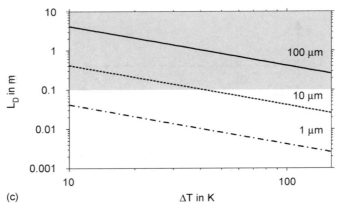

(c)

with the effective thermal expansion coefficient α_{eff} of the plastic substrate with the thin film layers on it given as

$$\alpha_{\text{eff}} = N\alpha_s + \frac{E_l\, d_l}{E_s\, d_s} N\alpha_l, \tag{21.8}$$

where

$$N = 1 \bigg/ \left(1 + \frac{E_l\, d_l}{E_s\,, d_s}\right). \tag{21.9}$$

For α_{eff} we obtain from Equations (21.8) and (21.9)

$$\lim_{d_e \to 0} \alpha_{\text{eff}} = \alpha_s \tag{21.10}$$

and

$$\lim_{d_e \to \infty} \alpha_{\text{eff}} = \alpha_l \tag{21.11}$$

With $\alpha_s \ll 1/\Delta T$, Equation (21.7) simplifies to

$$\varepsilon_v = (\alpha_{\text{eff}} - \alpha_s)\Delta T. \tag{21.12}$$

The value $\varepsilon_v/\Delta T$ in ppm/K from Equation (21.12) is plotted in Figure 21.3(a) versus the thicknesses d_l and d_s in µm, with the values of the parameters listed in the figure caption. As expected, the distortion is largest for a small thickness d_s of the substrate and a large thickness d_l of the layer, and smallest in the opposite directions. Point 1 in Figure 25.3(a) belongs to a $d_s = 200$ µm thick PES substrate, and to 300 nm SiN$_x$, 250 nm a-Si : H and 50 nm n$^+$a-Si : H, resulting in a $d_l = 0.6$ µm thick layer. Point 2 belongs to the same layer, but a $d_s = 100$ µm thick substrate. Lines of constant distortion are shown as full lines together with their distortion numbers in Figure 21.3(a). The stress σ at the interface of substrate and layer at room temperature is given by

$$\sigma = \sigma_l + \sigma_s \tag{21.13}$$

with

$$\sigma_l = E_l(\varepsilon_x - \alpha_l\, \Delta T) \tag{21.14}$$

as stress in the layer, and

$$\sigma_s = E_s(\varepsilon_x - \alpha_s\, \Delta T) \tag{21.15}$$

◀──

Figure 21.3 The expansion of a PES substrate versus d_l and d_s for $\alpha_s = 44$ ppm/K, $\alpha_l = 5$ ppm/K $E_s = 2.6$ kN/mm^2 (PES) and $E_t = 86$ kN/mm^2 in Equation 21.3.(Grimsdilch *et al.*, 1978); (b) the tangential tension σ versus d_l and d_s for the same parameters as in Figure 21.3(a); (c) diagonal L_D of a square versus ΔT for a constant distortion in µm as parameters

as stress in the substrate, where

$$\varepsilon_x = \frac{L_{1x} + \Delta L_x - L_{0x}}{L_{0x}} = \frac{L_{Rx} - L_{0x}}{L_{0x}}, \tag{21.16}$$

where Equation (21.5) was used.

ε_x is the distortion at room temperature reduced by the length L_{0x}. L_{0x} is the length of substrate and layer at the process temperature T_p, as shown in Figure 21.2, and is given by

$$L_{0x} = \frac{L_{1x}}{1 + \alpha_l \Delta T}. \tag{21.17}$$

L_{1x} and ΔL_x are the length and its distortion in Figure 21.2 at room temperature.

The stress σ is plotted in Figure 21.3(b), again versus d_s and d_l using the same parameters as in Figure 25.3(a). For a large thickness d_s and a small thickness d_l, we obtain the largest stress of $450\,\mathrm{N/mm^2}$. So, to avoid cracking of the layer, d_s has to be decreased and d_l increased. The stress at which the Si-based layers crack was determined as $\sigma_c = 400\,\mathrm{N/mm^2}$. The region in Figures 21.3(b) and 21.3(a) in which cracking occurs for $\sigma > \sigma_c$ is indicated as a hatched area. On the other hand, in the area bordered by the dashed line in Figure 21.3(b), the distortion at room temperature is larger than 5 ppm/K. Above 120°C the distortions in this region become so large that they become plastic, which must be avoided as they are then irreversible. Thus, point 1 but not point 2 meets both requirements of lying outside the crack area and in an area with a distortion below 5 ppm/K.

So far the selection of d_s and d_l has been based on purely mechanical considerations. However, the choice of $d_{SiN_x} + d_{a-Si} = d_l$ has to meet requirements imposed by the TFT. If d_{SiN_x} is chosen too thin in the range $d_{SiN_x} < 200\,\mathrm{nm}$, an electric breakdown of the dielectric will occur. If d_{SiN_x} is too thick the electrical charge generated in the channel will be too low. A total thickness $d_l \in [0.3\,\mu\mathrm{m}, 0.8\,\mu\mathrm{m}]$ is desirable. The thickness of the a-Si : H-layer being around 50 nm is small in comparison to d_{SiN_x}. Point 1, selected for mechanical reasons, also meets the electrical constraint. In fact, $d_s = 150\,\mu\mathrm{m}$ at point 1 could be chosen even closer to $100\,\mu\mathrm{m}$. The substrate thickness $d_s = 200\,\mu\mathrm{m}$ often selected for plastic substrates also meets the mechanical requirements, and has the advantage of easier handling than a thinner one. The result in Figure 21.3(a) has been evaluated in Figure 21.3(c) to give the diagonal L_0 of the largest square versus $|\Delta T|$, in which a given distortion in μm is not exceeded. This distortion lies on straight lines in Figure 21.3(c), at which the pertinent distortion is given. The shaded area for $L_0 > 10\,\mathrm{cm}$ indicates the area for which, so far, no fabrication equipment for displays with plastic substrates is available. As the distortion can be predicted from Figure 21.3(c), the layout and hence cell masks can be predistorted such that a geometrically correct display is obtained after the unavoidable elasto-mechanical distortion has occurred.

21.5 Fabrication of TFTs and MIMs at low Process Temperatures

21.5.1 Fabrication of a-Si : H TFTs at low temperature

Conventional PECVD used for the deposition of layers, made of a-Si : H and SiN$_x$, required in TFTs works at process temperature of 250–280°C. Plastics can only tolerate temperatures up to 180°C in the case of PES, therefore low temperature processes have to be developed for TFTs.

Low temperature PECVD below 200°C provides a-Si : H-layers that are too rich in H_2. The reason is that stoichiometry is largely determined by thermally activated mechanisms at the film growth surface. On the other hand, sputtering is a cold process. For this reason, sputtering was tried first for the low temperature fabrication of a-Si : H-TFTs.

Reactive magnetron sputtering from a single undoped target in an crystalline Si $Ar + H_2$ working gas provides good a-Si : H, n^+-a-Si : H and SiN_x layers (McCormick *et al.*, 1997). Anode biasing minimizes ion bombardment of the substrate. The partial pressure of Ar and H_2 is adjusted to the selected film composition. At a UHV background pressure of $< 10^{-9}$ Torr, a total pressure of $Ar + H_2 < 3$ mTorr, with partial pressure of H_2 at 0.4 mTorr and of Ar at 1.5 mTorr, and a temperature of 125°C an a-Si : H-layer providing an effective mobility in the TFT of 0.3 cm^2/Vs, a switching ratio of $5 \cdot 10^5$ and a threshold voltage of 3 V, was obtained. For the n^+-a-Si : H Layer a doped target was used.

In spite of the H_2-richness, low temperature PECVD can also provide good a-Si TFTs (Polach, 1999). The following feed gases were used: Monosilone for a-Si : H, a mixture of monosilone and ammonia for SiN_x, with a phosphine dopant for N_+ a-Si : H. On an a-SiN_x reed layer, 330 nm of SiN_x, 200 nm of l-a-Si : H and 50 nm of N_--a-Si : H are deposited at a low rate of 7 nm/min at 180°C. The pertinent power density is 5 mW/cm^2 for a-Si : H and 12 mW/cm^2 for SiN_x. The low deposition rate at elevated temperature may allow H_2 to leave the films by providing a large enough diffusion length of the reactive components on the surface. This favours hydrogen elimination and reconstruction of molecules after deposition (Luft and Tsou, 1993).

The properties of the a-Si : H-TFTs are a mobility from 0.3 cm^2/Vs to 0.45 cm^2/Vs, a switching ratio of 10^6, an off-current < 0.6 pA and a threshold voltage of 2 V (Polach, 1993).

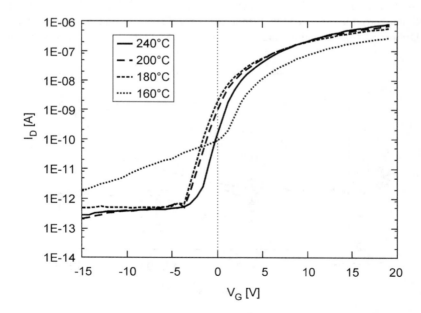

Figure 21.4 Input characteristics of a-Si : H TFTs fabricated at different process temperatures

Figure 21.4 shows the input characteristics of a-Si TFTs prepared at PECVD process temperatures of 240°C to 160°C. Only the TFT deposited at 160°C exhibits a noticeably degraded behaviour in particular an increased off-current. The shift ΔV_{th} of the threshold voltage during a bias-temperature test is shown versus the test time in Figure 21.5 (Ikeda et al., 1999). Again, a-Si TFTs prepared at very low temperatures, in this case at 120°C, differ from the others, albeit here in a positive sense.

21.5.2   Fabrication of low temperature poly-Si TFTs

The preparation of poly-Si TFTs is based on the same PECVD a-Si:H layers that are deposited for a-Si:H TFTs. The additional step is laser annealing. The best results were achieved by annealing with a scanning krF excimer laser with an energy density of 200–260 mJ/cm^2 per laser pulse (Young et al., 1996). The pulse frequency is 10 Hz, the scan lines have about a 10 percent overlap, while the scan speed is 0.5 mm/s, leading to 100 pulses hitting each point. Lower energy laser pulses at the edge of the Gaussian energy distribution gently dehydrogenate the a-Si:H layer which is too rich in H$_2$. No noticeable heating of the substrate by the laser pulses is observed. To limit the off-current, a gate-drain overlapped lightly doped drain is used. The input characteristics of an n-channel poly-Si-top gate TFT without LDD are shown in Figure 21.6. The TFT was annealed at 250°C and 200°C. The 200°C anneal provides a mobility of 60 cm^2/Vs, a minimum off-current of 10^{-10} A and a threshold of 8 V.

Among different recrystallizations such as Solid Phase Crystallization (SPC) and Metal-Induced Lateral Crystallization (MILC), the Excimer Laser Anneal (ELA) has so far provided the best results, especially the lowest off-currents (Young et al., 1999). With a similar process but a maximum temperature of only 100°C, top-gate poly-Si TFTs on PET were fabricated (Carey et al., 1997). Excimer laser anneal with 35 ns pulses did not heat the substrate, and provided TFTs with a mobility of 7.5 cm^2/Vs and a good switching ratio of 10^6. Drain and source regions are doped by a gas immersion laser at 308 nm and up to

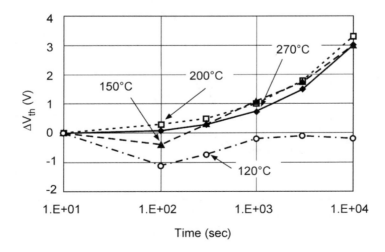

Figure 21.5   Variation of V_{th} during BT stress of a-Si:H TFTs fabricated at various temperatures

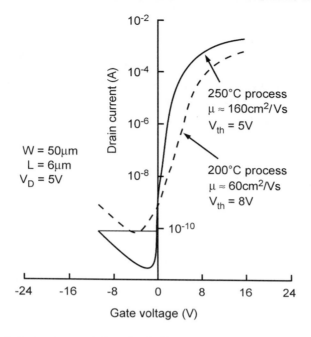

Figure 21.6 Input characteristics of poly-Si TFTs fabricated at 250°C and 200°C

$350\,mJ/cm^2$. By melting the surface, p-type doping was also performed in a BF_3 ambient and n-type doping in PF_5.

21.5.3 Fabrication of MIMs at low temperature

The MIM devices based on Ta_2O_5 as the dielectric, and presented in Chapter 17, only use sputtering for deposition of layers. As sputtering is a relatively cold process not exceeding 120°C, all process steps can be used for plastic substrates. The only difference stems from the fact that Ta and Ta_2O_5 are brittle materials which tend to crack when the plastic substrate is bent. The remedy consists in adding Al as a more ductile material to Ta (Klette et al., 1995). The cosputtered Ta/Al alloy has a Ta/Al ratio of about 6 : 1 and can be anodized as Al is also anodizable.

Thin film diodes with a PECVD generated a-SiN$_x$: H dielectric were fabricated at temperatures below 150°C, and possessed properties suitable for LCDs with plastic substates (Shannon et al., 1996).

21.5.4 Conductors and transparent electrodes for plastic substrates

Conductors and transparent electrodes with a high modulus of elasticity, such as Ta, Cr and ITO, tend to crack, and have to be replaced by ductile materials with a low modulus of elasticity. For conductors the remedy is to use Ti with a modulus of $107\,kN/mm^2$, or the very ductile Al (Klette *et al.*, 1995). ITO could be replaced by a polymer transparent conducting layer consisting of the organic material PEDOT (Tahon *et al.*, 1999). PEDOT can be spin-coated, and exhibits a resistance of 400 $\Omega/\square$ at a thickness of 200 nm and a transparency of >80 percent. It is thermally stable for 24h at 80°C.

In a bending test, PEDOT lines were compared to ITO lines. After 100 bends with a radius of 3 cm, 10 percent of the ITO lines were interrupted by a crack, whereas all PEDOT lines were still intact.

22

Printing of Layers for LC Cells

Replacing the conventional vacuum processes for the manufacture of LC cells by printing offers the potential of lower fabrication cost and lower energy consumption, as well as a smaller investment in equipment. Suitable printing processes for thinner films in the 1 μm range are flexographic printing, knife coating and ink jet printing, as well as silk screen printing for thicker layers in the 10 μm range. These processes will be described and their application outlined for orientation layers and liquid crystal films (Randler *et al.*, 2000; Lueder, 2000).

22.1 Printing Technologies

22.1.1 Flexographic printing

In the flexographic printer in Figure 22.1, the doctor blade distributes a layer of the printing ink on the anilox roll. This roll carries a pattern of gravures which are filled with a volume of ink corresponding to a layer 3.5 times thicker than the desired printing thickness. This makes good for the transfer loss. The printing ink is transferred to the letter press of compressable polymer with a relief containing the pattern to be printed. The letter press finally prints the pattern onto the substrate. This process is also called gravure printing.

22.1.2 Knife coating

In knife coating in Figure 22.2, a heated steel blade is used to spread the ink over the substrate area. Etched spacers on the substrate serve as a thickness control. The blade must be flexible in order not to damage the spacers. For the same reason, the blade pressure should be chosen as low as is allowed by the viscosity of the ink.

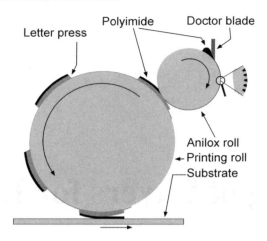

Figure 22.1 Flexographic printing

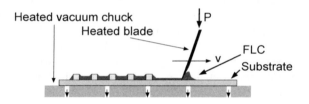

Figure 22.2 Knife coating

22.1.3 Ink jet printing

An ink jet printer is shown in Figure 22.3. A piezo-electric transducer compresses a glass capillary, and thus causes the ejection of a droplet of ink. The diameter of the nozzle is in the range of 30–100 μm, which is also approximately the diameter of the droplet ejected. The ejection frequency can reach 10 kHz. The printing head has to be heated if the viscosity of the ink is high.

To cover the area on the substrate completely and ensure it is free of defects, a droplet of ink on the surface in Figure 22.4 has to exhibit a low contact angle. This is reached with a cleaned grease free surface. A low contact angle indicates a high surface tension.

22.1.4 Silk screen printing

The squeegee in Figure 22.5 presses the ink through the meshes suspended on a frame onto a substrate. The meshes carry a photoresist layer with openings defining the pattern to be printed. High resolution frames have more than 130 meshes per cm, enabling them to print lines with a width down to 50 μm. A typical thickness of printed films is 10 μm. An alternative to framed meshes is a metal mask with openings for the pattern. Lifting off this mask

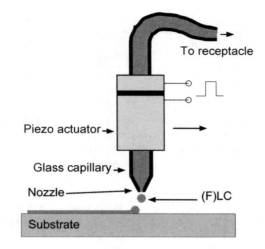

Figure 22.3 Inkjet printing

Figure 22.4 Contact angle of a drop on a surface

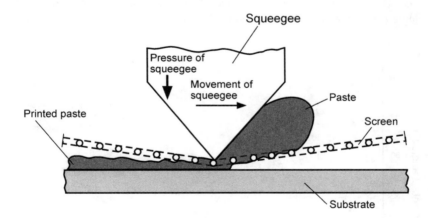

Figure 22.5 Silk screen printing

after printing may degrade the line definition, whereas the meshes being pressed down during printing jump back up while the ink still has a low viscosity. After printing the ink has to be burnt, driving out the ingredients providing a low viscosity and thereby solidifying the printed layer.

Table 22.1 lists some properties of the four printing methods (Randler *et al.*, 2000).

Table 22.1 Properties of four printing methods

	Flexographic printing	Ink jet printing	Knife coating	Silk screen printing
Minimum line	50 μm ... 60 μm		only for areas	60 μm ... 70 μm
Thickness of layer				
wet	1 μm ... 2 μm	defined by	defined by	15 μm ... 25 μm
dry	20 μm ... 2 μm	volume of drop	spacers	5 μm ... 25 μm
Required viscosity	10 ... 300 mPas	10 ... 20 mPas linear region	—	10 ... 100 Pas plastic
Required cleanliness of substrate	good	very good	good	moderate
Defects on substrates	no	no	possible	no
Printing speed	high	low	medium	moderate
Application	polyimide glue LC-material	LC material glue	LC material	etch resist glue

22.2 Printing of Layers for LCDs

Printing starts with a thorough cleaning of the substrate by a UV, plus ozone treatment removing organic contaminants. An indication of the cleanliness of the surface reached is a high surface tension indicated by a low contact angle in Figure 22.4 for a drop of water on that surface. Figure 22.6 shows the influence of the time for the surface treatment on the contact angle if unprocessed ITO is put on PES (Randler *et al.*, 1999, 2000; Lueder, 1999b).

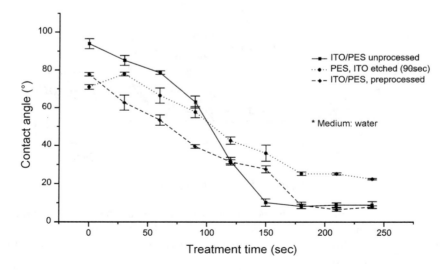

Figure 22.6 Influence of UV+Ozone cleaning time on contact angle

An unmodified polyimide orientation layer cannot be printed rapidly, volatile solvents have to be added to it. Such solvents are N-methyl-2-pyrrolidone (NMP), together with butyl-cellose (BC). The 20 nm thick polyimide layers printed flexographically possess a standard deviation of 2.5 nm, whereas a spin coated layer has a standard deviation of only 1 nm. The surface roughness of the printed polyimide is 5 nm.

The first FLC layers were printed in a polymer stabilized version (Okoshi *et al.*, 1998; Furue *et al.*, 1998). In printing FLCs without this stabilization, best results are achieved by printing them in the isotropic phase (Randler *et al.*, 2000). In this phase, mobility of the LC molecules is high, allowing them to align in the grooves of the rubbed polyimide. The proper alignment is enhanced by a printing direction antiparallel to the rubbing direction. The proper anchoring direction manifests itself in a low addressing voltage. Misaligned molecules are anchored in the wrong direction. As a consequence, the electric field necessary to align the molecules as required in the field-on state is larger in the misaligned areas, resulting in an increased voltage. The switching voltage versus pulse duration for FLCs

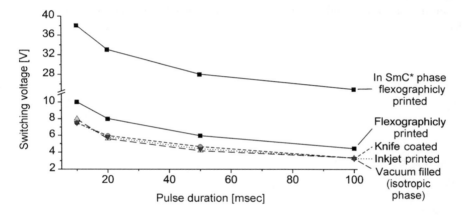

Figure 22.7   Switching voltage of printed FLCDs versus pulse duration for various printing processes

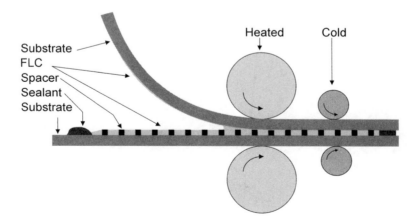

Figure 22.8   Cell building by lamination of plastic substrates

printed in the isotropic state with four methods and in the SmC* state by flexographic printing is shown in Figure 22.7 (Randler *et al.*, 2000). The SmC* state has a higher viscosity which hampers adjustment. This can be overcome by strong directional forces during printing, *e.g.* as in knife coating. In flexographic printing, those directional forces are weak, reflected in a high addressing voltage.

22.3 Cell Building by Lamination

After printing of the FLC, the two plastic substrates are aligned and fixed at one end. The sealing frame is either dispensed or silk screen printed. Contacts between the two substrates are made with a conductive adhesive. Then the substrates are laminated as shown in Figure 22.8 . A pair of heated rolls is required to allow the surplus material to flow outside the cell area. This way the cell gap defined by the etched spacers is realized. The cold rolls in Figure 22.8 ensure that the cell gap does not change after lamination. The sealant is cured by UV-light. Plastic substrates and printing are emerging display technologies (Randler *et al.*, 1999).

Appendix 1: Formats of Flat Panel Displays

Common sizes (no. of columns × no. of rows)		Aspect ratio (vertical:horizontal)
VGA	640 × 480	3:4
SVGA	800 × 600	3:4
XGA	1024 × 768	3:4
wide XGA	1366 × 768	9:16
SXGA	1280 × 1024	4:5
UXGA	1600 × 1280	3:4
NTSC	450 × 484	3:4
US HDTV	1280 × 720 (F1)	9:16
	1920 × 1080 (F2)	9:16
QSXGA	2560 × 2048	4:5

Appendix 2: Optical Units of Displays

Physics related parameters are:

Radiant flux of a light
source: Φ_e in Watt, W

Radiant intensity: $I_e = \dfrac{\Phi_e}{sr}$ in Watt per steradian, W sr^{-1}

Steradian: $sr = \dfrac{\text{area } A_0 \text{ on a sphere of radius R exposed to radiation}}{R^2} = \Omega$

$$sr = \frac{A_0}{R^2} = \frac{a \cdot 4\pi R^2}{R^2} = \alpha \cdot 4\pi$$

$\alpha 4\pi R^2 = $ portion of entire area $4\pi R^2$ of a sphere
exposed to light; sr is also called solid angle

Spectral radiant flux: $\dfrac{d\Phi_e}{d\lambda} = \Phi_{e\lambda}\dfrac{W}{nm}, \lambda = $ wavelength

*Efficiency of an electro-
optical light source*: $\eta_e = \dfrac{\text{power of light output } \Phi_e \text{ in W}}{\text{power of electrical input in W}}$ in %

Physiological parameters are:

*Spectral perception sensitivity
of the human eye*: $V(\lambda) = $ dimensionless weighting function
of the eye

Luminous flux:
$$\Phi = \int_{\lambda_1}^{\lambda_2} \Phi_{e\lambda} V(\lambda) d\lambda \quad \text{lumen}, \ell m$$

Luminous intensity:
$$I = \frac{\Phi}{\Omega} \text{ in cd} \quad cd = \frac{\ell m}{sr}, \quad cd = \text{candela}$$

Luminance:
$$L = \frac{I}{\text{area A}} cd/m^2 = \frac{\ell m}{sr \ m^2}$$
$$\text{often called brightness}$$

Illuminance:
$$E = \frac{\Phi}{A} \ell m/m^2 = \text{lux}, lx = cd \ sr \ m^{-2}$$

Efficiency in physiological terms:
$$\eta = \frac{\text{luminous flux } \Phi \text{ of light output}}{\text{power of electrical input}} \quad \frac{\ell m}{W}$$

Contrast C of a display
$$C = \frac{\text{highest luminance in a picture}}{\text{lowest luminance in a picture}}$$

Since ambient light affects contrast of a transmissive display, C is measured in darkness.

Aperture ratio A of a pixel:
$$A = \frac{\text{area of a pixel transmitting light}}{\text{entire area of a pixel}} \text{in } \%$$

Older units: $1 \text{ Nit} = 1 \text{ nt} = 1 \text{ cd}/m^2$
$1 \text{ Stilb} = 1 \text{ sb} = 10^4 \text{ nt} = 1 \text{ cd}/cm^2$
$1 \text{ Lambert} = 1 \text{ L} = \frac{1}{\pi} \text{ cd}/cm^2$
$1 \text{ Footlambert} = 1 \text{ fL} = \frac{1}{\pi} \text{ cd}/ft^2$

Appendix 3: Properties of Polarized Light

The description follows the presentation in Born and Wolf (1980).
We start with Equations (3.42) and (3.43) for the components of the electric field

$$E_x(z, t) = A_x \cos (\omega t + \varphi - \delta_1) \tag{1}$$

and

$$E_y(z, t) = A_y \cos (\omega t + \varphi - \delta_2), \tag{2}$$

with

$$\delta_1 = k_x z, \quad \delta_2 = k_x z - 2\pi \frac{\Delta n}{\lambda}, \quad A_x = E_{\xi 0} \cos \alpha \quad \text{and} \quad A_y = E_{\zeta 0} \sin \alpha.$$

The locus of the vector $E = x_0 E_x + y_0 E_y$ in the x-y-plane as time t evolves is calculated by eliminating $\tau = \omega t + \varphi$. Rewriting Equations (1) and (2) yields

$$\frac{E_x}{A_x} = \cos \tau \cos \delta_1 + \sin \tau \sin \delta_1 \tag{3}$$

and

$$\frac{E_y}{A_y} = \cos \tau \cos \delta_2 + \sin \tau \sin \delta_2, \tag{4}$$

resulting in

$$\frac{E_x}{A_x}\sin\delta_2 - \frac{E_y}{A_y}\sin\delta_1 = \cos\tau(\cos\delta_1\sin\delta_2 - \cos\delta_2\sin\delta_1) = \cos\tau\sin(\delta_2 - \delta_1) \quad (5)$$

and

$$\frac{E_x}{A_x}\cos\delta_2 - \frac{E_y}{A_y}\cos\delta_1 = \sin\tau\sin(\delta_2 - \delta_1). \quad (6)$$

Squaring and adding Equations (5) and (6) yields

$$\left(\frac{E_x}{A_x}\right)^2 + \left(\frac{E_y}{A_y}\right)^2 - 2\frac{E_x E_y}{A_x A_y}\cos\delta = \sin^2\delta, \quad (7)$$

with

$$\delta = \delta_1 - \delta_2 = 2\pi\frac{\Delta n}{\lambda}z. \quad (8)$$

This is the equation of the ellipse drawn and described in Figure 3.5. We rotate the coordinates in Figure 3.5 by an angle Ψ into the new coordinates ξ and η, which coincide with the main axis of the ellipse. The rotation matrix in Equation (3.28) gives the following relation between the components E_ξ and E_η in the $\xi-\eta$-coordinates and E_x and E_y:

$$E_\xi = E_x\cos\psi + E_y\sin\psi \quad (9)$$

$$E_\eta = -E_x\sin\psi + E_y\cos\psi. \quad (10)$$

The equation for an ellipse with the principal axis a and $b \leq a$ in the coordinates parallel to these axes is in parameters τ and δ_0

$$E_\xi = a\cos(\tau + \delta_0) \quad (11)$$

and

$$E_\eta = \pm b\sin(\tau + \delta_0). \quad (12)$$

The sign of E_η distinguishes the two possible rotations of the vector E with time τ along the ellipse. For the positive sign the rotation is counter clockwise, or left-handed in Figure 3.6 and for the negative sign clockwise or right-handed. We determine a and b by comparing Equations (9) and (10) with Equations (11) and (12) by using Equations (3) and (4):

$$a(\cos\tau\cos\delta_0 - \sin\tau\sin\delta_0) = A_x(\cos\tau\cos\delta_1 - \sin\tau\sin\delta_1)\cos\psi$$
$$+ A_y(\cos\tau\cos\delta_2 - \sin\tau\sin\delta_2)\sin\psi$$

and

$$\pm b(\sin \tau \cos \delta_0 + \cos \tau \sin \delta_0) = -A_x(\cos \tau \cos \delta_1 - \sin \tau \sin \delta_1)\sin \psi$$
$$+ A_y(\cos \tau \cos \delta_2 - \sin \tau \sin \delta_2)\cos \psi.$$

We equate the coefficients of $\cos \tau$ and $\sin \tau$, and obtain

$$a \cos \delta_0 = A_x \cos \delta_1 \cos \psi + A_y \cos \delta_2 \sin \psi \tag{13}$$

$$a \sin \delta_0 = A_x \sin \delta_1 \cos \psi + A_y \sin \delta_2 \sin \psi \tag{14}$$

$$\pm b \cos \delta_0 = A_x \sin \delta_1 \cos \psi - A_y \sin \delta_2 \cos \psi \tag{15}$$

$$\pm b \sin \delta_0 = -A_x \cos \delta_1 \sin \psi + A_y \cos \delta_2 \cos \psi. \tag{16}$$

Squaring and adding Equations (13) and (14) (resp. (15) and (16)) and using Equation (8) provides

$$a^2 = A_x^2 \cos^2 \psi + A_y^2 \sin^2 \psi + 2A_x A_y \cos \psi \sin \psi \cos \delta \tag{17}$$

$$b^2 = A_x^2 \cos^2 \psi + A_y^2 \sin^2 \psi - 2A_x A_y \cos \psi \sin \psi \cos \delta, \tag{18}$$

and hence

$$a^2 + b^2 = A_x^2 + A_y^2. \tag{19}$$

To obtain ψ, we multiply Equations (13) and (15) (resp. (14) and (16)), and add, resulting in

$$\mp ab = A_x A_y \sin \delta. \tag{20}$$

The division of Equation (15) by (13) (resp. (14) by (16)) yields

$$\pm \frac{b}{a} = \frac{A_x \sin \delta_1 \sin \psi - A_y \sin \delta_2 \cos \psi}{A_x \cos \delta_1 \cos \psi + A_y \cos \delta_2 \sin \psi} = \frac{-A_x \cos \delta_1 \sin \psi + A_y \cos \delta_2 \cos \psi}{A_x \sin \delta_1 \cos \psi + A_y \sin \delta_2 \sin \psi},$$

from which the following equation for ψ results:

$$(A_x^2 - A_y^2)\sin 2\psi = 2A_x A_y \cos \delta \cos 2\psi. \tag{21}$$

With $A_y/A_x = \tan \gamma_0$, Equation (21) becomes

$$\tan 2\psi = \frac{2A_xA_y}{A_x^2 - A_y^2} \cos \delta = \frac{2 \tan \gamma_0}{1 - \tan^2 \gamma_0} \cos \delta. \tag{22}$$

or

$$\tan 2\psi = (\tan 2\gamma_0) \cos \delta. \tag{23}$$

From Equations (19) and (20), we obtain

$$\mp \frac{2ab}{a^2 + b^2} = \frac{2A_xA_y}{A_x^2 + A_y^2} \sin \delta = (\sin 2\gamma_0) \sin \delta. \tag{24}$$

The additional auxiliary angle Ω with $-(\pi/4) \leq \Omega \leq (\pi/4)$ defines the ellipticity e as

$$e = \pm \frac{b}{a} = \tan \Omega. \tag{25}$$

For left-handed polarization we assign the upper sign to e, and for right-handed polarization the lower sign as in Equation (15). Thus, $\Omega \in [-(\pi/4), 0)$ stands for left-handed and $\Omega \in (0, \pi/4)$ is right-handed polarization. Equation (24) is rewritten using Equation (25) as

$$\sin 2\Omega = (\sin 2\gamma_0) \sin \delta. \tag{26}$$

As a result, the three parameters A_x, A_y and δ describing the polarization are transformed into three equivalent ones, namely $a^2 + b^2$, ψ and Ω by the following equations: from Equation (19):

$$a^2 + b^2 = A_x^2 + A_y^2, \tag{27}$$

from Equation (23):

$$\tan 2\psi = (\tan 2\gamma_0) \cos \delta, \tag{28}$$

and from Equation (26):

$$\sin 2\Omega = (\sin 2\gamma_0) \sin \delta, \tag{29}$$

with

$$\tan \gamma_0 = A_y/A_x, \tag{30}$$

from Equation (20). $a^2 + b^2$ stands for the intensity of the light, ψ is the angle of the ellipse to the x-y-coordinates in Figure 3.5 and Ω is related to the ellipticity of the ellipse in Figure 3.5. The orientation is given by the sign of b or, as shown in Chapter 3, by the sign of $\sin \delta$. The two special cases of the polarization, the linear and the circular polarization are discussed in Equations (3.52) through (3.55).

The three parameters A_x, A_y and δ on the other hand, can be replaced by three additional equivalent parameters called the Stokes parameters:

$$S_0 = A_x^2 + A_y^2 \tag{31}$$

$$S_1 = A_x^2 - A_y^2 \tag{32}$$

$$S_2 = 2A_xA_y \cos \delta, \tag{33}$$

and

$$S_3 = 2A_xA_y \sin \delta, \tag{34}$$

at has to be proven.

One of the four parameters is dependent on the others as

$$S_0^2 = S_1^2 + S_2^2 + S_3^2. \tag{35}$$

The proof of the Stokes parameters with the parameters $a^2 + b^2$, ψ and Ω in Equations (27), (28) and (29) with Equation (30) starts with Equations (31) and (27), stating that

$$S_0 = a^2 + b^2 = A_x^2 + A_y^2, \tag{36}$$

which is the intensity of the light.

From Equation (20) follows $\sin^2 \gamma_0 = A_y^2/(A_x^2 + A_y^2)$ and $\cos^2 \gamma_0 = A_x^2/(A_x^2 + A_y^2)$, and further $\sin \gamma_0 \cos \gamma_0 = (1/2) \sin 2\gamma_0 = A_xA_y/(A_x^2 + A_y^2) = (A_xA_y/S_0)$.

Inserting $A_xA_y = (1/2)S_0 \sin 2\gamma_0$ into Equations (34) provides $S_3 = S_0 \sin 2\gamma_0 \sin \delta$, which is (with Equation (29))

$$S_3 = S_0 \sin 2\Omega. \tag{37}$$

With Equations (22) and (32) we obtain from Equation (33)

$$S_2 = S_1 \tan 2\psi \tag{38}$$

Substituting S_3 in Equation (37) and S_2 from above into Equation (35) results in

$$S_1 = S_0 \cos 2\psi \cos 2\Omega, \tag{39}$$

and with Equation (38) in

$$S_2 = S_0 \cos 2\Omega \sin 2\Omega. \tag{40}$$

Equations (36), (37), (39) and (40) express the Stokes parameters in the already known parameters $a^2 + b^2$, ψ and Ω, and allow a geometrical interpretation by the Poincaré sphere in Figure 4.5. The radius of the sphere is $S_0 = a^2 + b^2$, which stands for the intensity of the light. A point P on the sphere is given by the two angles 2Ψ and 2Ω in Figure 4.5, resulting in the Cartesian coordinates of point P, as in Equations (37), (39) and (40).

References

Alt, P. M. and Pleshko, P. (1974) Scanning limitations of liquid crystal displays. *IEEE Electron Devices*, ED-21.

Azzam, R. M. A. and Bashara, M. M. (1972b) Simplified approach to the propagation of polarized light in anisotropic media–application to liquid crystals. *J. Opt. Soc. A*, **62**(11), 1252.

Azzam, R. M. A. and Bashara, N. M. (1972a) Polarization transfer function of an optical system as a bilinear transform. *J. Opt. Soc. A*, **62**, 222, 336.

Baeuerle, R. (2000) Technologie und Bau MIM-adressierter Fluessigkristall-Bildschirme mit Kunststoff- und Glassubstraten. Dissertation, Universität Stuttgart.

Bahadur, B. (1990) *Liquid Crystals, Applications and Uses, vol. 1*. World Scientific.

Ban, A. *et al.* (1996) A simplified process for SVGA TFT-LCDs with single-layered ITO source buslines. *SID 96 Digest*.

Beresnev, L. A. *et al.* (1989) Deformed helix ferroelectric liquid crystal display: a new electro optic mode in ferroelectric chiral smectic C liquid crystals, *Liquid Crystals*, **5**(4).

Berman, A. L. and Oh, C. S. (1980) *The Physics and Chemistry of Liquid Crystal Devices*. Plenum Press.

Berreman, D. W. (1972) Optics in stratified and anisotropic media: 4×4 matrix formulation. *J. Opt. Soc. A*, **62**.

Bleha, W. P. (1995) Development of ILA projectors for large screen display. *Int. Displ. Res. Conf. Record*.

Bleha, W. P. *et al.* (1978) Application of the liquid crystal light valve to real-time optical data processing. *Optical Eng.*, **17**(4), July/August.

Bodewig, E. (1959) *Matrix Calcules, 2nd ed.*, North Holland, Amsterdam.

Booth, C. J. (1998) In: *Handbook of Liquid Crystals, vol. 2A*. Wiley-VHC.

Borkan, H. and Weimer, P. K. (1963) An analysis of the characteristics of insulated gate thin film transistors. *RCA Review*, **24**(2).

Born, M. and Wolf, E. (1980) *Principles of Optics, 6th ed.*, Pergamon Press, Oxford.

Bos, P. J. *et al.* (1999) Status and trends in nematic bistable liquid crystal displays. *Proc. Eurodisplay 99*, Berlin.

Bos, P. J. and Koehler, K. R. (1984) The pi-cell: A fast liquid crystal optical switching device. *Mol. Cryst. Liq. Cryst.*, **113**.

Bos, P. L. and Rahman, J. A. (1993) An optically 'self-compensating' electro-optical effect with wide angle of view. *SID 93 Digest*.

Brennesholtz, M. S. (1997) Fundamentals of projection displays. *SID Short Course S-4*.

Brody, T. P. *et al*. (1973) A 6" × 6" 20 lines per inch liquid crystal display panel. *IEEE Trans. El. Devices* **ED 20**.

Brotherton, S. D. *et al*. (1997) Influence of melt depth in laser crystallized poly-Si TFTs. *J. Appl. Phys.*, **82**(8).

Bryan-Brown, G. P. *et al*., Grating aligned bistable nematic device. *SID 97 Digest*.

Buerkle, R. (1997) Ferroelektrische Fluessigkristall-Bildschirme als Lichtmodulatoren für die optische Bildverarbeitung und als bistabile Folienanzeigen. PhD Thesis, University of Stuttgart.

Bunz, R. (1999) Bistabile cholesterische Fluessigkristalldisplays mit Glas- und Kunststoffsubstraten. Diss. Universität Stuttgart.

Bunz, R. *et al*. (1997) Improved contrast ratio bistable cholesteric displays with an obliquely sputtered SiO_2 aligned layer. *SID 97 Digest*.

Buzak, T. S. (1990) A new active matrix technique using plasma addressing. *SID 1990 Digest*.

Buzak, Th. (1999) Plasma addressed liquid crystal displays. *Proc. Eurodisplay 99*, Berlin, Germany.

Cacharelis, P. *et al*. (1997) An 0.8 micron EEPROM technology modified for a reflective PDLC light valve application. *SID 1997 Digest*.

Carey, P. G. *et al*. (1997) Polysilicon TFT fabrication on plastic substrates. *SID 97 Digest*.

Castleberry, D. E. (1979) Varistor-controlled liquid crystal displays. *IEEE Trans. El. Devices*, **26**(8).

Chiang, Ch. and Kaniki, J. (1996) AMLCD lifetime evaluation based on an electrical instability of amorphous silicon thin film transistors. *Proc. Eurodisplay 96*.

Chien, L. C. *et al*. (1995) Multicolor reflective cholesteric displays. *SID 95 Digest*.

Clark, N. A. *et al*. (2000) FLC microdisplays, *Proc. Ferroel. LCD Conference*, Boulder, CO.

Clark, N. A. and Lagerwall, S. T. (1980) Submicrosecond bistable electro-optic switching in liquid crystals. *Appl. Phys. Lett.*, **36**.

Clifton, B. *et al*. (1993) Optimum row functions and algorithms for active addressing. *SID 93 Digest*.

Clifton, B. and Price, D. (1992) Hardware architecture for video-rate active addressed STN-displays. *Proc. Japan Display (12th Int. Displ. Res. Conf.)*.

Coates, D. *et al*. (1996) High-performance wide-bandwidth reflective cholesteric polarizers. *SID 96 Applications Digest*.

Colgan, E. G. and Uda, M. (1998) On-Chip metallization layers for reflective light valves. *IBM J. Res. and Dev.*, **42**(2−4), May/July.

Colos, H. (1998) In: *Handbook of Liquid Crystals, vol. 2A*. Wiley-VHC.

Conner, A. R. and Scheffer, T. J. (1992) Pulse-Height Modulation (PHM) grey shading methods for passive matrix LCDs. *Proc. Japan Display*.

Davies, D. *et al*. (1997) Multiple color high resolution reflective cholesteric liquid crystal displays. *SID 97 Digest*.

De Smet, H. *et al*. (1999) The design and fabrication of a 2560 × 2048 pixel microdisplay. *Proc. Eurodisplay 99* (http://www.elis.rug.ac.be/EL/Sgroups/tcg/projects/microdis.html).

De Vleeschouwer, H. *et al*. (1999) Image sticking theories and interpretation of measurements. *Proc. Eurodisplay 99*, Berlin, Germany.

Degen, W. H. (1980) *Physical Properties of Liquid Crystalline Materials*. Gordon and Breach, London.

Demus, D., Goodby, J., Gray, G. W., Spiess, H.-W. and Vill, V. (1998a) *Handbook of Liquid Crystals, vol. 2A*. Wiley-VCH.

Demus, D., Goodby, J., Gray, G. W., Spiess, H.-W. and Vill, V. (1998b) *Handbook of Liquid Crystals, vol. 2B*. Wiley-VCH.

De Gennes, P. G. and Prost, J. (1993) *The Physics of Liquid Crystals, 2nd ed.*, Clarendon, Oxford.

Dijon, J. (1990) Liquid crystals applications and uses. In: B. Bahadur (ed.), *Ferroelectric LCDs*. World Scientific, **1**.

Doane, J. W. *et al.* (1986) *Appl. Phys. Lett.*, **48**.

Doane, J. W. *et al.* (1992) Front lit flat panel displays from polymer stabilized cholesteric textures. *Japan Display 1992*.

Dozov, I. *et al.* (1997) Fast bistable nematic display from coupled surface anchoring breaking. *Proc. El. Imaging 97* (3015 – Liquid Crystal Materials, Devices and Applications V, February 9–14 1997, San Jose, USA).

Drzaic, P. S. *et al.* (1989) High brightness and color contrast displays constructed from nematic droplet/ polymer films incorporating pleochroic dyes. *SPIE*, **1080**(41).

Eaton Corp. (1997) *Flat panel equipment. High throughput ion implanter for AMLCD*. March.

Eblen, J. P. *et al.* (1994) Birefringent compensators for normally white TN-LCDs. *SID 94 Digest*.

Eblen, J. P. *et al.* (1997) Advanced gray scale compensator for TN-LCDs for avionics applications. *SID 97 Digest*.

Egelhaaf, J. *et al.* (2000) A super high aperture ratio 4 in. diagonal reflective TFT-LCD with overlapping pixel electrodes. *Proc. IDW 2000*, Kobe, Japan.

Entenberg, A. *et al.* (1987) Stress measurement in sputtered copper films of flexible polymide substrates. *J. Vac. Sci. Technol.*, **A5**(6).

Erhart, A. and McCartney, D. (1997) Charge conservation implementation in an ultra low power AMLCD column driver utilizing pixel inversion. *SID 97 Digest*.

Escher, C. *et al.* (1991) The SSFLC switching behaviour in view of Chevron layer geometry and ionic charges. *Ferroelectrics*, **113**.

Fünfschilling, J. and Schadt, M. (1994) Physics and electronic model of deformed helix ferroelectric liquid crystal displays. *Jap. J. Appl. Phys.*, **33**.

Fergason, J. (1966) Liquid crystals. *Proc. 2nd Kent Conf.*, Gordon and Breach, NY.

Fergason, J. L. *et al.* (1986) Polymer encapsulated liquid crystals for use in high resolution and color displays. *SID 86 Digest*.

Fréedericksz, V. and Zolina, V. (1933) Forces causing the orientation of an anisotropic liquid. *Trans. Farday Soc.* **29**.

Frey, V. (2000) Herstellung ferroelektrischer Folienanzeigen und Untersuchungen zum Schaltverhalten ferroelektrischer Fluessigkristalle. PhD thesis, University of Stuttgart.

Fryer, P. M. *et al.* (1998) A six mask TFT-LCD process using copper-gate metallurgy. *SID 96 Digest*.

Fuhrmann, J. *et al.* (1995) Improving the image quality of MIM-LCDs by compensative addressing. *Asia Display 95*, Hamamatsu, Japan.

Furue, F. *et al.* (1998) Characteristics and driving scheme of polymer stabilized monostable FLCD exhibiting fast response time and high contrast ratio with gray-scale capability. *SID 98 Digest*.

Ginter, E. *et al.* (1993) Optimized PDLCs for active-matrix addressed light valves in projection systems. *Proc. Eurodisplay 93*, Strasbourg, France.

Glueck, J. (1995). Mit a-Si:H-Dünnschichttransistoren angesteuerte flache Fluessigkristall– Bildschirme für Direktsicht und Projektion. PhD-Thesis, University of Stuttgart.

Glueck, J. *et al.* (1992) Color TV projection with fast switching reflective HAN-mode lightvalves. *SID 92 Digest*.

Glueck, J. *et al.* (1994) A 14-in.-diagonal a-Si-TFT AMLCD for PAL television. *SID 94 Digest*.

Gooch, C. H. and Tarry, H. A. (1974) Optical characteristics of twisted nematic Liquid-Crystal Films. *Electronic Lett.*, **10**, January.

Gore, J. M. (1990) *Mechanics of Materials*. PWS-Kent, Boston.

Gove, R. J. (1994) DMD display systems: The impact of an all digital display. *SID 1994 Digest*.

Grimsdilch, M. *et al.* (1978) Brillouin scattering from hydrogenated amorphous silicon. *Solid State Com.*, **26**.

Grinberg, J. and Jacobson, A. D. (1976) Transmission characteristics of twisted nematic liquid crystal layer. *J. Opt. Soc. A*, **66**(10).

Haas, G. (1999) Angular dependency of liquid crystal displays. *Workshop Eurodisplay 99*, VDE Verlag, pp. 5–24.

Harada, T. *et al.* (1985) An application of chiral nematic-C liquid crystal to a multiplexed large area display. *Digest SID 85*.

Harrer, Th. (to appear) Dissertation, University of Stuttgart.

Harrer, Th. *et al.* (1999) Single area excimer laser crystallization for the fabrication of bottom gate and top gate TFTs. *Proc. Eurodisplay 1999*, Berlin, Germany.

Heilmeier, G. H. and Zanoni, L. A. (1968) Guest-Host interactions in nematic liquid crystals. A new electro-optic effect. *Appl. Phys. Lett.*, **13**(3).

Hirai, Y. *et al.* (1995) An improved architecture of the multiple line addressed (MLA) LCD. *Proc. Asia Display*, p. 237.

Hochholzer, V. *et al.* (1994) A full color 14″ MIM-LCD with improved photolithography. *SID 94 Digest*.

Hoke, C. D. and Bos, P. J. (1998) A bistable twist cell exhibiting long term bistability suitable for page-sized displays. *Proc. IDW 98*.

Howard, W. E. (1995) Active matrix LCDs. *Seminar M5 at SID*.

Huang, X. Y. *et al.* (1995) Dynamic drive for bistable reflective cholesteric displays: a rapid addressing scheme. *SID 1995 Digest*.

Huang, X. Y. *et al.* (1996) High performance dynamic drive scheme for bistable reflective cholesteric displays. *SID 1996 Digest*.

Iberaki, N. (1990) Low temperature poly-Si-TFT-technology. *SD 99 Digest*.

Ihara, S. *et al.* (1992) A color STN-LCD with improved contrast, uniformity and response times. *SID Digest of Techn. Papers XXIII*.

Iimura, Y. and Kobayashi, S. (1994) Electrooptic characteristics of amorphous super-multidomain TN-LCDs prepared by a non rubbing method. *SID 94 Digest*.

Ikeda, M. *et al.* (1999) Characteristics of low-temperature-processed a-Si TFT on plastic substrates. *Int. Display Workshops*.

Jones, J. C. *et al.* (1998) Novel configuration of the zenithal bistable nematic liquid crystal device. *SID 98 Digest*.

Jones, R. C. (1941) New calculus for the treatment of optical systems. *J. Opt. Soc. Am.* **31**, 488.

Kakizaki, T. *et al.* (1996) Development of 25 in. active-matrix LCD using plasma addressing for video rate light quality displays. *SID 1996 Digest*.

Kaneko, E. (1987) *Liquid Crystal TV Displays*. KTK Scientific Publishers, Tokyo.

Kaneko, E. (1998) Active matrix addressed displays. In: Demus, D. *et al.* (ed.), *Handbook of Liquid Crystals*, vol. 2A. Wiley-VCH.

Kanoh, H. *et al.* (1998) A 28 cm diagonal 177 dpi UXGA color reflective Guest-Host TFT-LCD. *Asia Display 98*.

Kats, E. I. (1971) *Soc. Phys. JETP*, **32**.

Kawakami, E. *et al.* (1980) Brightness uniformity in liquid crystal displays. *Digest SID*.

Kawakami, H. (1976) US Patent 3976362.

Kawakami, H. *et al.* (1976) Matrix addressing technology of twisted nematic liquid crystal display. *SID-IEEE Record of Biennial Display Conf.*

Khakzar, K. (1991) Modellierung von a-Si-Dünnschichttransistoren und Entwurf von Ansteuerschaltungen für flache Fluessigkristall-Bildschirme. Diss. Univ. Stuttgart.

Kim, S. G. and Hwang, K. H. (1999) Thin-film micromirror array (TMA) for information display systems. *Proc. Eurodisplay 99*.

Kim, S. S. (1996) High aperture-ratio fault tolerant TFT-LCD using a-Si TFTs. *SID 96 Digest*.

Kim, S. T. *et al.* (1997) A novel method of charge-recycling TFT-LCD source driver for low power consumption. *Proc. IDW 1997*.

Kitazawa, T. *et al.* (1994) A 9.5-in. TFT-LCD with an ultra high aperture ratio pixel structure. *IDRC 94 Digest*.

Klette, R. *et al.* (1995) A reflective MIM-addressed PDLC-display with plastic substrates. *Asia Display 95*, Hamamatsu.

Koisnai, H., Iimura, Y. and Kobayashi, S. *et al.* (1995) Polarizer-free reflective amorphous chiral guest-host LCDs. *SID 95 Digest*.

Kuijk, K. E. *et al.* (1999) Minimum voltage driving of STN LCDs by optimized multiple row addressing. *Proc. Eurodisplay 99*.

Kuo, C. *et al.*, Reflective MTN-mode TFT-LCD with video rate and full color capabilities. *SID 97 Digest*, p. 79.

Kuo, C., Wie, C., Wu, S. and Wu, C. (1996) Reflective Displays using mixed-mode twisted nematic cells with wide viewing angle and high contrast. *SID-Proc. Eurodisplay*.

Labrunie, G. and Robert, J. (1973) *J. of Appl. Phys.*, **44**, 487.

Lambda Physik (2000) *Lambda Physik Industrial Report*, April, No. 11.

Lauer, H. U. (1996) Ansteuerung von TFT- und MIM-adressierten Fluessigkristall-anzeigen mit Farbfernsehsignalen. Diss. Univ. Stuttgart.

Lauer, H.-U. *et al.* (1990) A frame sequential color-TV projection display. *SID 90 Digest*.

Lee, M. J. *et al.* (1991) High mobility cadmium selenide transistors. *Proc. IDRC*.

Lee, S. W. *et al.* (1998) A low power poly-Si TFT-LCD with integrated 8-bit digital data drivers. *Asia Display 1998*.

Lowe, A. C. and Cox, R. J. (1981) Order parameter and the performance of nematic guest-host displays. *Mol. Cryst. Liq. Cryst.*, **66**.

Lowe, A. C. and Hasegawa, M. (1997) High reflectivity double cell nematic guest-host display. *SID 1997 Digest*.

Lu, K. and Saleh, B. E. A. (1990) Theory and design of the Liquid Crystal TV as an optical spatial phase modulator. *Opt. Eng.*, **29**.

Lueder, E. (1994a) Enhancement of luminance in LC-projectors. *Int. Workshop on AMLCs 94*, Shinjuku, Japan.

Lueder, E. (1994b) Fabrication of CdSe-TFTs and implementation of integrated drivers with poly-crystalline TFTs. *Proc. IDRC*.

Lueder, E. (1994c) Simplified manufacture of a-Si-TFTs with four or three masks. *2nd LCD Seminar*, (Nikkei) Chiba, Japan.

Lueder, E. (1998a) Fundamentals of Passive and Active Matrix Liquid Crystal Displays. Short Course S-1, SID, May.

Lueder, E. (1998b) Tendenzen in Forschung und Produktion bei Fluessigkristall-Bildschirmen. *Fernseh- und Kino-Technik*, **51**(3).

Lueder, E. *et al.* (1998c) Reflective FLCDs and PECVD-generated a-Si-TFTs with plastic substrates. *Asia Display 98*, p. 173.

Lueder, E. (1999a) Passive and active matrix liquid crystal displays with plastic substrates. *El.-Chem. Soc. Proc.*, **98–22**.

Lueder, E. (1999b) Passive and MIM-driven LCDs with plastic substrates. *Int. Display Workshop 99 Digest*.

Lueder E. (2000) SSFLCDs with plastic substrates fabricated by printing. *IDW 2000*, Kobe, Japan.

Luft, W. and Tsou, Y. S. (1993) *Hydrogenated Amorphous Silicon Alloy Deposition Processes*. Marcel Dekker, New York.

Luo, F. C. (1990) Active matrix LC displays, in liquid crystals applications and uses. B. Bahadur (ed.), World Scientific.

Möbus, D. (1990) Algorithmen zur Optimierung von Schaltungen und zur Loesung nichtlinearer Differentialgleichungen. Dissertation, Universitaet Stuttgart.

Maier, G. (1997) Technologie und Ansteuerung von spatialen Fluessigkristall-Lichtmodulatoren mit amorphen Silizium Dünnschichttransistoren. Dissertation, University of Stuttgart.

Maier, G. *et al.* (1997) Manufacturing process for a-Si-TFTs using ion implantation through the drain and source electrodes. *IDW 97 Proc.*, Nagoya, Japan.

342   REFERENCES

Manazawa, Y. *et al.* (1999) A 202 ppi TFT-LCD using low temperature poly-Si technology. *Eurodisplay 1999*, Berlin, Germany.

Maresch, S. *et al.* (1999) Lacrasil, A large area crystallization of amorphous silicon using a mesa-shaped laser. *Eurodisplay 1999*, p. 127.

Martinot-Lagarde, Ph. *et al.* (1997) Fast bistable nematic display using monostable surface anchoring switching. *SID 97 Digest*.

Matsueda, Y. *et al.* (1996) Low temperature poly-Si-LCD with integrated G bit digital data drivers. *SID 96 Digest*.

McCormick, C. S. *et al.* (1997) An amorphous silicon thin film transistor fabricated at 125°C by dc reactive magnetron sputtering. *Appl. Phys. Lett.*, **70**(2) 13 January.

Melcher, R. L. *et al.* (1998) Design and fabrication of a prototype projection data monitor with high information content. *IBM J. Res. and Dev.*, **42**(2–4), May/July.

Meyer, R. B. (1974) *V. Int. Liq. Cryst. Conf.*, June.

Miyachi, K. and Fukuda, A. (1998) *Antiferroelectric liquid crystals in Handbook of Liquid Crystals*, vol. 2B. Wiley-VCH.

Miyata, Y. *et al.* (1998) Dynamic characteristics of LCD addressed by a-Si-TFTs. *SID 98 Digest*.

Miyata, Y. *et al.* (1998) Two mask step-inverted staggered a-Si TFT-addressed LCDs. *SID 98 Digest*.

Mori, H. *et al.* (1997) Novel Optical compensation film for AMLCDs using a discotic compound. *SID 97 Digest*.

Nakamura, S. *et al.* (1997) Room-temperature continuous wave operation of InGaN multi-quantum-well-structure laser diodes with a long lifetime. *Appl. Phys. Lett.*, **70**.

Nehring, J. and Kmetz, A. (1979) Ultimate limits for matrix addressing of RMS-responding liquid crystal displays. *IEEE Trans. Electron Dev.*, **ED-26** (Also Ultimate limits for rms-addressing. In: *The Physics and Chemistry of Liquid Crystal Devices*, Plenum Press, NY, 1980).

Nehring, J. and Scheffer, T. J. (1998) On the electric field induced stripe instability threshold of twisted nematic layers. *Z. Naturforsch*, **45a**.

Nishimura, M. *et al.* (1998) A 21.3 in. UXGA TFT-LCD with new ringing method. *SID 98 Digest*.

Oh-e, M. *et al.* (1995) Principles and characteristics of Electro Optical behavior with in-plane switching mode. *Asia Display 95*.

Okabe, M. (1996) Low temperature poly-Si TFT-LCDs with monolithic drivers. *IDW 1996*.

Okada, M. *et al.* (1997) Reflective multicolor display using cholesteric liquid crystals. *SID 97 Digest*.

Okoshi, K. *et al.* (1998) FLC polymer and plastic-substrates for use in a large area optical shutter for 3D-TV. *SID 98 Digest*.

Polach, S. (1999) Matrix of light sensors addressed by a-Si:H TFTs on a flexible plastic substrate. *SPIE*, **364**, pp. 31–39.

Polach, S. (2000) Die Entwicklung von a-Si:H-TFT-adressierten Fluessigkristall-Displays und von Arrays von Lichtsensoren. Dissertation, Universität Stuttgart.

Polach, S. *et al.* (2000) A transmissive TN display addressed by a-Si:H TFTs on a plastic substrate. *Proc. IDW 2000*.

Prat, C. *et al.* (1999) A 200 ns-excimer laser pulse: a way toward optimization of the crystallization of a-Si for flat panel displays. *Eurodisplay 1999*, Berlin, Germany.

Priestley, E. B., Wojtowicz, P. J. and Sheng, P. (1979) *Introduction to Liquid Crystals*. Plenum Press, New York, London.

Randler, M. *et al.* (1999) Vacuum free manufacture of FLC-displays with plastic substrates, printed FLCs and laminated cells. *Proc. Eurodisplay 1999*, Berlin, Germany.

Randler, M. *et al.* (2000) Printing process for vacuum free manufacture of liquid crystal cells with plastic substrates. *SID 2000 Digest*.

Richou, R. *et al.* (1992) The 2S TFT process for low cost AMLCD manufacturing. *SID 92 Digest*.

Ricker, T. P. *et al.* (1987) Chevron local layer structure in surface stabilized ferroelectric smectic C cells. *Phys. Rev. Lett.*, **59**(23).

Rosenbluth, A. E. *et al.* (1998) Contrast properties of reflective liquid crystal light valves in projection displays. *IBM J. Res. Dev.* May–July.

Sage, I. (1992) In: *Liquid Crystals. Application and Uses, vol. 3.* World Scientific.

Saito, S. and Yamamoto, H. (1978) Transient behaviors of field-induced reorientation in variously oriented nematic liquid crystals. *Jap. J. of Appl. Phys.*, **17**(2).

Sakamoto, M. *et al.* (1997) Half-column-line driving method for low power and low cost TFT-LCDs. *SID 97 Digest.*

Sampsell, J. B. (1994a) An overview of the performance envelope of digital-micromirror-based projection display systems. *SID 94 Digest.*

Sampsell, J. B. (1994b) Digital micromirror device and its application to projection displays. *J. Vac. Sci. Technol. B*, **12**(5), November/December.

Sanford, J. L. *et al.* (1998) Silicon light valve array chip for high resolution reflective liquid crystal projection displays. *IBM J. Res. and Dev.*, **42**(2–4), May/July.

Sato, F. *et al.* (1997) High resolution and bright LCD projector with reflective LCD panels. *SID 1997 Digest.*

Schadt, M. (1999) Photo aligned liquid crystal displays and LC-polymer optical films. *Proc. Eurodisplay 99.*

Schadt, M. *et al.* (1996) Optical patterning of multidomain liquid crystal displays with wide viewing angle. *Nature*, **381**, 16 May. (*SID 97 Digest*, 1996/1997.)

Schadt, M. and Helfrich, W. (1971) Voltage-dependent optical activity of a twisted nematic liquid crystal. *Appl. Phys. Lett.*, **18**.

Scheffer, T. J. *et al.* (1993) Active addressing of STN displays for high performance video applications. *Displays Technology and Applications*, **14**.

Scheffer, T. J. and Clifton, B. (1992) Active addressing method for high contrast video rate STN-displays. *SID Digest of Techn. Papers XXIII.*

Scheffer, T. J. and Nehring, J. (1998) Supertwisted nematic LCDs. SID Lecture notes to Seminar M-6, vol. 1, May 1998, SID-Symposium in Anaheim, CA.

Schmidt, T. C. (2000) Fundamentals of projection displays. *SID Short Course S-1.*

Schneider, U. (1997) Entwicklung und Herstellung aktiv adressierter Fluessigkristall-Bildschirme fuer Projektion und Direktsicht. Dissertation, Universitaet Stuttgart.

Schott, D. J. *et al.* (1999) Reflective LCOS light valves and their application to virtual displays. *Proc. Eurodisplay 1999*, Berlin, Germany.

Schwarz, F. (1990) Treiberschaltungen in Dünnschichttechnik zur Ansteuerung von flachen Fluessig-kristallanzeigen. Diss. Univ. Stuttgart.

Schweikert, K.-H. *et al.* (1993) A fast driving scheme for an FLC-display in a pattern recognition system. *Proc. Eurodisplay 93.*

Seki, H. *et al.* (1996) A new reflective Guest-Host Display using a light scattering film. *SID 96 Digest.*

Shannon, J. M. *et al.* (1993) Electronic properties of a-SiN$_x$:H thin film diodes. *MRS Proc.*, **297**.

Shannon, J. M. *et al.* (1996) TFDs for high quality AMLCD on low temperature plastics. *Int. Display Workshop 96.*

Shields, S. E. and Bleha, W. P. (1990) Light valve and projection mode LCDs. In: B. Bahadur, (ed.), *Liquid Crystals, Application and Uses*, Vol. 1. World Scientific.

Simmons, J. G. (1967) Poole–Frankel effect and Schottky effect in metal-insulator-metal systems. *Phys. Rev.*, **155**(3), March.

Specht, J. (2000) Reflektive TN-Displays mit einem Polarisator. *Master Thesis*, Universitaet Stuttgart.

Sperger, R. *et al.* (1993) High performance patterned all-dielectric interference color filters for display applications. Balzers product information.

Stein, C. R. and Kashnow, R. A. (1971) A two frequency coincidence addressing scheme for nematic liquid crystal displays. *Appl. Phys. Lett.*, **19**(9) November.

Stroomer, M. V. C. (1984) Design of thin film transistors for matrix addressing of television liquid crystal displays. *Proc. Eurodisplay 1984.*

Stupp, E. H. and Brennesholtz, M. S. (1999) *Projection Displays*. John Wiley & Sons, SID Series in Display Technology.

Sun, Y. *et al.* (1994) Excimer laser annealing process for polysilicon TFT AMLCD application. *10RC*, Monterey, CA.

Sunohara, K. *et al.* (1996) A reflective color LCD using three-layer GH-Mode. *SID 96 Digest*.

Sunohara, K. *et al.* (1998) Reflective color LCD composed of stacked films of uncapsulated liquid crystal (SFELIC). *SID 98 Digest*.

Surguy, P. W. H. *et al.* (1991) The JOERS/Alvey ferroelectric multiplexing scheme. *Ferroelectrics*, **122**.

Suzuki, K. (1987) Compensative addressing for switching distortion in a-Si TFT LCD. *Proc. Eurodisplay 1987*.

Suzuki, K. (1992) Pixel design of TFT-LCDs for high quality images. SID 1992 Digest.

Suzuki, K. (1994) High aperture TFT array structures. *SID 94 Digest*.

Suzuki, Y. *et al.* (1983) A liquid crystal image display. *Digest SID*.

Sze, S. M. (1981) *Physics of Semiconductor Devices*. John Wiley & Sons, New York.

Tahon, J. P. *et al.* (1999) Flexible polymer laminates with excellent gas-barrier properties. *Proc. Eurodisplay 99*.

Takahashi, S. *et al.* (1990) 10-in. Diagonal 16 gray level a-Si-TFT LCD. *SID 90 Digest*.

Tanaka, T. *et al.* (1993) An LCD addressed by a-Si:H TFTs with peripheral poly-Si TFT Circuits. *IED 1993 IEEE*.

Tang, S. T. *et al.* (1999) New bistable TN LCD modes in transmission and reflection. *Proc. Eurodisplay 99*, Berlin.

Tani, M. and Sugiura, T. (1995) LCD color filters, characteristics and future issues. *SID Seminar F5*.

Tew, C. *et al.* (1994) Electronic Control of a digital micromirror device for projection displays. *IEEE Int. Solid-State Circuits Conf.*

Timoshenko, S. P. and Goodier, J. N. (1970) *Theory of Elasticity*. McGraw-Hill, New York.

Togashi, S. (1992) Two terminal device addressed LCD. *Optrelectronics-Devices and Technologies*, **7**(2), December.

Togashi, S. *et al.* (1984) A 210×228 matrix LCD controlled by double stage diode rings. *Eurodisplay 1984*.

Tsvetkov, V. (1942) *Acta Physicochim.* (USSR) **16**, 132.

Uchida, T. (1984) *Opt. Eng.*, **23**.

Uchida, T. (1999) High performance reflective color LCDs. *Proc. Eurodisplay 99*, Berlin.

Uchida, T. *et al.* (1980) Bright dichroic guest-host LCDs without a polarizer. *SID 80 Digest*.

van der Witte, P. *et al.* (1999) Preparation of retarders with a tilted optic axis. *Jap. J. Appl. Phys.*, **38**, February.

Vaz, N. A. *et al.* (1987) *Mol. Cryst. Liq. Cryst.*, **146**(1).

Vithana, H. K. M. and Faris, S. M. (1997) Polymer stabilized Pi-cells as switchable phase retarders. SID 1997 Digest.

West, J. L. *et al.* (1989) Polymer dispersed liquid crystals incorporating isotropic dyes. *SPIE*, **1080**(48).

White, D. L. and Taylor, G. N. (1974) New absorptive mode reflective liquid crystal display device. *J. Appl. Phys.*, **45**(11).

Wortman, D. L. (1997) A recent advance in reflective polarizer technology. *SID 97 Digest*.

Yamaguchi, Y., Miyashita, T. and Uchida, T. (1993) Wide viewing angle display mode for the active matrix LCD using bend-alignment liquid crystal cell. *SID 93 Digest*.

Yamamoto, S. *et al.* (1992) A self scanned light valve with poly-Si TFT drivers by low temperature process below 600°C. *Japan Display 1992*.

Yang, D. K. *et al.* (1994) Cholesteric reflective display: Drive scheme and contrast. *Appl. Phys. Lett.*, **64**(15).

Yang, K. H. (1991) Two domain twisted nematic and tilted homeotropic LCDs for active matrix applications. *IDRC 91 Digest 68*.

Yang, O. K. *et al.* (1992) Cholesteric liquid crystal/polymer dispersion for haze-free light shutters. *Appl. Phys. Lett.*, **60** (25).

Yeh, P. (1988) *Optical Waves in Layered Media.* John Wiley & Sons, Chichester.

Yeh, P. and Gu, C. (1999) *Optics of Liquid Crystal Displays.* John Wiley & Sons, Chichester.

Yoo, J. S. *et al.* (1999) A novel polysilicon TFT with lateral body terminal for driving circuit application of AMLCDs. *Eurodisplay 1999*, Berlin, p. 365.

Young, N. D. *et al.* (1996) AMLCDs and electronics on polymer substrates. *Eurodisplay 96*.

Young, N. D. *et al.* (1999) LTPS for AMLCD on glass and polymer substrates. *Int. Display Workshop 99 Digest*.

Zeile, C. (2001) Fluessigkristall-Lichtventile in der kohärenten fourieroptischen Signalverarbeitung. Diss., Universitaet Stuttgart.

Zhu, Y.-M. and Yang, D.-K. (1997) High speed dynamic drive scheme for bistable reflective cholesteric displays. *Digest SID 1997*.

Index

γ-correction   240
A/D-converter   238
absorber   127, 150, 264–265
active matrix addressing   1, 136, 211
active matrix AM   1
active matrix liquid crystal
    display (AMLCD)   18, 341
addressing, active
  block parallel   232, 240, 267, 270
  dc-free   165, 172, 200, 205, 217, 270
  direct   122, 161
  dual line   236
  e-beam   18, 281, 286
  laser   211, 250, 252, 254–255, 285–286,
    307, 318
  multiple line   340
  optical   6–8, 39–40, 45, 47–48, 65, 88–89,
    105, 107–108, 110–111, 113–114, 117,
    151, 162, 172, 175, 180, 199, 281, 283,
    287, 293–294, 297, 303
  plasma   161–162, 245, 252, 284–285
  TFT   16–17, 189, 211, 216–217,
    220, 224, 228–231, 233, 235, 241,
    245–247, 250–252, 256, 258–259,
    261, 273–274, 276, 278, 312,
    317–318
  waveform   1, 19, 165, 174, 176, 190, 200
alignment, homeotropic
  homogeneous   78, 97, 108, 110, 125,
    131–132, 223, 245, 255, 297
  vertical   105, 107–108, 111, 205, 238

alternating signs, columnwise
  framewise   237
  linewise   223, 228
AM   163, 231, 290
AMLCD   284
analyzer
  crossed   14, 39, 43, 48, 66, 69, 73, 93, 106,
    111, 113, 116, 132, 139, 152, 283,
  parallel   4, 6–9, 12, 14, 18, 23, 27, 29–30,
    34, 37, 39–41, 43–46, 49, 55, 57,
    64–66, 69, 72, 74, 85–86, 89, 93,
    97–98, 101, 108, 110–111, 113, 114,
    116–117, 123, 126–131, 137–139,
    141, 146–147, 153, 155–156, 205,
    212, 227, 232, 240, 265, 267, 269,
    283, 297, 332
anchoring forces   141, 146
angular colour separation   300
anisotropic   1, 4–5, 21, 24, 88–89,
    91, 102
  dielectric   6, 49, 70, 97–98, 101, 125, 137,
    148, 194, 197, 201, 203,213–214, 219,
    227, 231, 241, 245, 250, 271–272, 278,
    283, 286, 289, 316, 319,
anisotropy, optic   7, 8
anodically oxidized   276
aperture ratio   117, 136, 241, 265, 278, 285,
    298, 303, 307
a-Si   211, 214, 241, 245–247, 249–250, 252,
    254–255, 281, 283, 317–318
azimuth angle   105

back channel etch 245–247
backlight 16–17, 243, 285
backplate 165
backscatter 127
bend 10, 49, 309
bias ratio 172
binder 123, 128, 290
birefringence 12, 46, 78, 106–108, 114
  negative 94, 106–108, 114, 126, 153, 155,
    157–158, 194, 197, 199–201, 203, 212,
    217–221, 225–226, 228, 231, 233, 252,
    258–259, 261, 272, 276, 332,
  positive 18, 25, 106–108, 114, 153,
    155, 157–158, 194, 199–200, 203,
    217, 220–221, 225, 228–229, 231,
    233, 261, 271–272, 276,
    318, 332
birefringent 8, 98, 108, 110–111, 114,
  119–120
bistability 5, 70, 152
bit rate 266–269
black matrix 17, 242–243, 290
buffer 176, 185, 194, 232, 250

capacitive coupling 175, 223, 241, 244
cell assembly
  π 40–41, 47–49, 67, 85, 95
  DAP 43,–46, 51–54, 66, 106
  Fréedericksz 26, 28–29, 31–32, 39–41,
    43–47, 51–52, 54, 57, 116
  HAN 44–47, 51–54, 66, 303
  Heilmeier 128
  mixed mode twisted nematic 72
  OCB 117
  TN 20, 46–47, 55, 64–65, 67, 69–71,
    73–75, 77–78, 91, 93, 95–96, 106, 108,
    111, 113–114, 127, 167–168, 170, 172,
    174, 176, 178, 180, 182, 184–186, 188,
    190, 192, 194, 196, 197, 205, 265, 269,
    303,
  twisted nematic 2, 12, 132, 153, 297
  vertically aligned 43, 106, 108
cells, bistable twist
  reflective 2, 37, 39, 44–46, 54, 74,
    77–79, 119–120, 123, 127,
    131–133, 149–151, 205, 207, 241,
    243, 263, 265, 270, 283, 289,
    293–294, 297, 303, 305–307, 309
  transmissive 2, 12, 31, 39, 45, 54, 123, 127,
    131, 139, 151, 241, 243, 294, 297, 302,
    307, 330

characteristics, input
  output 31–32, 37, 43, 47, 57–58, 72, 88–89,
    111, 120, 128–129, 173, 214, 232, 238,
    240, 265, 281, 307
  transfer 162, 232, 238, 252, 261, 266, 284,
    321
chiral components 146147
cholesteric host 130
chromaticity 291
clearing point 3
cold mirror 305
colour coding 238
  coordinates 25, 27, 40, 55, 58, 67, 75, 82,
    84–85, 87–91, 97, 139–140, 332, 336
  filter 17, 79, 127, 133, 238, 284, 290–291,
    293–294, 297, 299
  generation, 289
  grating 152–153, 300
  plate 12, 16–18, 45, 108, 110, 116, 129,
    219–220, 240, 242–243, 284, 290
column 20, 24, 39, 89, 105, 161–162, 167,
  171–173, 175, 177, 179, 182–185,
  187–190, 192–194, 199, 201, 205, 207, 217,
  222–223, 228, 231–233, 240–241,
  243–244, 249, 261
  drivers 172, 175, 180, 185, 222, 231–233,
    240, 249, 259, 274–275, 279
  voltage 1, 12, 14, 16–20, 32, 34, 39–42, 44,
    47, 49, 51–52, 66, 69–70, 72, 78–79, 105,
    113, 125, 130–133, 146, 153, 156,
    158–159, 161–162, 165, 167–168, 170,
    172, 174–175, 177, 180–190, 193–194,
    196, 199–201, 205–208, 212, 216–222,
    225–226, 228, 230, 232–236, 238, 240,
    250–251, 258, 261, 265–267, 270,
    271–276, 278–279, 281, 283–286,
    317–318, 325
  voltage generator 193
commutator 240
compensation film 111
compensation impulse 219–220, 222
conic 29, 146–147, 149–151, 205–210, 286,
consecutive frames 174
contact angle 322, 324
contrast ratio 16, 114, 157
controller 238, 240
crossover frequency 194
crosstalk 161, 168, 175, 201, 251, 275,
Cu 224, 247
cyan 134, 289, 293
DAC 185

decay time 49–50, 52–54
dehydrogenation 250
density 9, 125, 127, 203, 213, 216, 249, 255,
   265, 279, 298, 302, 317–318,
DGH 130, 132
dichroic 126, 130, 293,–294, 300, 302–303,
   305–307
   ratio 83, 88, 126, 130, 132, 170–172, 175,
      178–180, 186, 189, 268, 284, 312,
      317–319, 327, 330
dielectric constant 6
diodes 162, 211, 281, 319
disclination 153
dissociation 20
distortion 313, 315–316
doctor blade 321
doping 127, 251, 319
drain current 213
driver 172, 232–233, 240, 249, 261
   row- 18, 232
droplet 125, 322
dual scan driving 232
dye 126–127, 130, 132, 286
dynamic 9, 49, 210

ECB 79, 108
electronic addressing 5, 161, 281
electro-distortional curve 70
electro-optic effects 1, 2, 5, 21
electro-optic response 195
electro-optical processor 40
end point control 245
enthalpy 6
ethyl alcohol 44
evaporated 44, 46, 145–146, 149, 311
exclusive or gate 185
extraordinary beam 7, 98, 102, 105, 119

fabrication process 245, 250
field effect transistor (FET) 18
flicker 188
folded magnifier 270
frame response 188
   time 19–21, 29–30, 34, 39, 46,
      49–52, 86, 111, 117, 125, 130–131,
      153, 156–158, 161, 165, 167–168,
      170, 172, 174–188, 190, 192–194,
      198–204, 209–210, 216–217,
      223–225, 227–228, 231, 233, 236,
      238, 245, 265–270, 276, 284, 287,
      300, 318, 324, 331–332

FRC 174
free energy 152–153
Fresnel lens 306
functions, bilevel row
   orthogonal row 180, 185
   trilevel 179 41, 49180, 193
   Walsh 179-180, 190, 192

gate 18, 211–213, 217–219, 222–224,
   230, 232, 238, 240, 243, 245–247,
   250–252, 258, 263, 265, 312, 318
Gaussian 318
grain 211, 252, 254–256
grey level inversion 105
grey shades 1, 16, 21, 39, 113–114, 116,
   119, 128, 149, 156, 161–162, 167–168,
   174, 188–190, 192, 205, 217, 240, 258,
   265–267, 269–270, 298,
guard ring 246

Hadamard matrix 180
HDTV 11, 275
helix 12, 55, 57, 64, 120, 130–132, 138, 141,
   149, 155–156, 208, 283
hydrogenation 252
hysteresis 197, 205

image converter 281, 283–284
impact ionization 258
index 7–9, 14, 22, 27, 34, 42–45, 78, 97–98,
   102–103, 107–108, 111, 113–114, 117,
   123, 156
ink 321–323
intensity of light 197
interference 16, 148, 294
intermolecular forces 14, 117
inverse mode 151
ion implantation 250, 256
IPS 106, 116–117
irradiation 125, 252
isocontrast curve 111, 118
isotropic 3, 45, 88, 98, 111, 114, 119, 146,
   285–286, 325–326
ITO 17–18, 20, 39, 127, 136, 241–243,
   245–246, 276, 278, 283, 285, 290, 312,
   320, 324
Jones matrix 24
Jones vector 24–25, 27–28, 33, 40, 57–58, 66,
   83, 85–90, 139

knife coating 321

laser annealing
  crystallization   250, 252, 254
  scanning   167, 254, 318
latch   232, 238, 240, 269
law of Alt and Pleshko   179, 190
layer, adhesion
  anti-reflection   286
  barrier   137, 265, 271, 278, 311–312
  doped   126, 246, 252, 317–318
  etch stop   247
  overcoat   17, 290,
  passivation   245, 250, 252,
  PEDOT   320
LCD   1, 11, 18, 40, 64, 108, 111, 128, 155, 157,
    161–162, 165, 176–177, 188, 197, 240,
    283–286, 297–300, 302, 305
LCOS   263, 270, 297, 305
lift-off   245
light blocking layer   283
  shield   242–243, 245, 263–264
  valve   125, 281, 284, 286–287, 297–298,
    300, 302–303, 305–307
linearly polarized   12, 27, 31–34, 36–37, 39,
    41, 43, 45, 55, 64, 66, 72, 81, 83, 86, 88, 93,
    96, 111, 116, 119, 128, 139, 305
liquid crystal display (LCD)   1
liquid crystal display, chiral nematic
  antiferroelectric   157
liquid crystals, calamitic
  discotic   108, 111, 114
  ferroelectric   155, 157, 163, 197, 203
  guest-host   131–132, 163, 289
  lyotropic   4
  supertwist nematic   69
  thermotropic   4, 11
LSB   266–267
luminance   16, 19–20, 36, 50, 69, 71–72, 105,
    113–115, 119, 122, 132, 136, 161, 240, 281,
    283–285, 298, 307, 309,

magenta   131, 134, 136, 289, 293,
magnifying optics   270
mask count   245–246
Maxwell's theory   8
melting point   3
meshes   322–323
mesophase   3
microlenses   300, 302
midlayer tilt   70–71
MIM   271–279, 281, 298, 319
mirror   39, 45–46, 74, 132, 263–265, 312

modulation, amplitude   28, 39
  pulse width   231, 266
modulus of elasticity   312
Mohr's circle   313
MTN   73
multiplexer   240

NCAP   124–125
n-channel   212, 250–252, 256, 258,
    261, 318
normally black   14, 16, 33, 36–37, 39, 43,
    66–67, 69, 74, 77–78, 116, 167, 190
normally white   14, 16, 33, 36–37, 39, 43, 48,
    66, 69, 73–74, 78, 117, 153, 165
nozzle   322
NTSC   11, 20

oblique angles
  sputtering   44, 245, 252, 276, 279, 312, 317,
    319
off-current   20, 216, 317–318
off-pixel   170, 174, 178, 180, 194
off-resistance   19, 229
ohmic contact   245
on-current   20
on-pixel   170, 174–175, 178, 194
on-resistance   19, 216
optical axis   7, 8
  multiplier   40
  path length   297
order parameter   6, 127, 132
ordinary beam   7–8, 98, 119
orientation layer   12, 13, 43, 46, 115–116, 127,
    145, 149, 325
Ormocere   312

parasitic   20, 161, 168, 201, 207, 217, 222,
    228–229, 233, 238, 241, 243, 251, 265, 275
passive matrix addressing   20
passive matrix (PM)   20
patterning   245, 250, 252, 291
p-channel   251-252, 258, 261
PECVD   245, 247, 252, 272, 278, 316–319
permeation of humidity   310–311
  water   311–312, 324
phase, chiral smectic C
  cholesteric   4–5, 120, 137, 145–146, 197,
    210
  nematic   3–4, 6, 9, 11, 49, 51–54, 67, 72,
    107, 114, 123, 125, 131–132, 137,
    145–146, 149–153, 163, 197, 205, 283,
    289, 294,

separation   124–125, 204, 212, 284–285
smectic   3–6, 137, 286
smectic A   3
smectic C   3
smectic C*   4
transition   6, 41, 49, 69, 71, 74, 127, 132, 151, 209, 213, 215, 224, 285,
phasor   23
PHM   192
photopolymer   115, 291
photoresist   246, 252, 256, 322
photosensor   283–284
piezo-electric transducer   322
pigments   290
pitch   57, 64, 132, 134, 138, 142, 149, 151, 155, 175–176, 208–209, 232, 284,
pitch helical   6
pixel   1, 12, 14, 16–20, 79, 111, 114–115, 134, 150, 159, 161–162, 167–168, 170–171, 174–177, 180, 186, 189–190, 193–194, 196, 199–201, 205–206, 208, 216–217, 221–222, 225, 228, 233, 235–237, 240–241, 243–247, 249, 259, 263, 265–266, 268–269, 271–273, 275–276, 278–279, 281, 283–285, 298, 300, 302–303, 309
  layout   241–243, 278, 316
  switch   46,142, 161–162, 173, 199, 203, 240, 261, 284
plastic substrates   309, 312–313, 316, 326
polarity stages   240
polarization   14, 29–32, 36, 81, 83–86, 88–90, 92–96, 120, 137–138, 142–143, 147, 153, 155, 157, 197–198, 202, 205, 283, 309, 334–335
  circular   30, 85–88, 90–92, 94–96, 120, 149, 335
  elliptic   12
  flexoelectric   153
  linear   20, 30–32, 36–37, 45, 55, 69, 71, 84, 86, 89, 93–95, 119–120, 145–146, 156, 161, 215, 221, 224, 272, 274, 335,
  piezoelectric   155
  spontaneous   137, 157–158
polarized light   1, 12, 31, 37, 39, 46, 67, 69, 81, 83, 86, 88–89, 95–96, 119, 147–149, 155, 303
  circularly   12, 31, 39, 46, 81, 83, 86, 89–90, 94–96 120, 147–149
  elliptically   12, 29, 30, 75, 86, 89

polarizer   12, 14, 21, 34, 37, 39–41, 45–46, 72, 74, 76–78, 93, 110, 119–120, 123, 128, 130, 139, 153, 285
polarizing beam splitter   303
polymer   47, 114, 123, 125–128, 132, 146–147, 149, 151, 153, 291, 320, 321, 325,
polymer dispersed LCD   4
poly-Si   20, 162, 211, 249–250, 254, 259, 261, 263, 312, 318
Pool-Frenkel effect
  normal   44, 49, 141, 145–146, 149, 157, 272, 293–294
power recycling   233
precursor   250
predistortion   276
prepolymer   125
pretilt   12, 14, 44, 47, 52, 70, 143, 145–146, 152
printing   2, 321–323, 325–326
  flexographic   321, 326
  gravure   321
  silk screen   284, 321, 326
probability   182
projection systems   265
projector, colour field sequential
  scrolling   299–300
projectors   128, 297, 300
propagation of light   26, 42, 72, 91, 102, 111, 139
PWM   174, 188

reciprocity   75
rectifier   283
reflectance   131–133, 136, 148–149, 290–291
reflectivity   17, 205, 264
reflow   256
retardation   31–32, 44–47, 66, 107–108, 110, 113–114, 116–118, 310
rotation matrix   332
row   18–20, 161, 167–168, 170–173, 175, 177–182, 184–190, 192–194, 199–201, 206–207, 217–218, 220–224, 230–231, 234, 236, 240, 249, 265–266, 268, 276
  selection voltage   172, 206, 273–274
rubbing   55, 70, 127, 325
saturation   14, 17, 125, 196, 213, 215–216, 258
screen   162, 297–298, 303, 305–307, 322
segments   161, 165
selection intervals   180
self-aligned   246, 250–251, 276
sheet resistance   213, 224

shift register   172, 232, 261
shifted front-plate potential   222
silane   44, 146
slow axis   72, 111
spacer   150, 284
speed of light   8–9, 21, 40
spin coated   325
spin coating   291
sputtered   44, 46, 145–146, 245, 276
step response   223
storage capacitance   217
storage capacitor   18, 238, 241, 243, 263, 276, 278
subpixel   268
subpixellation   269
substrate   12, 43, 125, 133, 165, 245, 250, 258, 263, 265, 273, 281, 310, 312–313, 315–316, 318–319, 321–322, 324
  glass   12, 16, 20, 105, 108, 111, 113–114, 116, 127, 129, 136–138, 146–147, 211, 232, 242–243, 245, 247, 249–250, 271, 273, 283–284, 290, 309, 312, 322
  plastic   20, 132–133, 211, 232, 271, 276, 279, 311–313, 315, 319
superframe   174
supertwist nematic (STN)   41
susceptibility   155
switching angle   139, 143–144
  dynamics   49, 230, 251, 276

target   286, 317
  bookshelf   141–144
  Chevron   141, 143, 145
  planar   21, 55, 99, 146–151, 153–154, 205–207, 209–210
  quasi-bookshelf   141–143, 145, 205
thermal expansion   252, 311–312, 315
threshold   14, 16, 49, 51, 143, 161, 196, 197–199, 204–205, 208, 213, 230, 250–251, 272, 278, 317–318
torque   47, 49, 69, 117, 125, 143, 149, 155, 197, 201
transient   49, 201, 209–210
transmittance   14, 125, 151, 157, 188, 291, 293
transparent electrode   12, 13
trapped electron   271
tunnelling   272

twist   9, 10, 12, 14, 49, 55, 57, 64, 67, 69–74, 78, 91, 93–94, 113–114, 116, 131, 143, 148, 152–153, 157, 283

uniaxial   98, 108, 114
UV light   8, 276

varistor   281
VGA   11
via hole   248, 265,
video adapter   238
  source   37, 151, 172, 211–212, 214, 218–220, 231, 240, 245–248, 250, 252, 281, 297, 305, 318
  voltage   18
viewing angle   49, 69, 106, 114, 117, 119, 131, 149, 157, 303
virtual picture   270
viscosity   6, 9, 49, 51, 157, 202, 321–323, 326
  dynamic   9
  kinematic   9, 49
  rotational   9, 49, 149, 152, 199, 202, 209
voltage swing   172, 186, 217, 233–234
  data   167, 172–173, 185–186, 189, 192, 200–201, 206, 216, 230–231, 233, 237–238, 240, 259, 261, 266, 274–276
  gate-source   212, 220, 230–231, 233, 243, 252
  holding   47, 221, 241, 273–274, 276
  selection   167, 170–172, 175, 178–180, 182, 185–186, 189, 196, 199, 209–210, 232, 316
  video   18, 161–162, 167, 216–217, 219, 222–223, 231–233, 235, 238, 240, 244, 261, 272–273, 284,

wave guides   122
  vector   3, 7–9, 12, 14, 21–27, 29, 39, 41, 43, 55, 57, 60, 75, 86–87, 97–105, 116, 184–185, 188, 331–332
  harmonic   21, 27, 99
wavelength in vacuum   22
wavelength   8, 16, 22, 32–33, 36–37, 39, 42, 66, 69, 72, 78, 93, 95, 114, 119, 130, 147–148, 291, 300

yellow   16, 72, 134, 136, 149–150, 289, 293